Climate Change Management

Series Editor

W. Leal Filho, Hamburg, Germany

More information about this series at
http://www.springer.com/series/8740

Walter Leal Filho
Editor

Climate Change in the Asia-Pacific Region

Springer

Editor
Walter Leal Filho
Hamburg University of Applied Sciences
Faculty of Life Sciences
Hamburg
Germany

ISSN 1610-2010 ISSN 1610-2002 (electronic)
Climate Change Management
ISBN 978-3-319-14937-0 ISBN 978-3-319-14938-7 (eBook)
DOI 10.1007/978-3-319-14938-7

Library of Congress Control Number: 2015935652

Springer Cham Heidelberg New York Dordrecht London

Printed on acid-free paper

Springer International Publishing AG Switzerland is part of Springer Science+Business Media (www.springer.com)

Preface

The Asia-Pacific region is rich in resources on the one hand but quite vulnerable to climate change on the other.

The growing evidence of increases in the intensity and frequency of various extreme events (e.g., draughts, heat waves, tropical cyclones, intense rainfall, tornadoes, thunderstorms, etc.) suggests that urgent measures are needed, in order to increase the capacity of the countries in the region to adapt. However, when it is considered that the majority of the estimated 500 million rural poor in the Asia and Pacific region are subsistence farmers occupying mainly rain-fed land, it can be seen that coping with climate change in the region cannot be regarded as an easy task.

This state of affairs suggests that it is important that countries across the region actively engage in the preparation of strategies or action plans to cope with climate change. These tools may assist government agencies, nongovernmental organizations, and communities to enhance their capability to adapt and to possibly reduce the various negative impacts the various countries are subjected to. But apart from policies and action plans, the seriousness of the threats posed by climate change means that new ideas, strategies, and projects are needed, so that they can also better handle the various uncertainties associated with a phenomena which is global in nature, but whose impacts are mostly local.

This book is a contribution to address the above needs and is aimed at assisting countries in the Asia-Pacific region to better prepare themselves to cope with climate change. It is structured into 2 parts:

Part I—Climate Change and Socio-Economic Issues
Part II—Policy and Technology Responses

This book is also an attempt to offer a platform for a range of actors working in different sectors and systems to share their knowledge about their experiences, failures, and successes in pursuing climate change adaptation. The needs and interests of critical and neglected groups are highlighted throughout the book, alongside the need for improving knowledge management on climate change.

The case studies and experiences involving countries as varied as Australia, Bangladesh, China, Fiji, India, Mongolia, Nepal, and the Philippines, as well as regional analyses offered by some authors, offer a concrete view of the issues on the ground.

I want to thank all authors for sharing their knowledge and experience in this book. I hope that the body of information and knowledge amassed in "Climate Change in the Asia and Pacific Region" will serve the purpose of illustrating for various sorts of actions which are possible and needed to address a problem which impacts the region as a whole and individual countries in particular.

Hamburg, Germany Walter Leal Filho

Contents

Chapter 1
Introduction

Walter Leal Filho

Climate change in Asia should be regarded as an issue of major concern and urgency. It is already taking place and is especially dangerous to South Asian countries, where many of the poorest countries are seen. The scope and intensity of climate change in the region means that the pro-poor adaptive capacity needs to be emphasised, so as to assist them to become more resilient, if not less vulnerable.

It is widely acknowledged that the climate change is already having major impacts on the economic performance of South Asian countries, as well as of countries in the Pacific region. In addition, it also negatively affects the lives and livelihoods of millions of poor people. A key role is played here by extremes events, including intense floods, droughts, and storms (Source: Managing Climate Risk: Integrating Adaptation into World Bank Group Operations).

There is therefore a perceived need to improve the capacity of countries in the region to adapt, a need which is also acute in the Pacific region, where many small island States are located, and are quite vulnerable to the consequences of sea level rise.

The Intergovernmental Panel on Climate Change, on its fifth Assessment Report (AR5) has provided a set of specific information concerning the nature and the future impacts of climate change in the Asia-Pacific region. Some of the future impacts may include:

- Glacier melting in the Himalayas is projected to increase flooding and will affect water resources within the next two to three decades.
- Climate change will compound the pressures on natural resources and the environment due to rapid urbanization, industrialization, and economic development.

W.L. Filho (✉)
Faculty of Life Sciences, Hamburg University of Applied Science, Ulmenliet 21033, Hamburg, Germany
e-mail: walter.leal@haw-hamburg.de

W. Leal Filho (ed.), *Climate Change in the Asia-Pacific Region*, Climate Change Management, DOI 10.1007/978-3-319-14938-7_1

- Crop yields could decrease up to 30 % in South Asia by the mid-twenty-first century.
- Mortality due to diarrhea primarily associated with floods and droughts will rise in South Asia.
- Sea-level rise will exacerbate inundation, storm surge, erosion and other coastal hazards.

Moreover, climate change may lead to forced migration with all socio-economic risks and consequences it entails.

In order to address the problems the Asia-Pacific faces, there is a need for coordinated action at country level on the one hand, and at the regional level on the other. Also, there is a need to strengthen the capacity of scientists, organizations, decision-makers and other stakeholders, contribute towards adaptation to climate change.

This book is aimed at assisting countries in the Asia-Pacific region to better prepare themselves to cope with climate change. It is structured along two parts: Part I (Climate Change and Socio-Economic Issues) and Part II (Policy, Responses, Technological Approaches and Methods), with various national and regional case studies.

Part I handles the many socio-economic issues related to climate change. In this context, Chap. 2, "Livelihood Vulnerability and Displacement in Coastal Bangladesh: Understanding the Nexus" authored by Md. Mustafa Saroar (Bangladesh), Jayant K Routray (India) and Walter Leal Filho (Germany), looks at the situation of climate migrants in coastal.

The research assessed the linkages between people's livelihood vulnerability and their intention for outmigration from the coast, and identified vulnerability reduction measures, the implementation of which may significantly influence the trend of forced migration.

Chapter 3, whose title is "The Application of the macroeconomic analysis of a climate change model in China", was written by Mario Arturo Ruiz Estrada (Guatemala), Ibrahim Ndoma (Nigeria), and examines the possibility of measuring the economic impact of climate change through the macroeconomics analysis of climate change (MACC) model. This is a new model to evaluate the impact of climate change on GNP growth. To give light to the model, a test run is carried out on China to evaluate the impact of the country's natural disasters on its economy.

Chapter 4 "Climate change and Intergenerational Democracy: Experiences from Vanuatu" was written by Kirsten Davies (Australia) and describes the findings of a 2011 study involving 444 participants from the Island of Espiritu Santo, Vanuatu. A whole community engagement methodology, known as Intergenerational Democracy, developed by the author, was adopted to capture snapshots of human-environmental relationships through the voices of children through to the elderly. The study was designed to assist a deeper understanding of community perceptions and observations pertaining to the impacts of climate change.

Chapter 5, "Urban Poor Women and Climate Change in India: Enhancing Adaptive Capacity through Communication for Development", authored by Sakshi

Saini, Savita Aggarwal and Geeta Punhani (from India), presents a study which has been conducted to raise the level of awareness and knowledge of urban poor women, change their attitude and behaviour to deal with climate change and thereby enhance their adaptive capacity, through the use of systematic and strategic communication, on a statistically defined sample from five major regions of Delhi, India. A Communication for Development (C4D) module was produced, and texted in selected slums across the five regions of Delhi.

Chapter 6, whose title is "Local community perception of climate change and scientific validation: a review of initiatives and perspectives in the Indian region", was written by Prakash Rao (India) and Walter Leal Filho (Germany) and reviews the various efforts to document and articulate people's perceptions and knowledge of the impacts of climate change in diverse ecosystems in India. The study analyses the relationship between local community knowledge of climate change and the available scientific evidence and suggests that people's local knowledge of climate indicators conforms to existing scientific data. The paper also suggests various stakeholder-based policy measures to support mitigation and adaptation strategies for climate change.

Chapter 7 "Managing Environmental Migration to Improve Economic and Social Outcomes in Developing Asia and Pacific" by Bart W. Édes (USA) and François Gemenne (Belgium) assesses the problems faced by the displaced in the Asia and Pacific region, due to large concentrated populations residing in areas exposed to environmental risks. They look at the issue of migration influenced by environmental factors, and assess how it is contributing to urbanization trends in the region.

Chapter 8 "Urban environmental governance: Bangladesh as a case in point", was prepared by Nidhi Mittal (India), Luca Petrarulo (Italy) and Nipunika Perera (Sri Lanka) and maps the role and dimensions of urban environmental governance and the systemic barriers, opportunities and pathways for strengthening this at all levels in the Bangladesh context. The paper draws on the concept of multi-level governance to rationalise and define some enabling structures, mechanisms and processes to strengthen urban environmental governance, as a basis to build sustainable urban futures.

Chapter 9 "The Economic Value of Cyclonic Storm-Surge Risks: A Hedonic Case Study of Residential Property in Exmouth, Western Australia", written by Rebecca Roberts, Lynnath Beckley and Malcolm Tull, all from Australia, stated that unbridled development, growing coastal populations and injudicious land planning amplifies the predicted disaster risk due to climate change and extreme weather events. They used the town of Exmouth (NW Australia) used to investigate economic strategies for coastal risk mitigation.

Chapter 10, titled "Understanding the role of trust in network-based responses to disaster management and climate change adaptation in the Asia-Pacific region", was written by Victoria L. Ross (Australia), Elizabeth L. Malone (USA), and Susan Kinnear (Australia), and discusses the role and impact of trust on facilitating improved network co-ordination in emergency responses to climate-driven events. These findings may provide important insights for policy makers and network participants with regard to building network trust to better underpin disaster management and climate change adaptation in the Asia-Pacific region.

Chapter 11, titled "Climate Change and Adaptation Challenges in the Pacific" was prepared by Cyn-Young Park, David A. Raitzer, Jindra Nuella G. Samson and Paulo Rodelio M. Halili and describes trends in the region.

Part II of the book deals with policy, responses and technologies, and entails a set of contributions which look at the nexus between these three key areas.

The first contribution is seen in Chap. 12 "Managing "Climate Migration" in Mongolia: The Importance of Development Policies", written by Benoît Mayer (Singapore) shows that, in the context of Mongolia's internal migration, climate change adaptation is inseparable from domestic development policies that, it is argued, need urgently to be rectified.

Chapter 13, titled "Climate Change and Adaptation Challenges in the Pacific", was written by Cyn-Young Park, David A. Raitzer, Jindra Nuella G. Samson and Paulo Rodelio M. Halili, all of whom work at the Asian Development Bank in the Philippines. The chapter draws on the downscaling results of Global Climate Models (GCMs), and summarizes climate change effects such as changes in temperature, rainfall patterns, and sea levels. Assessment of GCMs suggests that sea level increases are likely to be substantially higher than previously accepted estimates, posing risks of inundation and salinization of freshwater in the Pacific island countries.

Chapter 14 "The Strategic Partnerships on Climate Change in Asia-Pacific Context: Dynamics of Sino-US Cooperation", written by Fang-Ting Cheng (Taiwan) argues that the cooperative progress between the U.S. and China on climate change has come about because the two countries have shifted their positions and are now treating climate change as a strategic issue which affects their vital interests. As a result, they are more willing to cooperate strategically with each other through bilateral dialogue and programs on innovation and technology transfer.

Chapter 15 "Strengthening climate change adaptation in Nepal: needs and perspectives", written by Bimal Raj Regmi (Nepal), Cassandra Star (Australia), as well as by Apar Paudyal and Ram Chandra Karki (also from Nepal), explores the existing challenges of managing climate change adaptation in Nepal. The research is based on a case study of three different locations of Nepal, representing different landscapes. The findings show that the impact of climate change differs based on the socio-economic characteristics of households and communities.

Chapter 16 "Climate change policy and sustainable energy development in Fiji: implications to Pacific Island States", written by Anirudh Singh and Atesh Gosai—both from Fiji-presents a brief overview of Fiji's National Climate Change Policy, followed by a review of the recent energy developments in the country and its new National Energy Policy (NEP). It is found that there is a clear mismatch between the NEP and the NCCP on the issue of energy considerations with regard to climate change mitigation.

Chapter 17 "Building resilience to climate change impacts and socioeconomic attributes of rural households in Solomon Islands", written by Michael Otoara Ha'apio (Solomon Islands) and Ricardo Gonzalez (Chile), performed a descriptive analysis of the socioeconomic attributes of rural households that participate in the Coral Triangle initiative (CTI) and Mangrove rehabilitation project (MRP) in selected rural villages of Solomon Islands. Household surveys were conducted in order to raise information on the socioeconomic attributes of participant households.

Chapter 18 "The significance of contextual vulnerability in effective adaptation to climate change on Tuvalu", prepared by Florent Baarsch (Germany) and Lan Marie Nguyen Berg (Norway), used data from a governmental household survey (2006) on rainwater catchment and storage facilities on Funafuti, along with a thorough analysis of funding schemes for water catchment between 2006 and 2010, to analyze the inventories of private and public rainwater catchment and storage equipment available in Funafuti in 2006 and in 2010. Moreover, a survey was conducted in one of the informal settlements to understand freshwater access for some of the poorest and most vulnerable populations on the island.

In a context of an increasing need for climate change adaptation, especially in Small Island Developing States and other developing countries, the study highlights the current barriers to effective adaptation for the 5,000 inhabitants of Funafuti.

Chapter 19 "Pathways for Climate Resilient Livelihoods: The Case of a Large Cardamom Farming in the Dzongu Valley of the Tista River Basin, Sikkim Himalaya" was written by Vimal Khawas (India). It discusses the fact that climate change may lead to changes in temperature and rainfall patterns, which in turn could be potential factors and declining productivity of large cardamom plantations across the Sikkim region in India.

Chapter 20 "Progress towards Low Carbon Fuels in the Pacific: Prospects and Challenges" was authored by Pritika Bijay and Anirudh Singh, both from Fiji. The paper chapter discusses the present status and future prospects of ethanol and biodiesel as possible transportation and power generation fuels for the region. The range of indigenously available feedstock, including sugar juice, molasses, cassava, for ethanol and coconut oil, *Pongamia* and *Jatropha* for biodiesel are considered, and the possibilities of the commercial realization of hydrotreated vegetable oils (HVO) explored. Barriers to the successful development of a biofuels industry are considered and possible solutions evaluated.

Chapter 21 "Towards Low Carbon Hotels in the Pacific Region: A study of Energy Consumption and efficiency in Hotels using models based on energy performance indicators", written by Krishneel Prasad and Anirudh Singh –from Fiji-, reports on an analysis of the energy consumption characteristics of Pacific Island hotels using Fijian hotels as a case study. The study employs the results of detailed energy audits and modelling based on Energy Performance Indicators (EPIs) that permit the prediction of the energy consumption of hotels within specific architectural groupings and star ratings. The models allow the investigation of energy savings possible through retrofitting and other energy-saving techniques.

Finally, Chap. 22, whose title is "Integrating technologies, measures and policies for climate resilient economy: A case analysis from emerging countries in the Asia-Pacific region", was written by Rajesh S. Kumar, Nity Nishant, Keshav Jha, N.K. Binu and Balachandran Thampi, from India. It describes the climate change related threats and associated developmental risks in the Asia-Pacific region, as well as analyze the sensitivity, potential resilience strategies, various adaptation and mitigation initiatives in the region in the broad context of the role to be played by the region to deal with climate change.

All in days, these contributions offer a good overview of the some of the outstanding work taking place across the Asia-Pacific region today, and outline some areas where some visible results can be seen.

Part I
Climate Change and Socio-Economic Issues

Chapter 2
Livelihood Vulnerability and Displacement in Coastal Bangladesh: Understanding the Nexus

Md. Mustafa Saroar, Jayant K. Routray, and Walter Leal Filho

Abstract Although numerous researches came up with various estimates about climate migrants from coastal Bangladesh, yet the connections between vulnerability and forced displacement are poorly understood. This research is aimed first, to assess the linkages between people's livelihood vulnerability and their intention for outmigration from the coast; second, to identify the vulnerability reduction measures, implementation of which may significantly arrest the likely trend of forced migration. Empirical part of this research was conducted in three coastal villages in Bangladesh. A total of 285 respondents were randomly interviewed using a semi structured questionnaire. Given the scenario of sea level rise (SLR), respondents were first asked—what they or their descendent would probably do if their most parcels of farmland gradually go under half-knee deep (20–25 cm) sea water by the middle of this century. Then they were asked to identify the nature of livelihood vulnerability for which, they or their descendent may permanently leave their current place of residences. Finally, they were advised to identify various vulnerability reduction measures, implementation of which will reduce their vulnerability as well as will arrest the likely trend of forced migration. Findings suggest that one in every three families will be forced to migrate. Therefore, about five million people may turn out as climate migrants. There emerged a clear linkage between mass displacement and sources of livelihood vulnerability. Finally, Binomial logistic regression model has identified six vulnerability reduction measures from a total of 18 which will significantly arrest the likely trend of forced migration by

Md.M. Saroar (✉)
Development Studies Discipline, Khulna University, Khulna 9208, Bangladesh
e-mail: saroar.mustafa@yahoo.com

J.K. Routray
Regional and Rural Development Planning, School of Environment, Resources and Development, Asian Institute of Technology, PO Box 4, Klong Luang, Pathum Thani 12120, Thailand
e-mail: routray@ait.ac.th

W.L. Filho
Faculty of Life Sciences, Hamburg University of Applied Science, Ulmenliet 21033, Hamburg, Germany
e-mail: walter.leal@haw-hamburg.de

W. Leal Filho (ed.), *Climate Change in the Asia-Pacific Region*, Climate Change Management, DOI 10.1007/978-3-319-14938-7_2

minimizing people's exposure and sensitivity and enhancing adaptive capacity. The policy implication is to avoid the occurrence of any large scale forced migration from the vulnerable coast of Bangladesh measures must be initiated in line with the findings without further delay.

Introduction

The Intergovernmental Panel on Climate Change (IPCC) in its several assessment reports has made strong assertion about the occurrence of accelerated sea level rise (SLR) and more frequent extreme events such as coastal flooding, cyclones, storm surges and salinity intrusion. Low-lying deltaic and small island countries (SICs) are particularly vulnerable to the SLR and its associated extreme events (IPCC 2001a, 2007). Bangladesh with a vast low-lying deltaic coast is one of the countries very much susceptible to SLR and its associated disastrous events (IPCC 2001a; GOB 2009; UNDP 2009). Coastal morphology, hydrology, soil, agriculture, horticulture, forestry, fisheries, infrastructures, settlements all will be severely impacted by these disasters (Mirza 2002; CARE 2003). Numerous studies have warned that the situation may aggravate more as Bangladesh might experience 30, 50 and about 100 cm SLR by the year 2030, 2050 and 2100 respectively (World Bank 2000). For the projected SLR many of these impacts will be felt more severe than any time in the past.

Vulnerability of people of Bangladesh to SLR and its associated events is attributed to both physical and socio-economic factors. Physical factor, for instance, low elevation from the mean sea level makes the entire coast highly exposed to cyclonic storms, tidal incursion, coastal flooding and salinity intrusion. Particularly the southwest coast experiences recurrent exposure to these hazards as this area is located only about one meter above the mean sea level (World Bank 2000; Ali Khan et al 2000; Cannon 2002; GOB 2009; Thomalla et al 2006; UNDP 2009). Some socio-economic factors, such as high population growth, extreme poverty, poor resource base, low adaptive capacity and high dependency on natural resources also make the livelihood of coastal people highly sensitive to these SLR and its associated events (Mirza 2002; CARE 2003; Agrawala et al. 2003; Adger et al. 2003; Wisner et al. 2004).

As the livelihood of about 35 million natural resource-dependent coastal population would be severely affected by SLR and its associated events, a significant proportion of them might turn out as climate migrants by the middle of this century (GOB 2009, 2010). The issues of climate change induced displacement and climate migrants already have been appeared as serious concern for Bangladesh and its neighbouring countries as well. Drawing from the evidences of previous migration patterns in the Indian subcontinent, many scholars have warned that huge displacement from coastal Bangladesh may eventually drag Bangladesh in violent conflict with neighbouring India, Pakistan and Myanmar as many climate migrants may arrive in these neighbouring countries following some historical links (i.e. cultural, religious and ethnic). Such massive displacement from the vulnerable coast of Bangladesh could be reduced by reducing the vulnerability of the coastal population (Saroar and Routray 2010).

Theoretically, there are two important ways of reducing vulnerability and livelihood insecurity. First, it can be reduced by lowering the people's exposure and sensitivity to these [SLR induced] extreme events; second, by enhancing their adaptive capacity (Chambers and Conway 1992; Ellis 2000). To minimize exposure usually various hard (i.e. engineering) measures such as new safe (e.g. cyclone, flood) shelters, sea walls, sluice gates, dykes etc. are undertaken. Sensitivity can be minimized by taking various managerial measures such as providing early warning, quickly mobilizing rescue units and operating community food security program al local level. Similarly to enhance adaptive capacity various other programs that support the socio-economic development of the coastal population are initiated (Kelly and Adger 2000; Grothmann and Patt 2005; Adger 2006). Drawing on the findings of earlier researches (Kelly and Adger 2000; Grothmann and Patt 2005; Pelling and High 2005; Tol et al. 2008; Schmidt-Thome and Klein 2013), it is hypothesised that some of these vulnerability reduction measures will encourage the coastal people of Bangladesh to adopt anticipatory adaptation in situ rather than turn out as climate migrants. However, how each of these vulnerability reduction measures may affect the livelihood security of coastal population in Bangladesh is not empirically studied yet. Therefore, the connection among vulnerability reduction measures, people's livelihood security and intension for displacement from the coast is poorly understood. Acknowledging this lack in research that examines the casual links among vulnerability reduction measures, livelihood security and mass displacement, this evidence based research is aimed to fill this gaps in knowledge and practice by addressing two research questions. First, what are the impacts of SLR and its associated events that people count as the key threats to their livelihood security which may cause forced outmigration? Second, what potentials various vulnerability reduction measures have to arrest the likely trend of outmigration from the coast? The policy implication of the research finding is- it might help devising public interventions for vulnerability reduction and to avoid the occurrence of mass displacement from the fragile coast of Bangladesh and elsewhere.

Materials and Methods

Selection of Study Area, Respondents and Survey Procedures

Coastal Bangladesh covers an estimated area of 47 thousand sq km which is divided in three distinct zones, i.e. south-western, south-central and south-eastern zone. Although whole coastal area of Bangladesh is susceptible to multiplicity of disasters, historical evidences and scientific studies indicate that people of the south-west part are more vulnerable to various hydro-meteorological disasters including cyclones, storm surges, salinity intrusion, and tidal flooding (Castro-Ortiz 1994; Ali and Chowdhury 1997; Ali Khan et al. 2000; World Bank 2000; Khan 2008; GOB 2010). The vulnerability of the population of south-west coast may increase further in the changing context of climate; especially the accelerated SLR might amplify the devastating effects of many of these already occurring disasters. Patuakhali District of this south-west zone, which is worst affected by these disasters selected for empirical part of the research.

A multistage sampling technique was employed to collect data and information from the study area. In first stage, from the six Upazila (sub-district) of Patuakhali District, "Kalapara Upazila" (sub-district) was selected. In second stage, from nine Union Parishad (lowest tier of local government) in Kalapara, three UPs which are flanked by the bay of Bengal and located about 0–30 cm (1 ft contour) above mean sea level was selected for this study. Tidal water inundates a significant part of the study area during high tide as the sea water travels upward from the bay of Bengal through a network of rivers, natural cannels and creeks (Fig. 2.1). People have been living here are historically prone to various natural calamities. In the final stage, a total of 285 households selected randomly were interviewed during January–April 2009. Interviews of usually the head of households were done through administering semi-structured questionnaire. Among the respondents 175 are male and 110 are female. The Bengali version of questionnaire was used to facilitate the survey process. Several focus group discussion sessions both formal and informal were conducted to get deeper insight about issues of particular interest.

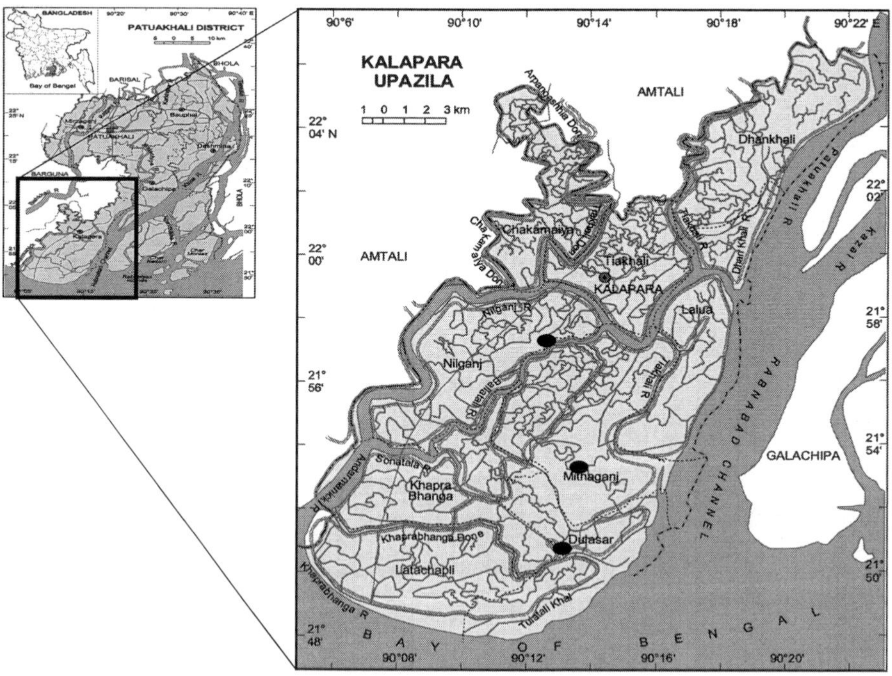

Fig. 2.1 Study sites (marked with *black circles*): in Dhulasar, Mithaganj and Nilganj "Union Parishad". *Source*. Islam (2003)

Approaches to Study and the Research Instruments

To address the research questions stated in the introduction part, a semi-structured questionnaire was prepared and tested in the study area before conducting the actual field survey. The questionnaire included among others various socio-economic and demographic information of the family, their exposure to various disasters, the damage that they experience due to recurrent disasters, their perception about the changing nature of climate, extreme events and sea level rise, and the likely impacts of these events on their live and livelihood. The following steps were followed to carry out the survey and data processing.

- First, the likely scenarios of SLR for the year 2020–2030 and 2050–2075 were narrated in brief before taking interview of the respondents. The direct impacts that might result from such SLR were also shared with them before the start of the interview. Respondents were asked to rate the impacts in a simple 3-point scale according to the level of severity of the impacts. They were further asked, what they or their decedents would probably do if their livelihood become insecure due to the impacts of future SLR and its associated events. From numerous responses, ultimately the likely occurrence of forced migration from the coastal tract was counted. The whole process of this step is discussed in some detail in respective section.
- Second, the respondents were asked to rank a maximum of five key causes/ sources of livelihood insecurity that might force them or their decedents (e.g. children or grandchildren) to migrate out from the fragile coast.
- Third, the respondents were asked to identify appropriate measures to reduce their livelihood vulnerability. Accordingly they were provided with three separate lists of measures, for instances, measures for minimization of exposure and sensitivity, and maximization of adaptive capacity. These lists of measures were prepared based on review of both hazard and climate literature. Broadly, these measures include—technological, managerial, policy, and other supportive measures that have the potential to ameliorate the negative impacts of SLR and its associate events on the livelihood of coastal people.
- Fourth, they were asked if appropriate measures are implemented whether they will still migrate out or change their intension of outmigration. It is expected that their responses will give a strong indication about the effectiveness of vulnerability reduction measures to stop mass displacement from the fragile coast.

Finally, to assess the effectiveness of each of the three kinds of vulnerability reduction measures in arresting the likely trend of mass displacement, a binomial logistic regressions (BLR) model was developed. This is discussed in some detail in respective section.

Result and Discussion

Socio-demographic Profiles of the Respondents

Among the respondents about 61 % are male and 39 % are female. The average age of the family head is 49 years and average duration of stay in the same locality is 44.45 years which means respondent's spatial mobility in terms of permanent migration is very less common. Of course, this interpretation does not consider the nearly permanent move out of female respondents from their parental house to husbands' house after their marriage. Almost 60 % of the respondents are illiterate; 20 % have completed grade-5 and 5 % have completed grade-8. Only about 2 % of them are college/university graduate. About 17 % of the respondents belong to different social and/or economic groups/organization but remaining 83 % are not member of any such groups.

The most dominant occupation of the respondents are crop agriculture (30.5 %) followed by casual labor (18.6 %) and fishing (17.2 %). Other available occupations are petty trade, business, transport-work, formal job, and various other on farm and off farm economic activities. However, almost every family is somehow engaged in natural resource-based livelihood which increases their vulnerability to climatic disasters. The annual average income of a family is about 141,438 BDT (US $2,065; 1 $ = 68.5 BDT). However, most of the families earn annually about 65,000 BDT (marginally below US $1,000). Average farm size among the surveyed household is only 0.36 ha and one fourth (24.6 %) of the families do not have any farmland; farmland holding is highly skewed.

Impacts of Hydro-meteorological Disasters on Livelihood Security

As the study area is historically exposed to multiple-hazards, the respondents are in general familiar with the impacts of cyclone, tidal surge, coastal inundation, salinity intrusion. They have a common perception that extreme events are more pronounced in recent time than these had been earlier. Respondents were asked to identify the likely impacts of various hydro-meteorological events including SLR especially if their farmlands gradually go under half of knee-deep water (20–25 cm) forever. Similarly they were asked about the impacts they will encounter if height of storm surges increase more few meters than they have had in the past. The use of this kind of plausible scenarios of climate change for studying the perception of common people is a widely accepted method (cf. Ford et al. 2010). To facilitate their responses, a total of 25 plausible impacts were included in the questionnaire. This list of 25 impacts of climate change induced hydro-meteorological events was prepared from the review of literature of climate and disaster science that have both global (see Smith 1997; Middleton 1999; IPCC 2001b; van Aalst 2006; Bosello et al. 2007; Wilbanks et al. 2007; Bunce et al. 2010) and regional focus, e.g. Asia pacific region (see Twigg and Bhatt 1998; Luna 2001; Ahmed 2005; Choudhury et al. 2005; Allen 2006; GOB 2009; De Silva and Yamao 2007). They have rated each of these impacts on their natural resource based livelihood in a simplified 3-point scale [low to high] instead of a 5-point Likert scale. Although a 5-point Likert scale is widely used, to facilitate the responses of rural illiterate/less educated respondents a 3-point simplified scale was preferred.

The Table 2.1 reports that respondent's livelihood security is impacted largely by physical damage of settlements, stock of foods, biomass fuels, cattle's fodders, and harvest failure. More than 50 % of the respondents mentioned that their

Table 2.1 Types of livelihood insecurity attributed to hydro-meteorological events in coastal Bangladesh

	High		Medium		Low	
Livelihood insecurity: type and nature[a]	%	f	%	f	%	f
Loss of crop production	48.77	139	36.49	104	14.74	42
Complete harvest failure	*51.93*	148	38.60	110	9.47	27
Increase cost of agricultural production	44.56	127	9.82	28	45.61	130
Degradation of pastureland	35.44	101	16.49	47	48.07	137
Seasonal shortage of fodder	43.86	125	7.37	21	48.77	139
Difficulty in animal/poultry husbandry	38.95	111	30.88	88	30.18	86
Over bank flow of fishponds/fish farm	3.86	11	8.07	23	88.07	251
Higher risk in offshore fishing	20.00	57	14.39	41	65.61	187
Limited scope of festival and social gathering	35.79	102	34.74	99	29.47	84
Increase number of non-fishing day	29.12	83	5.96	17	64.91	185
Decrease in fish catch per go	31.93	91	3.16	9	64.91	185
Difficulty in preserving fish	21.75	62	18.95	54	59.30	169
Physical damage of settlement	*62.46*	178	12.28	35	25.26	72
Damage of stock of food, biomass fuel and fodder	*62.11*	177	10.53	30	27.37	78
Cost of maintenance and rebuilding of private infrastructure	44.56	127	24.21	69	31.23	89
Damage of road infrastructure	22.46	64	43.16	123	34.39	98
Damage of social physical infrastructure, e.g. market, school etc.	38.25	109	32.28	92	29.47	84
Difficulty in physical mobility	18.25	52	44.56	127	37.19	106
Difficulty in carrying goods and commodities	32.28	92	39.30	112	28.42	81
Decrease number of earning/productive day	*61.40*	175	32.28	92	6.32	18
Fluctuation/decline in wage rate	48.42	138	42.46	121	9.12	26
Limited supply and stock of foodstuff in the market	42.81	122	43.86	125	13.33	38
Spread of contaminated water	28.42	81	54.04	154	17.54	50
Lack of saline free fresh water for drinking	*60.70*	173	20.00	57	19.30	55
Prevalence of waterborne diseases	26.32	75	54.39	155	19.30	55

[a]Percentage should be read row-wise. Figures in Italic imply high impacts

livelihood security will be highly impacted in each of these dimensions. Severe threat to these dimensions of livelihood security is acknowledged by a large number of population because these are the common dimensions of livelihood security of most occupational groups in the study area. Similarly 40–50 % of the respondents mentioned that their livelihood will be highly impacted by loss of crop production, high cost in production, seasonal shortage of fodders, fluctuation/decline in wage rate, and limited supply of food staff in the markets. Livelihood insecurities of many other families are highly related to damage of fish ponds/enclosures, higher risk in offshore fishing, decrease in fish catch per go. Some other impacts, for instances, damage of road infrastructure, difficulty in physical mobility, difficulty in carrying commodity and goods, spread of contaminated water, and prevalence of waterborne diseases also will cause the livelihood of coastal people insecure in the changing context of climate.

Contrary to general expectation, more than 50 % respondents acknowledged that their livelihood security is rather less impacts in the dimensions of damage of fish ponds/farms, higher risk in offshore fishing, increase number of non-fishing day, decrease in fish catch per go, and difficulty in preserving fish. More than 50 % of the respondents assigned a low score for each of these dimensions of livelihood insecurity. It is probably because the livelihood challenges in these dimensions are related to only a particular occupational group [e.g. artisanal fisher/fishing community]. Although, probably fisher-group expressed higher concern in these dimensions of livelihood security, however, as a whole due to the presence of higher number of respondents from other occupational groups who do not expressed higher concern, higher percentage of low scores are observed in these dimensions. About 30–50 % of the respondents mentioned that they will experience medium level of impacts in their livelihood security due The respondents perceived relatively medium level of livelihood insecurity from other impacts such as loss of crop production, fluctuation/decline in wage rate, difficulty in animal/poultry husbandry, damage of road infrastructure, difficulty in physical mobility, difficulty in carrying goods and commodities, spread of contaminated water, and prevalence of waterborne diseases.

Causes of Livelihood Insecurity and Mass Displacement

Not all impacts cited earlier in the Table 2.1 will cause livelihood insecurity that may trigger mass displacement from the coast. The respondents have identified and ranked five major causes of livelihood insecurity that might force them or their descendants to migrate out from the fragile coast. Table 2.2 reports that almost half of the respondents, i.e. 48 % (137 out of total 285), do not think about permanent displacement from their current place of living. Other half (148 out of total 285) of them, however, identified one or more causes of livelihood insecurity which may force them or their descendent to leave the fragile coast forever. About 36 % respondents from the latter category reported that they or their descendants may

Table 2.2 Reasons of livelihood insecurity which may cause forced migration from the fragile coast

Reasons of livelihood insecurity which may cause forced migration from the coast (*N = 148*[a], *total responses = 208*[b]):	Count	Percent of cases	Rank
i. If current main sources of income are likely to encounter irrecoverable loss due to SLR associated events	53	35.8	1
ii. If current physical accesses to services - local health care, market place, schooling are likely to be severely affected due to SLR associated events	52	35.1	2
iii. If current free/low cost accesses to potable water are likely to be diminished due to SLR associated events	45	30.4	3
iv. If food securities (production/availability) are likely to be severely affected due to SLR associated events	39	26.4	4
v. If most relatives are likely to quit/evacuate due to perceived threat of SLR associated events	19	12.8	5

[a]Remaining 135 (285–148) respondents do not consider outmigration as an adaptation strategy
[b]Multiple responses (adopted from Saroar and Routray 2013)

eventually leave the coast if their livelihood is severely affected due to complete loss of income from current/known sources. About same number of the respondents (35 %) think that people may leave the coast if their access to various services including local health care, market places, schools etc. are permanently disrupted. Another 30 % respondents believe that people may leave the coast because of severe crisis of salt free water for drinking and similar uses. Only about 25 % respondents have identified food insecurity (production loss/harvest failure) as the main cause in this regard (Table 2.2). It is a bit strange however to note that more people are willing to leave the fragile coast for severe scarcity of freshwater than loss of crop production due to harvest failure. About 10 % may prefer outmigration just because their close relatives will be doing so. The latter cause is not so serious from the point that normally about 5 % out migration is common from this part of fragile coast.

Linking Livelihood Vulnerability and Forced Migration

The likely scenarios of climate change induced sea level rise (CC-SLR) for the year 2020–2030 and 2050–2075 were presented to the respondents; these scenarios were adopted from the National Adaptation Programme of Actions (NAPA) for Bangladesh (GOB 2005). In general respondents are not familiar with the phrase "sea level rise". However, they are very much familiar with periodic/occasional inundation of their farmlands by tidal or cyclonic surges that originate from the bay of Bengal located nearby. They experience such inundation due to their recurrent exposures to tidal floods, storm surges, breach of embankments, and high tides. Considering their low level of familiarity with CC-SLR, the essence of SLR was rather communicated with them using practical means. Accordingly the respondents were asked- what they or their descendants would probably do, if their farmlands gradually go under "ankle-deep (10–15 cm)" and "half of knee-deep (20–25 cm)" salt water by the year 2020–2030 and 2050–2075 respectively? When the essence of SLR and its associated events were communicated using this kind of iconic images, the respondents were able to identify the likely impacts on their livelihood security and they replied accordingly. To deals with the future livelihood insecurity, the respondents may choose from eight courses of actions/responses. Their future responses will be: raise homestead and continue the same occupation anyway; raise homestead and continue agriculture (hoping salt tolerant varieties will be available); raise homestead and adopt brackish water aquaculture; switch to non-farm occupations; evacuate and settle nearby safer localities and towns; evacuate and arrive at major metropolises, especially in Dhaka; and evacuate the locality without predefined destination.

As it was aimed at identifying the links among SLR induced impacts, livelihood insecurity and forced migration of people from the fragile coast, therefore, the above responses concerning the people's spatial mobility and adaptive responses were grouped into two distinct and meaningful categories by collapsing closely related responses. The broad two categories of respondents are those who will not leave the coast (i.e. adaptation in situ) and those who will leave the coast (climate migrants/forced migrants) (Table 2.3). Finding suggests that no significant occurrence of forced migration will take place in the near future (by 2020–2030). However, almost 30 % respondents believe that they or their offspring may be forced to turn out as climate migrants in distant future (2050–2075) if appropriate measures are not taken beforehand.

Although about 35 million people inhabit in the entire coast of Bangladesh, half of them live in the low-lying parts alone. Based on the estimates of forced migrants (see last column: Table 2.3), one can expect that by the later part of this century (2050–2075) almost 5 million people will turn out as climate migrant (Table 2.3) from the low-lying part of the coast alone. However, if forced migration from the whole coastal tract is considered this number would be double, i.e. about 10 million population. Available estimates show that by 2050, 130 million more people will be

Table 2.3 Likely trend of forced migration for two plausible scenarios of SLR in coastal Bangladesh

Likely responses of respondent if farmlands permanently go below salt water	Below ankle-deep water (10–15 cm) by 2020–2030	Below half—knee deep water (20–25 cm) by 2050–2075
i. Adaptation in situ (no occurrence of forced migration)	100 % (285)	70.2 % (200)
ii. Permanent displacement (occurrence of forced migration)	00 % (0)[a]	29.8 % (85)
Total	100 % (285)	100 % (285)

[a]Figure in parenthesis indicates frequency/absolute response. Respondents do not considered 10–15 cm inundation as a problem and have rejected the idea of permanent evacuation

added in Bangladesh many of whom will be in this part of the fragile coast (GOB 2009). Therefore, it is reasonable to conclude that number of climate migrants will exceed 10 million from Bangladesh coast before the end of this century. These prospective climate migrants will evacuate their current localities and settle permanently elsewhere, mostly in the metropolises, and Dhaka—the capital city of Bangladesh. This kind of mass displacement would create numerous problems in the places of destination as well (Moser and Satterthwaite 2009). Offering healthy places of living in the cities for this huge number of forced migrants is really a critical concern (Table 2.3). Massive forced migration from the vulnerable coast in Bangladesh could only be avoided through careful anticipatory adaptation planning. For this there is a clear need of evidence based research which will examine the effectiveness of various anticipatory measures to arrest the likely trend of forced migration by reducing the people's livelihood insecurity (Stern 2006; Leal Filho 2009).

Preferred Measures to Reduce Livelihood Insecurity and the Occurrence of Forced Migration

Broadly three types of measures are suggested to address the livelihood security of natural resource-dependent coastal community against the threats of SLR and its associate events (Klein et al. 2001). These measures are initiatives to minimize people's exposure to and sensitivity against these SLR induced events; and also initiatives to enhance their adaptive capacity. Exposure is the nature and degree to which a system experiences environmental or socio-political stress (Klein et al. 2001). Sensitivity is the degree to which a system is modified or affected by perturbations of certain types (McCarthy et al. 2001; Adger 2006). On the other hand, adaptive capacity is the ability of a system to adjust to climate change to moderate potential damages, to take advantage of opportunities, or to cope with the consequences (Brooks et al. 2005; Smit and Wandel 2006).

From a total of 18 measures, the respondents have identified their preferred measures that will minimize their exposure and sensitivity and also maximize their adaptive capacity. They believe that implementation of these measures will discourage forced migration by ensuring their livelihood security. For exposure minimization their first priority is construction of more multipurpose cyclone/flood shelters. Maintenance of existing embankments/levees has been appeared as an important measure to discourage forced migration. Almost 75 % of them preferred operation of more new sluice gates. Similarly, about 65 % preferred construction of new flood walls/embankments/levees believing that construction of these will reduce their exposure to SLR and its associate events. Although everyone understood the roles of naturally occurring mangrove forest to minimize their exposure but only 30 % preferred reforestation of mangrove in their localities (Table 2.4).

Table 2.4 Respondent's preferred measures to discourage forced migration from Bangladesh coast

Measures preferred by the respondent to discourage forced outmigration from the coast	Count	Percent of cases
Measures to minimize exposure (N = 285, total responses = 1,187[a]):		
i. Construction of new flood wall/embankment/levee	187	65.6
ii. Maintenance of existing embankment/levee	226	79.3
iii. Operation of more new sluice gate	214	75.1
iv. Re-excavation of illegally occupied canals	192	67.4
v. Protection of mangrove and reforestation	83	29.1
vii. Construction and maintenance of cyclone/flood shelter	285	100.0
Measures to minimize sensitivity (N = 270, total responses = 810[a]):		
i. Early dissemination of warning information	169	62.6
ii. Emergency rescue and recovery unit at community level	197	73.0
iii. Quicker and better access to cyclone/flood shelter	204	75.6
iv. Community food security program	138	51.1
v. Others- as part of integrated coastal management	102	37.8
Measures to enhance adaptive capacity (N = 285, total responses = 1,187[a]):		
i. Awareness rising program on CC and SLR	285	100.0
ii. Strengthening community cohesiveness and bonding	140	49.1
iii. Help upgrading indigenous coping and adaptation	128	44.9
iv. Low-cost innovation for saline free water	170	59.6
v. Cash incentive for post disaster rehabilitation	285	100.0
vi. Special social safety-net for coastal communities	185	64.9
vii. Support for coastal resources based adaptive livelihood	221	77.5

[a]Multiple responses (adopted from Saroar and Routray 2013)

To minimize sensitivity about 75 % respondents preferred initiatives for quicker and better access to cyclone/flood shelters. Almost similar number (73 %) of them preferred establishment of emergency rescue and recovery units at community level. About 63 % preferred measures to disseminate warning information earlier than what is done now. About half of them preferred community food security program at community level (Table 2.4). To enhance their adaptive capacity about 77.5 % respondents preferred new initiatives for coastal resources based livelihood. About 65 % preferred special social safety nets for coastal area. Providing cash incentive for post disaster rehabilitation has been appeared as very popular initiative for adaptive capacity enhancement (Table 2.4).

Effectiveness of Measures to Discourage Forced Migration

To assess the effectiveness of both exposure and sensitivity minimization measures and adaptive capacity enhancement measures in lowering the likely trend of forced migration from the coast a Binomial Logistic Regression (BLR) model is built. This

BLR model only predicts the probability of arresting the likely trend of forced migration of those people whose avenues of income will be severely threatened by SLR and its associate events. Among the 18 vulnerability reduction measures, if a specific measure is considered by the respondents as an appropriate one then their response is coded with 1; but if that measure is considered inappropriate, then the responses is coded with 0. All the 18 measures are grouped under three major categories, such as measures for 'minimization of exposure (6 variables)', 'minimization of sensitivity (5 variables)', and 'enhancement of adaptive capacity (7 variables)'. Each of these 18 measures is used as an independent variable in the BLR model. Respondent's intension to migrate or not (when a specific measure will be implemented) is the dependent variable. Despite implementation of preferred measures if the respondents still intend to migrate out, their response is coded with 0. But if they intend not to migrate out any more, their response is coded with 1 (Table 2.5).

To fit the independent variables in the BLR model, the constant variables (i.e. 100 % responses are either 'yes' or 'no' or variables that have high colinearity/multicolinearity problem (i.e. correlation coefficient in excess of 0.70) (Field 2005) are excluded from the model. Following Field (2005), three independent variables, such as 'construction and maintenance of cyclone/flood shelter', 'awareness rising program on CC-SLR', and 'cash incentive for post disaster rehabilitation' are excluded from the model because these are constant variables (i.e. 100 % responses are same: see Table 2.4). Similarly, independent variable- 'quicker and better access to cyclone/flood shelter' is excluded from the model because it has strong colinearity (i.e. correlation coefficient is >80) with another variable 'emergency rescue and recovery unit at community level' (Table 2.6). Finally 14 independent variables are used in the model.

Respondents were asked if their preferred vulnerability reduction measures are implemented whether still they intend to migrate or not. Based on their responses, how far each measure will be effective in stopping forced migration is predicted with this BLR model. Models output are presented in Table 2.7. The result shows that exposure minimization measures (LR chi-square $= 46.80$, Pseudo $R^2 = 0.24$, $p < .01$), and adaptive capacity enhancement measures (LR chi-square $= 49.45$, Pseudo $R2 = 0.26$, $p < .01$) will have statistically significant influence than the sensitivity minimization (LR chi-square $= 8.53$, Pseudo $R^2 = 0.05$, $p < .10$) measures on the future migration behaviour of the respondents. It means if vulnerability reduction measures are implemented, prospective climate migrants may rethink and eventually may not migrate out from the coast.

Table 2.7 further reports that implementation of adaptive capacity enhancement measures will have highest influence on migration behaviour of coastal people. By contrast o implementation of "sensitivity minimization measures" will have least influence in this regards. As regards exposure minimization measures is it seen that the odds of preference for "not to out migrate (i.e. adaptation in situ)" is 24.1 times higher among the respondents who considered "construction of new flood walls/embankments/levees" as appropriate than who did not considered ($B = 3.18$, Exp $(B) = 24.10$, Wald $= 9.32$, $p < .01$). It means among the various measures of

Table 2.5 Variables used in BLR model and their coding

Variables	Coding
Independent variables:	
i. Exposure minimization measure:	
Construction of new flood wall/embankment/levee (dummy)	1, if think appropriate; 0, otherwise
Maintenance of existing embankment/levee (dummy)	1, if think appropriate; 0, otherwise
Operation of more new sluice gate (dummy)	1, if think appropriate; 0, otherwise
Re-excavation of illegally occupied canals (dummy)	1, if think appropriate; 0, otherwise
Protection of mangrove and reforestation (dummy)	1, if think appropriate; 0, otherwise
Construction and maintenance of cyclone/flood shelter (dummy)[a]	1, if think appropriate; 0, otherwise
ii. Sensitivity minimization measure:	
Early dissemination of warning information (dummy)	1, if think appropriate; 0, otherwise
Emergency rescue/ recovery unit at community level (dummy)	1, if think appropriate; 0, otherwise
Quicker and better access to cyclone/flood shelter (dummy)	1, if think appropriate; 0, otherwise
Community food security program (dummy)	1, if think appropriate; 0, otherwise
Others- as part of integrated coastal management (dummy)	1, if think appropriate; 0, otherwise
iii. Adaptive capacity enhancement measures:	
Awareness rising program on CC and SLR (dummy)[a]	1, if think appropriate; 0, otherwise
Strengthening community cohesiveness and bonding (dummy)	1, if think appropriate; 0, otherwise
Help upgrading indigenous coping and adaptation (dummy)	1, if think appropriate; 0, otherwise
Low-cost innovation for saline free potable water (dummy)	1, if think appropriate; 0, otherwise
Cash incentive for post disaster rehabilitation (dummy)[a]	1, if think appropriate; 0, otherwise
Special social safety-net for coastal communities (dummy)	1, if think appropriate; 0, otherwise
Support for coastal resources based livelihood (dummy)	1, if think appropriate; 0, otherwise
Dependent variables:	
If the preferred measured is implemented whether the respondent still intend to migrate out from the coast? (dummy)	1, no; 0, yes

[a]The responses are constant (i.e. 100 % respondents responded in the same way); cannot be used in final modelling

Table 2.6 Partial correlation matrix to identify colinearity among independent variables

		1	2	3	4	5	6	7
	Measures to minimize sensitivity[a]:							
1	Early dissemination of warning information	–						
2	Emergency rescue and recovery unit at community level	−0.16**	–					
3	***Quicker and better access to cyclone/flood shelter***[a]	−0.17**	***0.83****[a]	–				
4	Community food security program	0.10	0.24**	0.24**	–			
5	Others- as part of integrated coastal management	0.43	0.09	0.07	0.27**	–		
	Measures to minimize Exposure[a]:							
1	Construction of new flood wall/embankment/levee	–						
2	Maintenance of existing embankment/levee	0.54**						
3	Operation of more new sluice gate	−0.04	−0.11*	–				
4	Re-excavation of illegally occupied canals	0.33**	0.31**	0.01	–			
5	Protection of mangrove and reforestation	0.25**	0.33**	−0.11	0.10	–		
6	Construction and maintenance of cyclone/flood shelter’	c	c	c	c	c	–	
		1	2	3	4	5	6	7
	Measures to maximize Adaptive capacity[a]:							
1	Awareness rising program on CC and SLR	–						
2	Strengthening community cohesiveness and bonding	c	–					
3	Help upgrading indigenous coping and adaptation	c	0.09	–				
4	Low-cost innovation for saline free potable water	c	0.04	0.11	–			
5	Cash incentive for post disaster rehabilitation	c	c	c	c	–		

(continued)

Table 2.6 (continued)

		1	2	3	4	5	6	7
6	Special social safety-net for coastal communities	c	0.05	0.15*	−0.10	c	–	
7	Support for coastal resources based adaptive livelihood	c	0.11	0.06	−0.07	c	0.13*	–

*p < 0.05; **p < 0.01
[a]If a measure is considered appropriate by the respondent it is coded with 1; otherwise coded with 0
[b]Bold and Italicized item shows colinearity (e.g. Correlation coefficient value: 0.83)
[c]Cannot be computed because at least one variable is constant (i.e. 100 % responses are either yes or no)

exposure minimization, initiatives for "construction of new flood walls/embankments/levees" will have significantly higher influence to stop forced migration and to encourage adaptation in situ. In case of sensitivity minimization measures, it is found that the odds of preference for "not to out migrate (i.e. adaptation in situ)" is 1.9 times higher among the respondents who considered "community food security program" as appropriate sensitivity minimization measure than who did not (B = 0.64, Exp (B) = 1.90, Wald = 3.65, p < .10). It means, among the various measures of sensitivity minimization, initiatives for "community food security program" will have significantly higher influence to stop forced migration and to encourage adaptation in situ. In case of adaptive capacity enhancement measures, it is seen that the odds of preference for "not to out migrate (i.e. adaptation in situ)" is 5.17, 4.0 and 3.34 times higher among the respondents who considered "special social safety-nets for coastal communities" (B = 1.75, Exp (B) = 5.74, Wald = 12.01, p < .01), "strengthening community cohesiveness and bonding" (B = 1.38, Exp (B) = 4.0, Wald = 14.7, p < .01) and "support for coastal resources based adaptive livelihood" (B = 1.21, Exp (B) = 3.34, Wald = 4.52, p < .05) respectively as the appropriate measures (Table 2.7). It means implementation of adaptive capacity enhancement measures such as "special social safety-nets for coastal communities" and "support for coastal resources based adaptive livelihood" will have significantly high influence to stop mass displacement from the fragile coast of Bangladesh.

Analysis, Policy Implications and the Limitations of the Research

Results show that about 30 % families believe that they or their descendants will probably turn out as forced migrants because of loss of livelihood avenues. Considering the current population of fragile coast one can assume that about five

Table 2.7 Binomial logistic regression model to predict the influences of various vulnerability reduction measures on peoples' forced migration behaviour that may result from livelihood insecurity

	Model 1 Exposure minimization		Model 2 Sensitivity minimization		Model 3 Adaptive capacity enhancement	
Vulnerability reduction measure preferred by the respondent (1 = yes; otherwise, 0)	B: Coefficient (Exp (B): odds ratio)	Wald chi-square	B: Coefficient (Exp (B): odds ratio)	Wald chi-square	B: Coefficient (Exp (B): odds ratio)	Wald chi-square
Construction of new flood wall/embankment/levee	3.18*** (24.10)	9.32				
Maintenance of existing embankment/levee	1.47 (4.35)	1.86				
Operation of more new sluice gate	−0.39 (0.68)	1.19				
Re-excavation of illegally occupied canals	0.23 (0.33)	1.27				
Protection of mangrove and reforestation	−0.40 (0.67)	1.37				
Early dissemination of warning information			0.18 (1.2)	0.23		
Operate emergency rescue and recovery unit at community level			0.21 (1.23)	0.31		
Community food security program			0.64* (1.90)	3.65		
Others- as part of integrated coastal management			0.32 (1.37)	0.77		
Help upgrading indigenous coping and adaptation					0.41 (1.50)	1.43
Strengthening community cohesiveness and bonding					1.38 *** (4.0)	14.70
Low-cost innovation for saline free potable water					−0.49 (.62)	2.03
Special social safety-net for coastal communities					1.75*** (5.74)	12.01
Support for coastal resources based adaptive livelihood					1.21** (3.34)	4.52
Likelihood Ratio Chi-square	46.80***		8.53*		49.45***	

(continued)

Table 2.7 (continued)

Vulnerability reduction measure preferred by the respondent (1 = yes; otherwise, 0)	Model 1 Exposure minimization		Model 2 Sensitivity minimization		Model 3 Adaptive capacity enhancement	
	B: Coefficient (Exp (B): odds ratio)	Wald chi-square	B: Coefficient (Exp (B): odds ratio)	Wald chi-square	B: Coefficient (Exp (B): odds ratio)	Wald chi-square
Pseudo R^2 (Nagelkerke)	0.24		.05		0.26	
N	285		285		285	

Note: 1. 'Construction and maintenance of cyclone/flood shelter', 'Awareness rising program on CC and SLR', and 'Cash incentive for post-disaster rehabilitation' are excluded from the models as the responses are constant (100 % same)

2. 'Quicker and better access to cyclone/flood shelter' is also excluded because of its co-linearity with 'Emergency rescue and recovery centre at community level' ($r = 0.83$, $p < 0.01$: see Table 2.6)

3. Dependent variable: Whether the people still intend to migrate, if their preferred vulnerability reduction measures are implemented (no response is coded with 0; yes with 1)

*significant at 0.10; **significant at 0.05; ***significant at 0.01

million people will be forced to be climate migrant by the middle of this century (see Table 2.3). If the projected population are taken into consideration this number could be double by the end of this century. However, results further show that if various vulnerability reduction measures are initiated well before reaching the tipping point, a significant portion of the 'would be forced migrants' will adopt anticipatory adaptation instead of leaving the coast forever. In this respect implementation of exposure and sensitivity minimization measures and adaptive capacity enhancement measures will play very crucial role to hold back the prospective climate migrants. For instance construction of hard measures such as new sea walls and embankments will minimize exposure and introduction of community food security programs, special safety-net for coastal region, and coastal resources based adaptive livelihood will enhance people's adaptive capacity which eventually will slow down the likely trend of forced migration from the coastal tract. It is worth noting that the contributions/influences of exposure minimization measures have outweighed the total contribution of other measures in these respects. In fact, people put too much emphasis on these hard structural measures primarily because most peoples are engaged in agriculture and allied occupations; for them if production unit/place is safe from inundation of sea water they are comfortable with dealing with others anomalies.

Among various exposure minimization measures 'construction of new flood walls/embankments/levees' have been appeared as most effective to reduce the occurrence of forced migration. It is because for certain occupation groups who are willing to migrate because of fear of loss of income from higher exposure to disastrous events, they believe implementation of these hard measures will minimize their exposure and thus they will not require to leave the coast. For instances, agricultural farming communities, are more prone to saline water where as people engaged in aquaculture are more concerned about coastal inundation. For these both occupational groups minimization of their exposures to surges, salinity intrusion, and coastal inundation through construction of new sea walls or embankments having sufficient height will discourage them to leave the coast. Despite having huge potential to minimize people's exposure to disastrous events implementation of these hard measures such as construction of new sea wall, operation of new sluice gate and re-excavation of illegally occupied canals are really capital intensive as costs are substantial. If public funds are not allocated implementation of these exposure minimization measures from mere community or private initiative is hardly possible. Therefore there is need for planned public initiatives where other actors including community and individuals will participate proactively to avoid the occurrence of any large scale forced migration.

The Results further reveal that a significant portion of the respondents believe adaptive capacity enhancement measures will be effective to lower the likely trend of forced migration. Accordingly they have identified- 'special safety-net for coastal communities' and 'community food security program' as top priority initiatives in this regards. In fact, implementation of these two measures along with 'initiative for coastal resource based adaptive livelihood' will strengthen the very foundation of adaptive capacity of the prospective climate migrants. As the key sources of livelihood in the coastal area are agriculture and allied activities, and fishing it is not unusual that they have also prioritized introduction of coastal resources based adaptive livelihood as a long term strategy to avoid mass displacement. In the same vein they give heavy emphasis on measures of social insurances and cushion against temporary/seasonal loss of income as short-term or medium-term coping and adaptation strategy. This is possibly the case for people, for instance day labourer, off-farm workers, subsistence fishers who really lack year-round permanent sources of income.

Finally, it has also been appeared that some people think policy makers and planners should concentrate more on adaptive capacity enhancement of people as implementation of exposure minimization measures will require long time and allocation of huge public fund which the government of Bangladesh may not be able to afford. However, it should be noted that although implementation of adaptive capacity enhancement measures involve fewer technicalities, require less budgetary allocation and execution time, paradoxically in absence of exposure minimization measures such as construction of new sea walls/embankments people might rarely invest for anticipatory adaptation no matter how strong their adaptive capacity are. Thus there is a need for forward planning where a balance among exposure and sensitivity minimization and adaptive capacity enhancement measures will be maintained to arrest the likely trend of forced displacement from the fragile coast of Bangladesh. While this study has identified important avenues of intervention to bring down the likely trend of forced migration from coastal Bangladesh but the major limitation of this work is it has only considered those people who may migrate due to the loss of avenues of income. Earlier analysis shows that people might be forced to migrate for many other reasons associated with climatic disasters. This model does not consider those factors' induced mass displacement. Incorporation of those factors' induced forced migration in the model would have helped designing a more pragmatic plan to address the critical linkages among climate change induced vulnerability, livelihood insecurity and forced migration.

Conclusion

In the context of rural Bangladesh circular migration or seasonal displacement of people is not a new phenomenon especially in the poverty stricken and flood affected areas. However, permanent displacement is the last resort for the people who lose their farmlands and settlements due to natural calamities, for instance massive riverbank erosion. As the coast-lying areas are very fertile and resourceful, and riverbank erosion is not very severe, peoples rarely leave their place of living permanently only because of coastal hazards. However, given the context of future climate change induced vulnerability a significant number of people from the coastal tract might turn out as climate migrants. Because their perceived threat on livelihood security due

(continued)

to SLR and its associated disastrous events are very deep rooted in their mind-set. Unless they are assured with substantial visible initiatives for building a more resilient coastal community, it will be very hard to encourage anticipatory adaptation and to stop forced migration. Experience shows that anticipatory adaptation in situ rarely take place automatically when there is uncertainty. Therefore it is strongly suggested that a balanced and coordinated effort must be made to minimize exposure and sensitivity, and enhance adaptive capacity of people which will eventually bring down the likely trend of forced migration from the coastal tract of Bangladesh.

References

Adger WN (2006) Vulnerability. Glob Environ Change 16:268–281

Adger WN, Huq S, Brown K, Conway D, Hulme M (2003) Adaptation to climate change in the developing world. Prog Dev Stud 3(3):179–195

Agrawala S, Ota T, Ahmed AU, Smith J, van Aalst M (2003) Development and climate change in Bangladesh: focus on coastal flooding and Sundarbans. OECD, Paris

Ahmed AU (2005) Adaptation options for managing water related extreme events under climate change regime: Bangladesh perspectives. In: Mirza MMQ, Ahmad QK (eds) Climate change and water resources in South Asia. Balkema, Leiden, pp 255–278

Ali Khan TM, Singh OP, Rahman MS (2000) Recent sea level and sea surface temperature trends along the Bangladesh coast in relation to the frequency of intense cyclones. Mar Geod 23:1–14

Ali A, Chowdhury JU (1997) Tropical cyclone risk assessment with special reference to Bangladesh. MAUSAM Former Indian J Meteorol Hydrol Geophys 48:305–322

Allen KM (2006) Community-based disaster preparedness and climate adaptation: local capacity building in the Philippines. Disasters 30(1):81–101

Bosello F, Roson R, Tol RSJ (2007) Economy-wide estimates of the implications of climate change: sea level rise. Environ Res Econ 37(3):549–571

Brooks N, Adger WN, Kelly PM (2005) The determinants of vulnerability and adaptive capacity at the national level and the implications for adaptation. Glob Environ Change 15:151–163

Bunce M, Rosendo S, Brown K (2010) Perceptions of climate change, multiple stress or sand livelihoods on marginal African coasts. Environ Dev Sustain 12(3):407–440

Cannon T (2002) Gender and climate hazards in Bangladesh. Gend Dev 10(2):45–50

CARE (2003) Report of a community level vulnerability assessment conducted in Southwest Bangladesh. Reducing Vulnerability to Climate Change (RVCC) Project. CARE Bangladesh, Dhaka

Castro-Ortiz CAC (1994) Sea level rise and its impact on Bangladesh. Ocean-Coast Manage 23 (3):249–270

Chambers R, Conway G (1992) Sustainable rural livelihoods: practical concepts for the 21st century, IDS Discussion paper 296. Institute of Development Studies, University of Sussex, Brighton, UK.

Choudhury AM, Neelormi S, Quadir DA, Mallick S, Ahmed AU (2005) Socio-economic and physical perspectives of water related vulnerability to climate change: results of field study in Bangladesh. Sci Cult (Spec Iss) 71(7–8):225–238

De Silva DAM, Yamao M (2007) Effects of the tsunami on fisheries and coastal livelihood: a case study of tsunami-ravaged southern Sri Lanka. Disasters 31(4):386–404

Ellis F (2000) Rural livelihood diversity in developing countries. Oxford University Press, Oxford

Field AP (2005) Discovering statistics using SPSS, 2nd edn. Sage, London

Ford JD, Pearce T, Prno J, Duerden F, Ford LB, Beaumier M, Smith T (2010) Perceptions of climate change risks in primary resource use industries: a survey of the Canadian mining sector. Reg Environ Change 10:65–81

Government of Bangladesh (GOB) (2005) National Adaptation Programme of Action (NAPA), Ministry of Environment and Forests, Government of Bangladesh.

Government of Bangladesh (GOB) (2009) Bangladesh Climate change Strategy and Action plan 2008. Ministry of Environment and Forests, Dhaka, Bangladesh

Government of Bangladesh (GOB) (2010) National Plan for Disaster Management 2010–2015. Disaster Management Bureau, Disaster Management and Relief Division Government of Bangladesh, Bangladesh

Grothmann T, Patt A (2005) Adaptive capacity and human cognition: the process of individual adaptation to climate change. Glob Environ Change 15:199–213

IPCC (2001a) Climate Change 2001: the scientific basis. Contribution of working group I to the third assessment report of the Intergovernmental Panel on Climate Change. Cambridge University Press, Cambridge, p 881

IPCC (2001b) Climate Change 2001: summary for policymakers, a report of working group I of the Intergovernmental Panel on Climate Change. Cambridge University Press, Cambridge, p 20

IPCC (2007) Climate Change 2007: impacts, adaptation and vulnerability. Contribution of Working Group II to the Fourth Assessment Report of the Intergovernmental Panel on Climate Change. Cambridge University Press, Cambridge

Islam S (2003) Banglapedia: the national encyclopedia of Bangladesh. Asiatic Society of Bangladesh, Dhaka, Bangladesh

Kelly PM, Adger WN (2000) Theory and practice in assessing vulnerability to climate change and facilitating adaptation. Clim Change 47:325–352

Khan MSA (2008) Disaster preparedness for sustainable development in Bangladesh. Disaster Prevention and Management 17(5):662–671

Klein RJT, Nicholls RJ, Ragoonaden S, Capobianco M, Aston J, Buckley EN (2001) Technological options for adaptation to climate change in coastal zones. J Coast Res 17(3):531–543

Leal Filho W (2009) Communicating climate change: challenges ahead and action needed. Int J Clim Chang Strateg Manag 1(1):6–18

Luna EM (2001) Disaster mitigation and preparedness: the case of NGOs in the Philippines. Disasters 25(3):216–226

McCarthy JJ, Canziani OF, Leary NA, Dokken DJ, White KS (eds) (2001) Climate change 2001: impacts, adaptation and vulnerability, intergovernmental panel on climate change (IPCC), Work Group II Input to the Third Assessment Report. Cambridge University Press, Cambridge

Mirza MMQ (2002) Global warming and changes in the probability of occurrence of floods in Bangladesh and implications. Glob Environ Change 12(2):127–138

Moser C, Satterthwaite D (2009) Towards pro-poor adaptation to climate change in the urban centers of low-and middle-income countries. Paper presented at the Conference on Urban Poverty and Climate Change Infrastructure and Development, BRAC Center, Dhaka, Bangladesh, 26–28 January.

Pelling M, High C (2005) Understanding adaptation: what can social capital offer assessments of adaptive capacity? Glob Environ Change 15(4):308–319

Saroar MM, Routray JK (2010) Adaptation in situ or retreat? A multivariate approach to explore the factors that guide the peoples' preference against the impacts of sea level rise in Bangladesh. Local Environ 15(7):663–686

Saroar MM, Routray JK (2013) 'Climate Refugee' is not a hoax. But we can avoid it. Empirical evidence from the Bangladesh coast. In: Schmidt-Thome P, Klein J (eds) Climate change adaptation in practice: from strategy development to implementation. Wiley-Blackwell, UK, pp 283–301

Schmidt-Thome P, Klein J (eds) (2013) Climate change adaptation in practice: From strategy development to implementation. Wiley-Blackwell, UK

Smit B, Wandel J (2006) Adaptation, adaptive capacity, and vulnerability. Glob Environ Change 16(3):282–292

Smith JB (1997) Setting priorities for adapting to climate change. Glob Environ Change 7 (3):251–264

Stern N (2006) The stern review on the economic effects of climate change (Report to the British Government). Cambridge University Press, Cambridge

Thomalla F, Downing T, Spanger-Siegfried E, Han G, Rockström J (2006) Reducing hazard vulnerability: towards a common approach between disaster risk reduction and climate adaptation. Disasters 30(1):39–48

Tol RSJ, Klein RJT, Nicholls RJ (2008) Toward successful adaptation to sea-level rise along Europe's coast. J Coast Res 24(2):432–442

Twigg J, Bhatt M (1998) Understanding vulnerability: South Asian perspectives. IT Publications and Duryog Nivaran, London and Colombo

UNDP (2009) 2009 Global assessment report on disaster risk reduction: risk and poverty in a changing climate. UNISDR, United Nations, New York

van Aalst MK (2006) The impacts of climate change on the risk of natural disasters. Disasters 30 (1):5–18

Wilbanks TJ, Leiby P, Perlack R, Ensminger JT, Wright SB (2007) Toward an integrated analysis of mitigation and adaptation: some preliminary findings. Mitig Adapt Strateg Glob Change 12 (5):713–725

Wisner B, Blaikie P, Cannon T, Davis I (2004) At risk: natural hazards, people's vulnerability and disasters, 2nd edn. Routledge, London

World Bank (2000) Bangladesh: climate change and sustainable development. Report No. 21104-BD, Rural Development Unit, South Asia Region, Dhaka, Bangladesh

Chapter 3
The Application of the Macroeconomics Analysis of Climate Changes Model (MACC-Model) in China: Floods

Mario Arturo Ruiz Estrada, Ibrahim Ndoma, and Donghyun Park

Abstract Global climate change has a potentially large impact on economic growth but measuring such economic impact is subject to a great deal of uncertainty. Hence, this chapter examines the possibility of measuring the economic impact of climate change through the macroeconomics analysis of climate change (MACC) model, a new model to evaluate the impact of climate change on GNP growth. To give light to the model, a test run is carried out on China to evaluate the impact of the country's natural disasters on its economy.

Introduction

The issue of climate change and its ravaging impact on the socio-economic, environmental and political wellbeing of people and societies across the world is obviously well acknowledged. In fact, given the importance to avert its deleterious effects to the world we live in today, the debate on climate change is a front burner in the agenda of many world leaders. Apparently, climate change is inextricably linked to economic activities that drive economic growth and development.

Hence, with the economic benefits, comes the unavoidable depredation of the ecosystem that in turn affects the different facets of human livelihood. This is testament to the unprecedented and hostile nature of the climate in many parts of the world particularly over the last decade that witnessed record number of lives lost, destruction of property and a significant strain on the economy, all resulting from environmental disorders most often linked to climate change.

M.A.R. Estrada (✉) • I. Ndoma
Faculty of Economics and Administration (FEA), University of Malaya, 50603 Kuala Lumpur, Malaysia
e-mail: marioarturoruiz@gmail.com; marioruiz@um.edu.my; http://www.econonographication.com; ibrahim.ndoma@yahoo.com

D. Park
Principal Economist, ADB, 6 ADB Avenue, Mandaluyong City, Metro Manila 1550, Philippines
e-mail: dpark@adb.org

W. Leal Filho (ed.), *Climate Change in the Asia-Pacific Region*, Climate Change Management, DOI 10.1007/978-3-319-14938-7_3

Literally, climate change can be considered a natural disorder occurrence (cause) that is generated by natural evolutionary elements often linked to the high and competitive demand and extraction of natural resources in the production and consumption of goods and services. This process poses a significant strain on the ecosystem, which reacts with imbalances and irregular climatic conditions (effect) in different ecospheres (Ruiz Estrada 2013). In the same vein, it is without doubt that climate change has a significant impact on economic growth in light of the fact that it imposes both direct and indirect costs on the economy of a country, and those costs change and evolves over time.

Moreover, climate change adversely affects economic activities in the short run through a number of channels. For example, the record occurrence of floods in many parts of China severely curtailed the country's agriculture sector output following the destruction of plantations, forestry, fisheries, livestock, water resources, transportation systems, telecommunication systems, private and social infrastructure, and housing.

Beyond the very short term, however, the negative economic impact of climate change tends to fade when there is massive spending towards human and infrastructural development. The deadly floods and earthquakes that brought about the huge human and material losses in Central South China (Henan, Hubei, Hunan, Guangdong, Guangxi, Hainan) and Southwest China (Chongqing, Sichuan, Guizhou, Yunnan, Tibet) and the recovery efforts that followed between 1992 and 2012 is a good reference point. The government's reconstruction spending spearheaded a robust recovery in private investment and consumption. As a result, macroeconomic indicators recovered slowly and stabilized after its initial plunge.

Apparently, developing countries in particular are more vulnerable to the effects of climate change given their huge quest for economic growth and development, which is often accompanied by huge pollution levels and almost uncontrolled depredation of natural resources. In fact, developing Asia accounted for 55 % of global fatalities and 30 % of all persons affected globally by climate change between 2000 and 2012 (Asian Development Bank 2012). According to Table 3.1 shows the fatalities and estimated damages from various types of climate change in China between 2000 and 2012. The estimated damages imply a sizable negative economic impact on the region.

Given the effects of climate change on economic growth in developing countries, it is important that policymakers have reasonable and accurate estimates of its effects on a country's Gross Domestic Product (GDP) (Kunreuther and Rose 2004). However, measuring such impact to get an acute understanding of the degree of its effects on a country's GDP is subject to a great deal of uncertainty (Loayza et al. 2009). In essence, this chapter aims to examine and shed light on the effect of climate change on the economy of a country. Moreover, the chapter sets out to showcase a novel contribution to the study and economics of climate change by introducing the macroeconomics analysis of climate change (MACC) model, an economic instrument to evaluate the impact of climate change on GNP growth.

Hence, as means to illustrate and illuminate the MACC model, the climate change occurrences in China is dissected using the model to assess and evaluate

Table 3.1 The climate change growth rates (αi) and National climate change vulnerability rate (ΩT)

No.	Country	$\alpha 1$	$\alpha 2$	$\alpha 3$	$\alpha 4$	$\alpha 5$	$\alpha 6$	$\alpha 7$	$\alpha 8$	$\alpha 9$	$\alpha 10$	$\alpha 11$	$\alpha 12$	$\alpha 13$	$\alpha 14$	$\alpha 15$	$\alpha 16$	ΩT
1	China	0.95	0.35	0.99	0.75	0.15	0.99	0.35	0.25	0.3	0	0.25	0.95	0.25	0.1	1	0.95	0.54

$\boldsymbol{\Omega_i}$ = The climate change growth rates
$\alpha 1$ = Mean temperature
$\alpha 2$ = Temperature extremes
$\alpha 3$ = Mean precipitation
$\alpha 4$ = Precipitation extremes
$\alpha 5$ = Snow and ice
$\alpha 6$ = Carbon cycle
$\alpha 7$ = Ocean acidification
$\alpha 8$ = Sea level
$\alpha 9$ = El Niño
$\alpha 10$ = Monsoons
$\alpha 11$ = Sea level pressure
$\alpha 12$ = Radiative forcing
$\alpha 13$ = Tropical cyclones
$\alpha 14$ = Hailstorms
$\alpha 15$ = Sandstorm
$\alpha 16$ = Hurricanes and typhoons
High level of risk: 1 = Mean temperature, 2 = Mean precipitation, 3 = Carbon cycle
Source: Intergovernmental Panel on Climate Change (IPCC)

the economic impact of climate change on the country to give us a sense of the degree to which the problem affects the country's economic wellbeing. The model is based on five basic indicators viz. (i) the climate change growth rate (α_i); (ii) the national climate change vulnerability rate (Ω_T); (iii) the climate change magnitude rate (Π); (iv) the economic desgrowth rate (δ); and (v) the CC-Surface. In addition, the model is based on elements from an alternative mathematical approach analysis framework from a multidimensional perspective.

Economic Modeling in the Analysis of Climate Change

The economic dimension to understanding and analyzing the effects of climate change on economic growth is obviously not new. In fact, this approach has taken an evolutionary development over the years with significant improvement to the tools and analytical approach to the economic study of climate change. Hence, in this section, we review the classical and modern economic contributions to the study of climate change from where we find a place for the MACC model.

Classic Economic Modeling in the Analysis of Climate Change

The study and origins of the economics of climate change could well be credited to Cline's (1992) publication entitled "The Economic of Global Warning" and Reilly and Thomas's (1993) "Toward Economic Analysis of Climate Change Impacts: A Review and Analysis of Studies of the Impact of Climate Change". In these two publications, we locate the very earliest yet fundamental analyses on the impact of climate change from both microeconomic and macroeconomic perspectives. Subsequently, Harris and Roach (2002) in their book further expanded the discourse and debate by providing comprehensive analyses on the causes and consequences of climate change from a broader economic angle. In fact, Harris and Roach (2002: 25–50) summarized their main economic arguments on climate change as follows.

> Concern has grown in recent years over the issue of global climate change. The problem, frequently called global warming, is more accurately referred to as global climate change. A basic warming effect will produce complex effects on climate patterns—with warming in some areas, cooling in others, and increased climate variability. In terms of economic analysis, greenhouse gas emissions, which cause planetary warming, represent both environmental externalities and overuse of a common property resource. If indeed the effects of climate change are likely to be severe, it is in everyone's interest to lower their emissions for the common good. But where no agreement or rules on emissions exist, no individual firm, city, or nation will choose to bear the economic brunt of being the first to reduce its emissions. In this situation, only a strong international agreement binding nations to act for the common good can prevent serious environmental consequences.

While we are very much in consonance with Harris and Roach's (2002) contributions particularly with regards to policies and implications that usher in some fundamental points about climate change, Cline (1992) and Reilly and Thomas's (1993) contributions, however, remain the cornerstone in the study of economics of climate change. This so given their comprehensive analysis of the short and long term recovery model that made reference to the climate stabilization process involving the community back to the past economic level. Climate stabilization is defined and seen by these three main authors as "this should be the goal, rather than economic optimization of costs and benefits. Stabilizing greenhouse gas emissions is not sufficient, since at the current rate of emissions carbon dioxide and other greenhouse gases will continue to accumulate in the atmosphere.

Hence, stabilizing the accumulations of greenhouse gases will require a significant cut below present emission levels". It is important to mention that the short and long term recovery model formulation is based on the use of the cost benefit by using the equilibrium general circulation model (GCM), which runs at 2xC02 and gives different levels of C° by Manabe and Kirk (1969) to estimate the annual damages to any economy from global climate change. Also worth mentioning and useful to our study are research on "CETA: A Model for Carbon Emissions Trajectory Assessment" by Peck and Teisberg (1992) and "The Economics of Controlling Stock Pollutants: An Efficient Strategy for Greenhouse Gases" by Ita and Mendelsohn (1993). Hence, analyses of the above papers by Reilly and Thomas (1993: 15–35) brought about the following submission deemed critical to the study of economics of climate change.

"These two models developed provide more applicability in representing damages as non-linearly related to a single climate change indicator and they study the implications of damages that are linear, quadratic, and cubic in the climate variable. Peck and Teisberg also evaluate the case where damages are related to the rate of change rather than the level of climate change. If damages are related to the rate of climate change, the economically optimal level of control is less. If climate stops changing at any level, no more damages occur. In contrast, if the level of change matters, then the flow of damages accruing during each period continues to accumulate even if climate change is halted. To stop the flow of damages, climate change must actually be reversed. Viewing damages as related to the rate of change is consistent with a view that damages are due largely to adjustment, where slow climate change may have negligible effects even if the rate persists over many years.

In considering these different possibilities, Peck and Teisberg do not provide evidence for any particular damage function relationship. Their work only illustrates the importance of further research to clarify how damages can best be represented". In our opinion, building a model of this magnitude back in 1992 is obviously a significant effort and one deserving huge commendation indeed particularly in light of the limitation of database confined to simple observations. Without doubts, all the authors mentioned have made remarkable contributions given their futuristic views about climate change and its impacts.

The Macroeconomics Analysis of Climate Change Model

The macroeconomics analysis of climate change (MACC) model assumes that any country is vulnerable to the effects of climate change anytime and anywhere. Additionally, each climate change occurrence has its own level of potential damage and impact on the final GNP of any country. Hence, our world is in a constant dynamic imbalanced state. This means, at anytime and anywhere, there is the possibility of a climate change occurrence, which could generate different magnitudes of climate change effects. Climate change in the context of this model refers to any occurrence beyond human control that can generate massive destruction anytime, anywhere, without any advance warning. The quantification and monitoring of climate change is inherently difficult, and we cannot evaluate and predict them with any degree of accuracy, but we can compute series of climate change within a fixed period of time (per year or decades). In addition, the MACC model is useful for demonstrating how the GNP growth rate is directly connected to the presence of climate change.

Intrinsic to the MACC model are five new key indicators viz. the climate change growth rates ($\boldsymbol{\alpha_i}$), the national climate change vulnerability rate ($\boldsymbol{\Omega_T}$), the climate change magnitude rate ($\boldsymbol{\Pi}$) the economic desgrowth rate ($\boldsymbol{\delta}$) and the CC-Surface. These five indicators aim to simultaneously show the different levels of vulnerability and devastation arising from different climate change occurrence. The five indicators are determined by the collection of historical data of different climate change occurrence that have impacted any country, where climate change is defined according to certain intervals of time and magnitude. Based on our model, the analysis of any climate change from an economic point of view must take into account the production reduction (national output) and human capital mobility (labor) simultaneously. With this, we introduce a new concept called "economic desgrowth (δ)" (Ruiz Estrada 2010).

The economic desgrowth rate (δ) is defined as a leakage of economic growth due to any climate change. The main objective of this concept is to determine the ultimate impact of any climate change on the final GNP growth rate behavior over a certain period of time. The basic data used by the MACC model is based on the use of sixteen different possible climate change events. These include mean temperature; temperature extremes; mean precipitation; precipitation extremes; snow and ice; carbon cycle; ocean acidification; sea level; El Niño; monsoons; sea level pressure; radiative forcing; tropical cyclones; hailstorms; sandstorm; hurricanes and typhoons.

The National Climate Change Vulnerability Rate (Ω_T)

Based on the MACC model, we assume an irregular oscillation into different climate change events all the time. We do so by applying the climate change growth rates (αi), which is equal to the total sum of the same type of climate change event in the present year ($\Sigma\lambda_o$) minus the total sum of the same type of climate change event in the past 10 years ($\Sigma\lambda_{n\text{-}1}$) divided by the total sum of the same type of climate change event in the past 10 years ($\Sigma\lambda_{n\text{-}1}$) (see Expression 3.1).

$$\alpha_i = \Sigma\lambda_o\text{-}\Sigma\lambda_{n\text{-}1}/\Sigma\lambda_{n\text{-}1} \tag{3.1}$$

It means that our world is going to be in a permanent dynamic imbalanced state under high risk of having a climate change event at anytime. The MACC model allows for different magnitudes of climate change. Therefore, we have different climate change events growth rates (α_i) as described in Expression 3.2. Also, we assume that the national climate change vulnerability rate (Ω_T) is directly connected to time (T_j). At the same time, T_j is affected directly by different climate change growth rates (α_i). In our case, "j" is a specific period of time and "i" represents the type of climate change, which according to our classification and usage comprise of 16 different types. Hence, the national climate change vulnerability rate (Ω_T) includes a total of 16 possible climate change events that are as follows: mean temperature (α_1); temperature extremes (α_2); mean precipitation (α_3); precipitation extremes (α_4); snow and ice (α_5); carbon cycle (α_6); ocean acidification (α_7); sea level (α_8); El Niño (α_9); monsoons (α_{10}); sea level pressure (α_{11}); radiative forcing (α_{12}); tropical cyclones (α_{13}); hailstorms (α_{14}); sandstorm (α_{15}); hurricanes and typhoons (α_{16}) respectively. Each global climate change has its magnitude of intensity according to the geographical position and environmental problems.

We assume that if any climate change occurrence was observed to have had a wide interval before a subsequent occurrence, then it is not possible to be predicted with accuracy as in Expression 3.4. Hence, we can calculate the national climate change vulnerability rate (Ω_T), which is equal to the total sum of all α_i divided by the total of climate change in analysis (i_{total}) (see Expression 3.3). In our case, we are making use of sixteen different climate change variables.

$$\Omega_T = (\Sigma\alpha_i)/i_{total} \text{€} \, [0 < \Sigma\alpha_i < 1] i_{total} = 16 \tag{3.2}$$

$$\Omega_T e = \text{Ln}\left[(\alpha_i)_{Tj} - (\alpha_i)_{Tj\text{-}1}\right]/(\alpha_i)_{Tj}] \quad \forall \Omega_T e \neq 0 \tag{3.3}$$

$$\Omega_T p = \text{Ln}\left[(\alpha i_{max})_{Tj}\right] - \left[(\alpha i_{min})_{Tj})\right] \; 0 > \alpha i_{max} \leq 1 \text{ or } 0 \geq \alpha i_{min} < 1 \tag{3.4}$$

$$\Omega_T e \ddagger \Omega_T p \tag{3.5}$$

Expressions 3.3 and 3.4 show the effective national climate change vulnerability rate ($\Omega_T e$) and the potential national climate change vulnerability rate ($\Omega_T p$). The effective national climate change vulnerability rate ($\Omega_T p$) is based on compare the past and present climate change events growth rates. We assume that the present national climate change vulnerability rate Ω_T cannot be equal to zero (see Expression 3.3). However, the potential national climate change vulnerability rate ($\Omega_T p$) is based on the uses of a maximal and minimal climate change events growth rate into a determinate period of time (T_j) (see Expression 3.4).

Additionally, we need to assume that in the potential national climate change vulnerability rate ($\Omega_T p$) exist a random database which makes it possible for the MACC model to analyze unexpected results from different climate change events which cannot be predicted and monitored with the traditional methods of linear and non-liner mathematical modeling. Hence, the effective climate change events growth rate is identified in Expression 3.3.

Finally, our identity about the potential climate change event growth rate cannot be equal to the effective climate change events growth rate in the short run or long run (see Expression 3.5). This is because we assume at the very outset that our world is in a dynamic imbalanced state.

Thus Ω_T calculation in Table 3.2 is possible for observation by different countries by using different α_i and a single Ω_T. The analysis of the national climate change vulnerability rate (Ω_T) is applied to three different levels of vulnerability (see Expression 3.6)

$$\begin{array}{c} \text{Level 1}: \text{ High vulnerability (red color alert)}: 1 - 0.75 \\ \text{Level 2}: \text{ Average vulnerability (yellow color alert)}: 0.74 - 0.34 \\ \text{Level 3}: \text{ Low vulnerability (red color green)}: 0.33 - 0 \end{array} \tag{3.6}$$

However, in Fig. 3.1, it is possible to observe diminishing returns between the economic desgrowth rate (δ) and the national climate change vulnerability rate (Ω_T). We can have three possible scenarios of analysis in this relationship between the economic desgrowth rate (δ) and the national climate change vulnerability rate (Ω_T).

First scenario, if the national climate change vulnerability rate (Ω_T) is very high then the economic desgrowth rate (δ) will be high. Second scenario, if the national climate change vulnerability rate (Ω_T) is very low then the economic desgrowth rate (δ) will be low (see Figure 3.1). Finally, we assume that the national climate change

Table 3.2 Climate change magnitude rate of China floods in the year 1931 and 2010

Climate change magnitude rate of China floods in the year 1931 (Π_{1931})			(Π)
Φk 20b	80b	0.25	
ΨL 4 M	60 M	0.067	**−5**
Climate change magnitude rate of China floods in the year 2010 (Π_{2010})			
Φk 51b	375b	0.14	
ΨL 5000	30 M	0.000067	**−11**

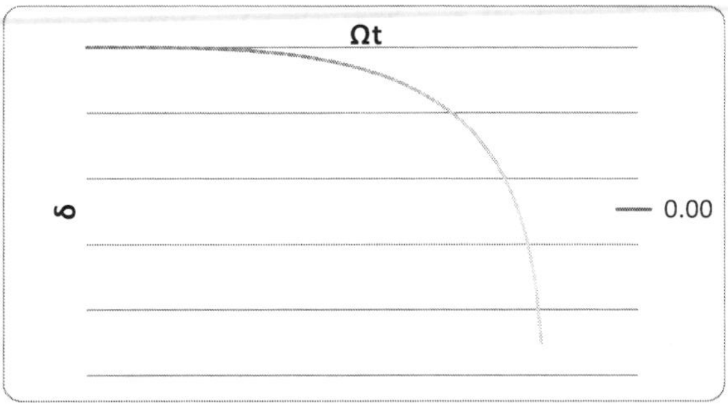

Fig. 3.1 The relationship between the national climate change vulnerability rate (Ωt) and the economic desgrowth (δ). *Source*: See Table 3.2

vulnerability rate (Ω_T) can never intercept the economic desgrowth rate (δ), because we use "The Dynamic Imbalanced State (DIS)". The DIS never keeps static but constantly keeps changing. Hence, we suggest the application of the Omnia Mobilis assumption to keep the DIS in the long run. It changes according to change in the national climate change vulnerability rate (Ω_T).

The Climate Change Magnitude Rate (Π)

Basically, we use two main variables to calculate the climate change magnitude rate (Π). The first main variable, which is capital devastation (Φk) is computed by dividing the area of infrastructure destroyed by the climate change (km^2) by total infrastructure area (km^2) in the same geographical space. The second main variable is human capital devastation (ΨL). We compute human capital devastation (ΨL) by dividing the number of people killed by or missing due to climate change by the total population in the same geographical space.

After calculating both main variables, we can then multiply the results to get our natural disaster magnitude rate (Π). In short, the climate change magnitude rate (Π) is equal to the product of the capital devastation (Φk) and the human capital devastation (ΨL). Finally, we generate the natural logarithm to calculate the final climate change magnitude rate (Π) that is expressed in the Expression 3.7.

$$\Pi = f(\Phi k, \Psi L) = \text{Ln}\left[(\Phi k) \times (\Psi L)\right] \tag{3.7}$$

We decide to apply the product rule of differentiation in Expression 3.7 to obtain the first derivative test to find the relative maximum and minimum in the capital devastation (Φk) and capital devastation (Φk) (see Expressions 3.8, 3.9, and 3.10).

$$\partial f/\partial(\Phi k) = \Phi'(k)\Psi L/\Phi(k)\Psi L \tag{3.8}$$

$$\partial f/\partial(\Psi L) = \Psi'(L)\Phi(k)/\Psi(L)\Phi(k) \tag{3.9}$$

$$\partial\Pi = \Phi'(k)\,\Psi(L) + \Phi(k)\,\Psi'(L) \tag{3.10}$$

Moreover, we can also observe that the climate change magnitude rate (Π) is directly proportional to the national climate change vulnerability rate (Ω_T).

The Economic Desgrowth (δ)

We define economic desgrowth (δ) as a macroeconomic indicator that shows the final impact of any climate change on the GNP (Ruiz Estrada 2010). We could say that the final GNP post-climate change effect is a function of the climate change magnitude rate (Π) (see Expression 3.11). At the same time, the climate change magnitude rate (Π) is directly dependent on the national climate change vulnerability rate (Ω_T) (see Expression 3.11) according to Figs. 3.1 and 3.2. In Expression 3.12, we calculate the preliminary GNP post-climate change effect (Q'). Hence, the Q' is in function of Π.

$$\Pi = f(\Omega_T) \tag{3.11}$$

$$Q' = f(\Pi) \tag{3.12}$$

The Climate Change Surface (CC Surface)

The construction of the CC-Surface is based on the climate change growth rates (Ωi) results and the mega-surface coordinate space (see Expression 3.13 and Fig. 3.2). The climate change vulnerability surface is a four by four matrix that contains the individual results of all 16 variables (taken from Table 3.1). However, the 16 variables are plotted in a four by four array with the vertical value on the CC-Surface.

The idea is to produce a surface for a quick pictorial representation of the overall propensities for any one country. The underlying idea here is to use the results of 16 variables in the climate change growth rates (Ωi) to build a symmetric surface.

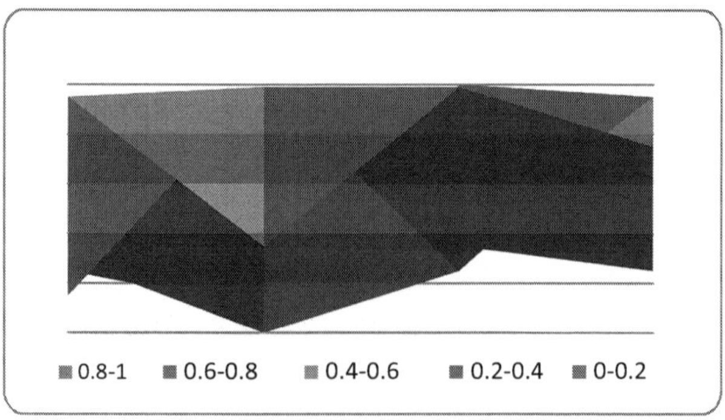

Fig. 3.2 CC-Surface. *Source*: Intergovernmental Panel on Climate Change–IPCC- data and MECC Model results

When the MD-coordinate system (η) has strictly the same number of rows as the number of columns, then the climate change growth rates (αi) can always be perfectly symmetric.

$$\eta = \begin{pmatrix} \alpha_1 & \alpha_5 & \alpha_9 & \alpha_{13} \\ \alpha_2 & \alpha_6 & \alpha_{10} & \alpha_{14} \\ \alpha_3 & \alpha_7 & \alpha_{11} & \alpha_{15} \\ \alpha_4 & \alpha_8 & \alpha_{12} & \alpha_{16} \end{pmatrix} \quad (3.13)$$

The final analysis of the CC-surface depends on any change that this surface can experience in a fixed period of time.

The Macroeconomics Analysis of Climate Change Model: The Case of China

Applying the MACC-Model to China's economy will give us a much better idea of how the model works. Before we do so, it is useful to have a closer look at the general local economic data about China. Such data includes the contribution of each of the country's region to the final GNP of the country as well as its geographical distribution of agricultural production. In terms of the geographical distribution of contribution to the country's GNP, we find that North China contributes around 12 % and East China with the highest share of 34 %, while Northeast China has the least contribution of 15 %. Therefore, the major contributors to the country's GNP are the Central South China and Southwest China regions that collectively account for 39 % of the country's overall output. While the Northeast region contributes 15 %, Central South China and East China account for about 57 % of the country's overall output. Similar trend could be observed with the country's agriculture output by regions where North China accounts for 12 % and Northeast China 10 %, East China 13 %, Central South China 30 %, and Southwest China 35 % respectively (see Fig. 3.3).

The Climate Change Growth Rates (α_i)

In this section, we first examine the natural disaster vulnerability propensity rate for me some countries around the world and then we take a closer look at China's natural disaster vulnerability propensity rate.

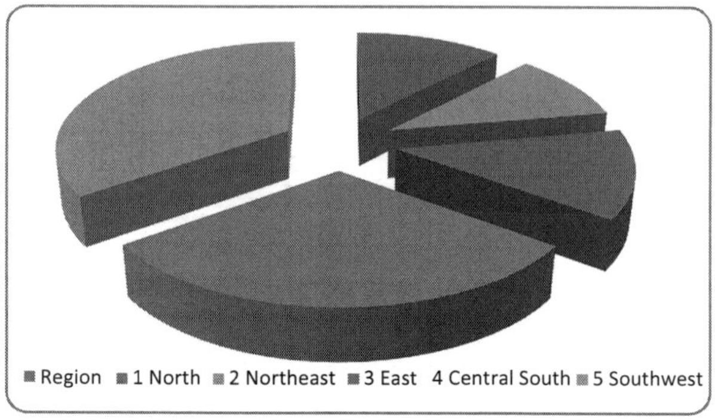

Fig. 3.3 The agriculture production concentration of China. *Source*: FAO and Ministry of Land and Natural Resources of China

China's Climate Change Vulnerability Rate (Ω_T): Max and Min

In the case of China, we find large differences between maximum and minimum climate change vulnerability rate (Ω_T). According to historical data of climate change, North has the lowest vulnerability, with a Ω_{Tmin} of only 0.15 and Ω_{Tmax} of 0.25. In the rest of China, the climate change vulnerability propensity rates are higher. More specifically, vulnerability rate ranges from 0.45 to 0.95 in East, from 0.35 to 0.95 in Southwest region, and from 0.25 to 0.85 in Northeast and Central regions (see Fig. 3.4).

The Climate Change Magnitude Rate (Π)

Here we would compare the climate change magnitude rate (Π) of floods in China between 1931 and 2010. Hence, the paper estimates and compares the magnitude of the impact of that climate change variable on China. According to our results, the devastation resulting from the China floods in the year 2010 was quite limited at −11. However, according to our computations below, the devastation caused by floods in 1931 were much larger at −5. Similarly, from a clear graphical perspective, we could observe in Table 3.2 and Fig. 3.5 that the China floods in 1931 caused much devastation several times larger than the floods in 2010.

The Economic Desgrowth (δ)

Finally, to measure the impact of the floods and temperature change on economic growth, we use the new concept of "economic desgrowth (δ)" introduced by Ruiz Estrada (2010). According to this concept, we try to discover possible leakages that can adversely affect GNP performance. Basically, this new concept assumes that in

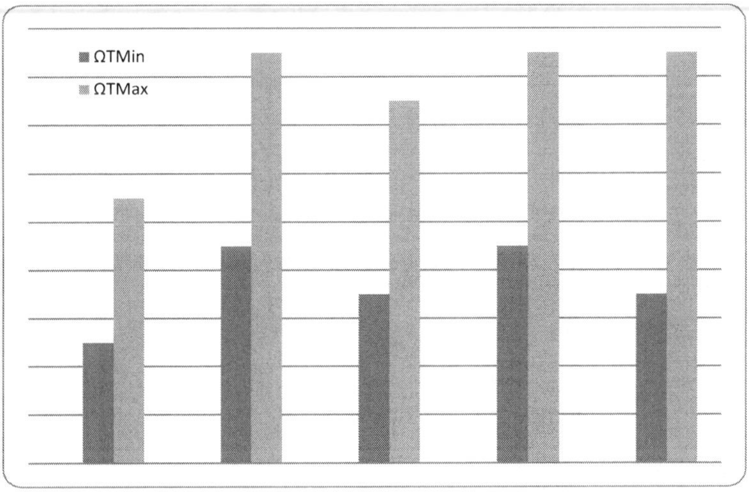

Fig. 3.4 The climate change vulnerability rate by region (China) (Ω_T). *Source*: MACC Model

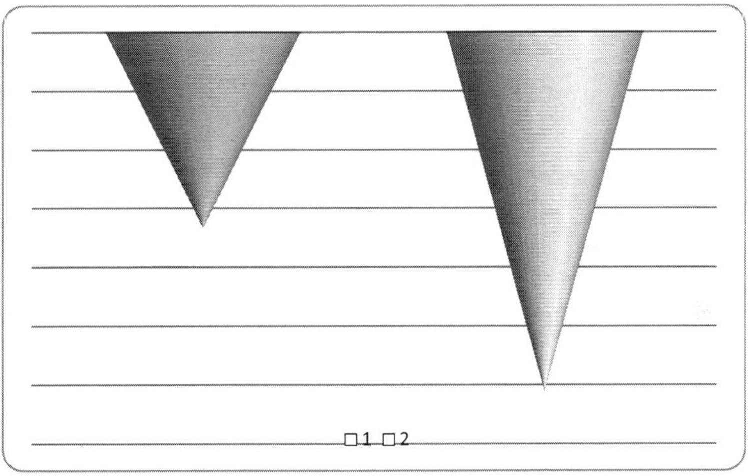

Fig. 3.5 Climate change magnitude rate (Π) between China floods 1931 and 2010. *Source*: See Table 3.2. Note: Final results from MACC Model

the process of the GNP formation, leakages may arise due to different factors, in our case, climate change. According to our estimates, the economic desgrowth caused by the Central South China floods in year 1931 has an impact of −1.51 on China's economic desgrowth (δ). Our estimates indicate that the economic desgrowth caused by the Central South China floods of 2010 has been much larger, at −2.8 in 2010 (see Table 3.3).

Table 3.3 GNP growth rates from China (1931 and 2010)

1	1931	3 % 1.49%	$\delta = -1.51$	$\Omega_T = 0.95$	$\Pi = -5$
2	2010	13.1 10.3%	$\delta = -2.80$	$\Omega_T = 0.99$	$\Pi = -11$
	Variables:				

δ = GNP desgrowth rate

ΩT = The national climate change vulnerability rate

Π = The climate change magnitude rate

Source: International Monetary Fund (IMF) and National Bureau of Statistics of China

Conclusions and Policy Implications

Climate change obviously has a significant negative impact on economic performance but measuring such impact with any degree of certainty is a herculean task. In this chapter, we propose a new model for evaluating the impact of climate change on economic performance. The macroeconomics analysis of climate change (MACC) model, which is based on five key five indicators viz. (i) the climate change growth rates (α_i); (ii) the national climate changes vulnerability rate (Ω_T); (iii) the natural disaster magnitude rate (Π); (iv) the economic desgrowth rate (δ); (v) and the CC-Surface.

The underlying intuition is that the economic impact of climate change depends on a country's vulnerability to temperature change and the floods devastation caused by climate change, which jointly determines the leakage from economic growth and hence, the impact on growth. We are of the belief that our model will contribute to a better and deeper understanding of measuring the economic impact of climate change.

A more useful measurement of the impact from climate change is conducive for appropriate policies, both for dealing with the effects of climate change and also for anticipatory policy measures, which seek to lessen the impact of climate change before they occur. For example, on the one hand, underestimating the impact may lead to the government allocating too few resources for addressing the impact of climate change—e.g. public investment in physical infrastructure and income support for households most affected by the climate change.

On the other hand, overestimating the impact may cause the allocation of too many resources, raising the risk of inefficiency and waste. By the same token, determining the appropriate level of anticipatory investments to limit the impact of future climate change would benefit from an accurate ex-ante assessment of their impact. The MACC Model can also help in determining the appropriate mix of climate change management and policies. For

(continued)

example, the model may allow policymakers to better estimate and compare the impact of different types of climate change occurrences.

The application of our model to two climate change occurrences in China – the floods of 1931 in Central South China and the Zhangshu and Jiangxi floods in year 2010 indicate that the Zhangshu and Jiangxi floods in 2010 will have a bigger impact than the Central South China floods of 1931. Nevertheless, the immediate implication for Chinese policymakers is that they need to support growth with stronger measures than as implemented in 2010.

In particular, they need to provide more fiscal resources for reconstruction efforts to re-build the region's devastated physical infrastructure, which in turn, will lay the foundation for the recovery of the region's productive activities, particularly manufacturing. In addition to rebuilding the infrastructure, the government should provide income support for the residents whose homes and livelihoods have been destroyed by natural disasters emanating from climate change. While China's high public debt level constrains the government's fiscal space, concerted fiscal support is nevertheless vital for China's floods recovery. At a broader level, our results confirm that climate change can have a significant economic impact even in advanced countries with good infrastructure and high level of preparedness.

Anticipatory measures can reduce the extent of climate change damage, loss of life and disruption to economic activity. Such measures include: (1) Good design and adherence to rigorous building codes; earthquake and storm proofing of buildings; floodplain and drainage designs; hillside stabilization, and other measures related to the natural and manmade environments, (2) Early warning system for floods, storms, epidemics, typhoons, tsunamis, and others. (3) Emergency response plans: evacuation systems; emergency response drills; equipment readiness; supplies storage - e.g. medicine and water. Given the high opportunity costs of using fiscal resources to mitigate the effects of climate change in developing countries, the MACC model's more accurate measurement of the economic impact of climate change is all the more valuable.

The failure of authorities to quickly and reliably inform the public led to widespread concerns and fear, which further dented consumer and business confidence. Therefore, more and better information is likely to reduce the impact of climate change, and looking at the role of information would contribute to a more accurate measurement of its impact.

References

Asian Development Bank (2012) Available at ADB: http://www.adb.org/data/statistics

Cline WR (1992) The economics of global warming. Peterson Institute for International Economics, Washington DC

Harris J, Roach B (2002) Environmental and natural resource economics: a contemporary approach. GDAE-Institute-Tufts University, Medford, MA
Ita F, Mendelsohn R (1993) The economics of controlling stock pollutants: an efficient strategy for greenhouse gases. J Environ Econ Manage 25(1):76–88
Kunreuther H, Rose A (2004) The economics of climate changes, vol I and II. Edward Elgar, Northampton, MA
Loayza N, Olaberria E, Rigolini J, Christiaensen L (2009) Climate change and growth: going beyond the averages. World Bank Policy Research Working Paper 4980
Manabe S, Kirk B (1969) Climate calculations with a combined ocean–atmosphere model. J Atmos Sci 26:786–789
Peck S, Teisberg TJ (1992) CETA: A model for carbon emissions trajectory assessment. Energy J 13(1):55–77
Reilly J, Thomas C (1993) Toward economic analysis of climate change impacts: a review and analysis of studies of the impact of climate change. MIT-CEEPR 93-009WP
Ruiz Estrada MA (2010) Economic desgrowth. Available at SSRN: http://ssrn.com/abstract=1857277 or http://dx.doi.org/10.2139/ssrn.1857277
Ruiz Estrada MA (2013) Economic vulnerability under the global climate changes analysis model (GCCE-Model): The case of China. Available at SSRN: http://ssrn.com/abstract = 2303707

Chapter 4
Kastom, Climate Change and Intergenerational Democracy: Experiences from Vanuatu

Kirsten Davies

Abstract The Republic of Vanuatu is a nation that is particularly vulnerable to the impacts of global climate change. It is a country with deep-rooted cultural practices, including customary (Kastom) Law that, throughout history, has protected the sustainability of its people, land, flora, fauna, rivers and oceans. This paper will describe the findings of a 2011 study conducted by Dr Kirsten Davies, involving 444 participants from the Island of Espiritu Santo, Vanuatu. A whole-of-community engagement methodology, known as Intergenerational Democracy, developed by the author, was adopted to capture snapshots of human-environmental relationships through the voices of children through to the elderly. The study was designed to assist a deeper understanding of community perceptions and observations pertaining to the impacts of climate change. As the Espiritu Santo community is becoming increasingly urbanised and less dependent on subsistence practices, valuable lessons can be learnt from these relational shifts. Focusing on adaptation, this paper discusses these shifts in a global context, as urbanised, transient communities become increasingly separated from the places of their ancestors. As the planet encounters the impacts of climate change, much can be learnt from the people of Vanuatu and their environmental and cultural connections.

Introduction

Global climate change (GCC) is threatening the stability of ecosystems and the maintenance of biodiversity across the world (Stern 2006; Costello et al. 2009; IPCC 2007a, 2012, 2013). It is widely agreed that human survival, in the face of the impacts of climate change, is dependent on mitigation and adaptation strategies

K. Davies (✉)
Faculty of Arts, Macquarie Law School, Centre for Environmental Law, Macquarie University, Balaclava Road, North Ryde, NSW 2109, Australia
e-mail: kirsty.davies@mq.edu.au

W. Leal Filho (ed.), *Climate Change in the Asia-Pacific Region*, Climate Change Management, DOI 10.1007/978-3-319-14938-7_4

(Costello et al. 2009; IPCC 2007a, 2012, 2013).[1] This is a challenging task, given the global community has reached unprecedented levels of urbanisation and migration (UNFPA 2007). People have increasingly become separated from the lands of their ancestors, resulting in chains of environmental knowledge and behaviour, once handed down through generations, being severed (Inda and Rosaldo 2002).

The Intergovernmental Panel on Climate Change (IPCC) has defined climate change adaptation as the "adjustment in natural or human systems in response to actual or expected climatic stimuli or their effects, which moderates harm or exploits beneficial opportunities" (IPCC 2007a). The IPCC noted that few plans "have explicitly included either adapting to climate change impacts or promoting adaptive capacity" (IPCC 2007b). Adaptation is underpinned by the understanding that human survival and natural ecosystems are intrinsically connected (Shiva 2006). As highlighted by Shiva (2006), it is important to learn from communities, such as those in Vanuatu, that have traditionally managed natural resources as 'living cultures'.

This paper explores coupled human and nature relationships, through the voices of the community of Espiritu Santo, Vanuatu, as they encounter and consider the impacts of GCC. A unique whole-of-community engagement method, known as Intergenerational Democracy (ID), developed by the author in 2006, was applied to this study. "Intergenerational Democracy is a whole-of-community method of engagement and participation that requires the inclusion of citizens representing all ages (from 8 to 100+ years). The foundation of ID is embedded in the principles of direct democracy and human rights. ID recognises that there are many quieter but equally legitimate voices, particularly those of children.... rarely heard in policy and planning forums. Through its age-based methodology ID enables the application of intergenerational equity which is at the heart of environmental sustainability. ID cuts through barriers of inequality, by engaging, connecting and motivating communities in planning and managing their sustainable futures" (Davies 2012 p. xxiii).

The attitudes, observations and traditional practices of citizens of all ages from Vanuatu, will add a depth of knowledge to the international discourse surrounding future adaptation approaches to the impacts of climate change.

The Urgent Problem

International concern surrounding GCC has been escalating over the past 20 years! The United Nations (UN) Framework Convention on Climate Change (UNFCCC) was adopted in 1994 in response to this challenge as an important tool for guiding the international community. The UNFCCC was concerned with setting targets to

[1] Threats to human survival resulting from the impacts of GCC on ecosystems include: changing patterns of disease, water and food insecurity, extreme climatic events, coastal erosion and flooding, world famines and population migration (Costello et al. 2009).

limit global temperature rises and mitigate the associated effects of GCC (UN 1992, 2012). The Kyoto Protocol, enacted in 2005, sought to address shortcomings identified in the original UNFCCC Agreement. Establishing its first commitment period from 2008 to 2012, the Protocol established legally binding targets to reduce Green House Gas (GHG) emissions to a global average of 5 % emissions compared with 1990 levels (UN 2012).

The IPCC was established in 1988 jointly by the United Nations Environment Program (UNEP) and the World Meteorological Organisation (WMO), with the aim to providing clear scientific updates on climate change and its impacts, both environmental and socio-economic, to the global community. In 2013, the IPCC released its Fifth Assessment Report (Cubasch et al 2013) which highlighted the need for strengthening climate change adaptation responses together with a reduction in greenhouse gas emissions. The Report described, since the previous IPCC Report in 2007, an increase to a 95 % confidence level, confirming the anthropocentric cause of GCC (Cubasch et al 2013). The Report has forecast within a century, the global surface temperature will increase between 0.3 and 4.8 °C, with a mean increase of two degrees. It predicted that a current reduction in carbon dioxide emissions will not reverse many of the expected changes in climate for centuries, and reinforced the urgent need for the international community to focus on adaptation.

Vanuatu and Climate Change

Brown Weiss (2008) noted "the impacts from warming are predicted to be long-term, widespread and severe... Developing countries will very likely suffer the worst effects from climate change because they have least resilience and capacity to adapt" (p. 616). Indeed the Vanuatu Archipelago and its inhabitants are already experiencing the effects of climate change. Rising sea levels and intensified extreme events, such as droughts and tropical cyclones are examples of environmental change occurring across the region. Incremental changes in the underlying climate are impacting on day-to-day living. Farmers are observing an increase in weed invasions and plant and human disease (e.g. vector borne disease) outbreaks are more prevalent. Both can be linked to increasing temperatures and shifts in precipitation regimes that now exceed sensitive thresholds (VNACCC 2009). More than 75 % of Vanuatu's settlements are located in coastal areas and are experiencing coastal erosion problems. Furthermore, the Vanuatu Department of Fisheries reported that outbreaks of fish poisoning due to *sigatora* are highly related to climate variability (BOM 2006).

Vanuatu is a nation committed to addressing the impacts of GCC. In 1993 The UN Framework Convention on Climate Change (UNFCCC) was ratified by Vanuatu. In 1999, Vanuatu's first Initial National Communication (INC) to the UNFCCC was submitted. Since then, Vanuatu has commenced the process of creating an institutional framework that positions climate change issues within national legal frameworks. Vanuatu's INC presents compelling evidence that it is one of the nation's most vulnerable to climate change and rising sea-level. Vanuatu's commitment to addressing climate change is also demonstrated by its being a Party to UN conventions. In 2012 the Government developed the National Advisory Board on Climate Change and Disaster Risk Reduction (NAB). Its key role is to: "Act as Vanuatu's supreme policy making and advisory body for all disaster risk reduction and climate change programs, projects, initiatives and activities" (NAB 2012).

Kastom

The framework in which environmental management has been historically practiced, and is still alive today, is known as 'Kastom' in Vanuatu, or more commonly as customary law in other jurisdictions. A fundamental principle of international law is customary law, which has its genesis in the principles of natural law. Both natural law and customary law are founded in human relationships with natural systems. International customary law is defined by the International Court of Justice Statute (Art. 38(1)(b)) as "evidence of general practice accepted as law" and is determined by the general practice of states and the acceptance of law by states.

The Republic of Vanuatu offers the critical attributes in which a study of customary law and global climate change adaptation, should be positioned. The nation has a deep history of recognising the significance of customary law as evidenced in the nation's (1980) Constitution, a combination of customary, French civil and English common law. Vanuatu's Constitution reflects the commonly adopted sustainability principle of intergenerational equity (Brundtland 1987), as evidenced by its aim: "to protect the Republic of Vanuatu and to safeguard the national wealth, resources and environment in the interests of the present generation and of future generations" (Chapter 2, Part II (7)(d)). 'Kastom', incorporates the religious beliefs and traditional Melanesian customs that play an integral part in the day to day lives of most Vanuatu citizens. Kastom informs environmental protection, attitudes, values, family and community structures, behaviour and participation (ESTA 2010; Cassidy 2004; Romulus and Lucas 2000). It is practiced within the Constitution through the nation's Chiefs, recognised as the National Council of Chiefs (RV 2007, Chap. 5).[2]

[2] Chapter 8 of the Constitution is concerned with the Justice System. It describes how the Judicial Service Commission includes a representative of the National Council of Chiefs and that "Parliament shall provide for the establishment of village or island courts with jurisdiction over customary and other matters and shall provide for the role of chiefs in such courts." (Art. 52). Chapter 12 of the Constitution is dedicated to land ownership and is particularly critical to the role of Traditional Law. Article 73 states "All land in the Republic of Vanuatu belongs to the Indigenous custom owners and their descendants." And "The rules of custom shall form the basis of ownership and use of land in the Republic of Vanuatu" (Art. 74).

The Tabwemasana Project

The Tabwemasana Research Project (Davies 2012) was conducted from 2010 to 2011 in the Republic of Vanuatu on the island of Espiritu Santo (Santo), which is the largest in the nation's archipelago of 83 islands. The study derived its name (with permission from the local Chiefs) from the highest mountain in Vanuatu, Mt Tabwemasana, located on the island of Santo.

The study was funded and supported by an Australian Government Endeavour Research Fellowship Award and conducted in partnership with Vanuatu Earth Care Association (VECA) and the University of the South Pacific (Luganville campus) after attaining a research permit from the Government of Vanuatu. The project had the endorsement and involvement of local Chiefs and Councillors.

At the time of the study the total population of Vanuatu was 243,304 and Santo 34,388 (VNSO 2009). The nation's population was largely comprised of young people, with 41 % of the population aged 0–15 years (VNSO 2009). Attaining snapshots of human–environmental relationships from the past and the present were key threads in this study. The social impacts of GCC were of special interest as the study examined community observations, attitudes and perceptions pertaining to GCC. Capturing the views of young people was prioritised. It was hypothesised that as the Santo community becomes increasingly urbanised and less dependent on subsistence practices, valuable data could be collected from these relational shifts, particularly in respect to Kastom Law (VCC 2007; Romulus and Lucas 2000; WHC 2010) and the impacts of GCC.

Methods

The study adopted a mixed method approach through individual interviews, focus groups and the distribution of a survey in the urban location of Luganville and rural village of Port Olry. Survey instruments and information materials were translated in to Bislama, English and French. Interpreter services were usually required to assist each focus group and interview. The whole-of-community method, titled Intergenerational Democracy (ID) was adopted, requiring the involvement of citizens from eight years of age to the elderly. Participants were recruited through purposeful selection, assisted by the Director of VECA, local chiefs and the University of the South Pacific. Young people participated in the study through local schools.

Quantitative and qualitative data was collated and analysed thematically, by age, gender and rural versus urban responses.

Limitations

The limitations of this study included language barriers, gender, engaging the elderly, literacy levels, 'island time' and challenges with difficult terrain to access. With the many language groups spoken in Espiritu Santo, the researcher was reliant on translators to interpret participant's responses. Some content and meaning may have been lost in translation. Issues relating to gender inequality in Vanuatu have previously been documented. When the researcher attempted to run a women's only focus group in one of the study locations, men insisted on being in attendance, limiting the women's freedom to speak out on issues that concerned them. There were difficulties accessing the elderly as aged care institutions do not exist in Santo. It was easier to contact elderly people in the rural village as they were more accessible. Many of the citizens participating in the study had low literacy levels especially the elderly. To ensure their inclusion in the study, students from the University of the South Pacific completed their survey forms on their behalf as they dictated their responses. While this was not a preferable approach, it did enable their inclusion in the survey component of study.

There are different concepts of time a comparison is made of a Western understanding of time with 'island time' in Vanuatu. This made it difficult for the researcher to make appointments for interviews and focus groups and added to the time required to conduct the research. Conversely 'Island time' also meant that once the researcher had engaged with participants they generously gave her ample time. Initially it was planned that one of the research localities would be on the western coast of Espiritu Santo. This area is only accessible by boat and high seas prevented this localities inclusion in the study.

Results

There was 444 participants in this study, of which 237 (53.4 %) were male and 191 (43.0 %) female (Table 4.1).

The majority of participants in this study were young people. A total of 158 participants (35.6 %) were aged from 11 to 20 years, 94 (21.2 %) were aged from 21 to 30 years, 63 (14.2 %) were aged from 31 to 40 years and 45 (10.1 %) were aged from 41 to 50 years (Table 4.2).

There was a relatively even distribution of participants from urban and rural localities. A total of 207 participants (46.6 %) came from rural areas, and 237 (53.4 %) were from urban areas (Table 4.3).

Table 4.1 Number and gender of participants

Gender	Interviewees	Focus group participants	Survey respondents	Total
Male	21 60.0 %	61 46.2 %	155 56.0 %	237 53.4 %
Female	13 37.1 %	61 46.2 %	117 42.2 %	191 43.0 %
No Response	1 2.9 %	10 7.6 %	5 1.8 %	16 3.6 %
Total	35 100.0 %	132 100.0 %	277 100.0 %	444 100.0 %

Table 4.2 Participants by age group

Age group in years	Interviewees	Focus Group participants	Survey respondents	Total
0–10	1 2.9 %	27 20.5 %	5 1.8 %	33 7.4 %
11–20	12 34.3 %	84 63.6 %	62 22.4 %	158 35.6 %
21–30	4 11.4 %	4 3.0 %	86 31.0 %	94 21.2 %
31–40	8 22.9 %	2 1.5 %	53 19.1 %	63 14.2 %
41–50	5 14.3 %	0 0.0 %	40 14.4 %	45 10.1 %
51–60	3 8.6 %	0 0.0 %	12 4.3 %	15 3.4 %
61–90+	2 5.7 %	0 0.0 %	11 4.0 %	13 2.9 %
No Response	0 0.0 %	15 11.4 %	8 2.9 %	23 5.2 %
Total	35 100.0 %	132 100.0 %	277 100.0 %	444 100.0 %

Table 4.3 Participant distribution based on locality

Type of locality	Interviewees	Focus group participants	Survey respondents	Total
Rural	9 25.7 %	96 72.7 %	102 36.8 %	207 46.6 %
Urban	26 74.3 %	36 27.3 %	175 63.2 %	237 53.4 %
Total	35 100.0 %	132 100.0 %	277 100.0 %	444 100.0 %

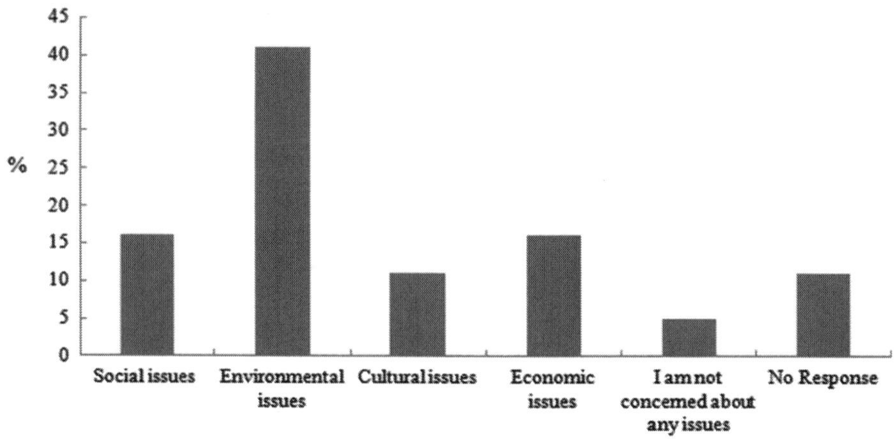

Fig. 4.1 Participant prioritisation of areas of concern ($N = 444$)

Area of Highest Concern

In the quantitative data, when asked to identify the area they were most concerned about relating to Santo, the majority of respondents nominated 'the environment' (Fig. 4.1). 180 respondents (40.5 %) ranked environmental issues as their highest area of concern. Economic and social issues were ranked equally second (16.0 %). Cultural issues were ranked third (11.5 %).

Respondents aged from 11 to 50 years had the highest levels of environmental concern when compared to other age groups (Fig. 4.2). Those aged from 51 to 60 years were the most concerned group regarding social issues. Children aged 0–10 years were equally concerned about cultural and environmental issues.

Notable differences in responses by gender related to cultural and economic issues as men were more concerned than women about both of these areas. Men and women were equally concerned, (40.7 % females, 40.4 % males) about 'environmental issues' as their main concern. Urban respondents (43.9 %) were more concerned about environmental issues than rural citizens (36.7 %). Urban respondents (17.7 %) were also more concerned about social issues when compared with rural citizens (14.0 %).

In the qualitative data, climate change was prioritised as the most important environmental issue for Espiritu Santo. Interviewees and participants described elevating levels of concern surrounding extreme weather events such as: tidal waves, cyclones, earthquakes, tsunamis and changing weather patterns. They described the indirect effects of less defined seasons and how these had impacted on food security and erosion. An interviewee noted that "Planting seasons have changed and now they don't know when to plant various crops so find it really hard to grow sufficient crops. Laplap is dry and dying so that we cannot use it anymore and some insects are eating them. Sometimes it rains a lot so new rivers and creeks

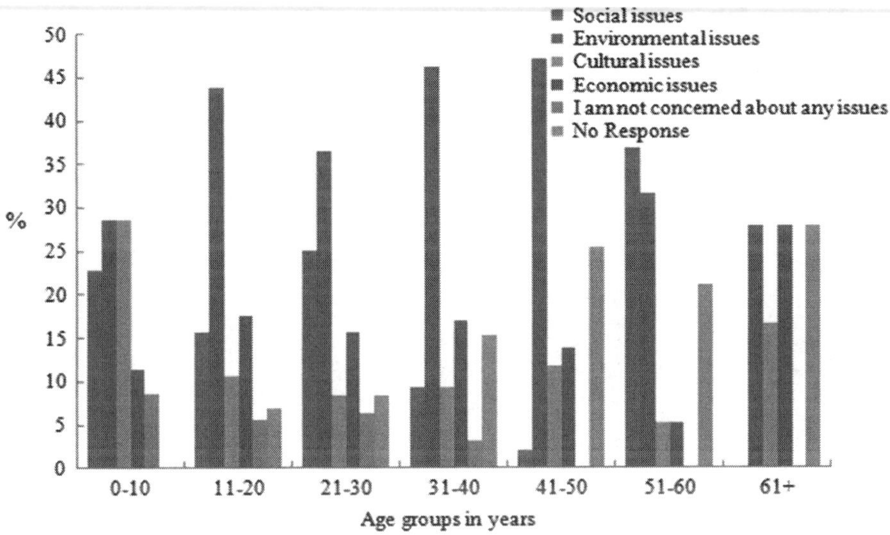

Fig. 4.2 Prioritisation of participant concerns by age group ($N = 444$)

form from nowhere." And "I think the weather pattern is quite unpredictable. It is an issue. . .. spices such as vanilla was a very good crop for communities in Sanma, but with the irregular climate patterns they are not producing and this affects the communities."

Community Concern About Climate Change

The survey measured respondents' levels of concern surrounding climate change (Fig. 4.3). Almost one third of respondents (29.3 %), described their attitude as 'very concerned'. Ranked second was 'concerned' (23.9 %) and third 'extremely concerned' (19.1 %). Very few people were either 'marginally' (11.0 %) or 'not concerned' (7.2 %) about climate change. More older respondents, particularly those aged upward from 51 years of age, were 'very concerned' about climate change. It was notable that children aged from 0 to 10 years were the highest 'extremely concerned' group.

More males were 'very concerned' or 'extremely concerned' when compared to females. A significantly larger cohort of urban respondents was either 'extremely concerned' or 'concerned' about climate change when compared with rural responses. One interviewee said, "I have already seen the impacts of climate change. These are all negative impacts. It affects people, animals and the environment. I am very concerned."

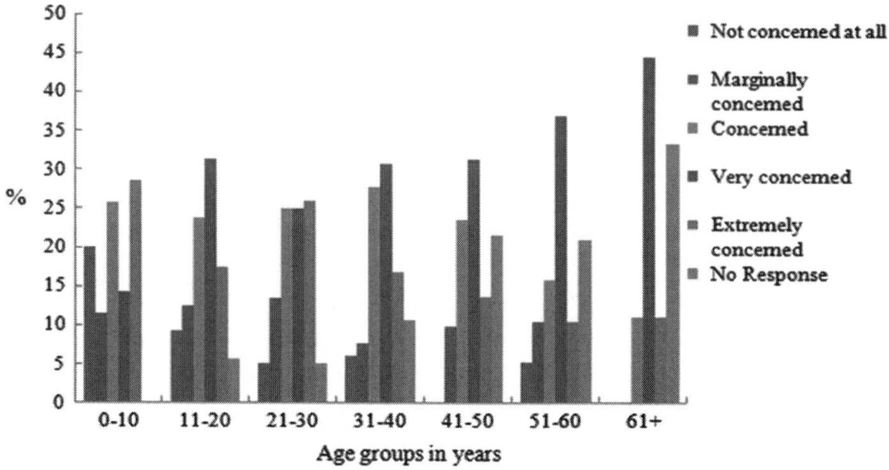

Fig. 4.3 Participant levels of concern about climate change by age group ($N = 444$)

Timeframe of Climate Change Effects

Respondents were asked if they thought the effects of climate change were a problem 'now', or into the future (Fig. 4.4). More than half (54.1 %) responded that they believed it was a problem 'now'. The next highest response (17.8 %) was 'in the next 50 years'. A small group of respondents (7.7 %) believed that climate change wouldn't be a problem. The immediacy of the problem was prioritised by respondents from all age groups.

Responses according to gender were equitably spread across most options with 110 (55.3 %) females and 130 (53.1 %) males believing that climate change was a concern 'now'. When comparing urban and rural responses, more urban residents believed that climate change was a problem 'now'. 59.5 % of urban people said that climate change was a problem 'now', compared with 47.8 % of rural residents. An interviewee said, "Now when we sleep we barely have our blankets on. Before we used to have thick blankets [at this time of the year]."

Changes Attributable to Climate Change

When asked if they had noticed any changes attributable to climate change 73.9 % of people responded 'yes', 11.3 % said 'no' and 4.7 % 'I don't know' (Fig. 4.5). These responses were consistently high across all age groups with the strongest representation from 0 to 50 years. Affirmative responses were found to be less in those aged from 51 to 61+ years and it was notable that these age groups had a significantly larger representation of 'no response' to this question.

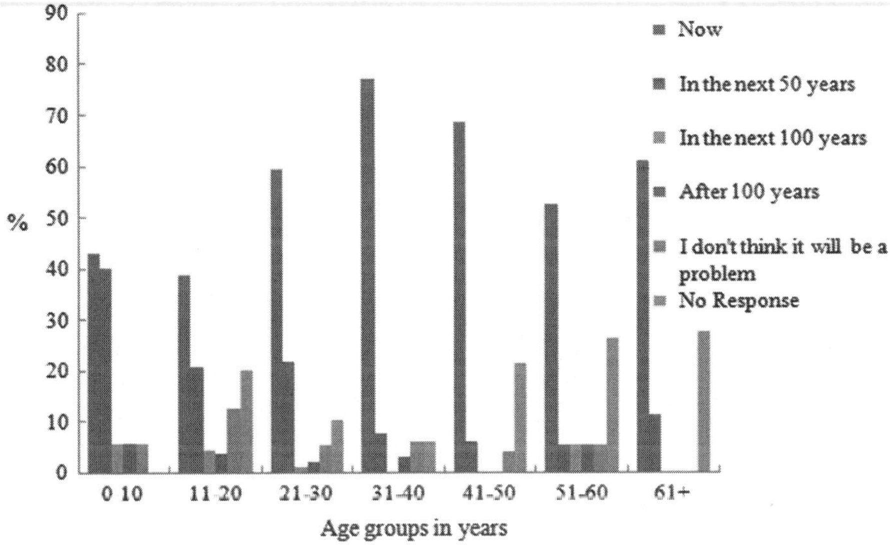

Fig. 4.4 Participant concern about timeframe of climate change effects by age group ($N = 444$)

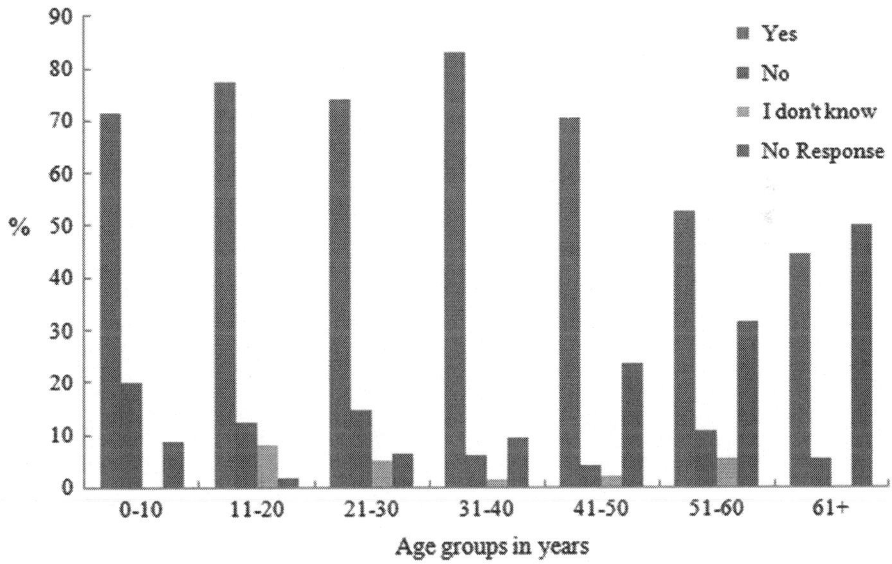

Fig. 4.5 Changes perceived to be attributed to climate change by age group ($N = 444$)

84.4 % of urban residents compared to 61.8 % of rural residents responded that they had noticed changes which they attributed to climate change. A notable number (34.8 %) of rural residents did not respond to this question compared to 19.0 % of urban residents.

Areas of Impact

Survey respondents were asked to describe aspects of Santo they thought had changed due to climate change (Fig. 4.6). 44.8 % of people ranked 'the sea' as their top priority. A close second was 'climate and weather patterns', which was nominated by 34.5 % of respondents. The third area was 'the land' which was nominated by 21.2 %. More people in older age groups (31–61+) observed changes in 'the sea' and 'climate and weather patterns' when compared with responses from younger respondents. A notably high number (26.8 %) of people did not respond to this question. More men (43.7 %) prioritised 'the sea' than women (39.7 %). More males (24.1 %) identified changes in 'the land' than females (17.6 %).

Extreme weather conditions and changes in climatic patterns were of significant concern to interviewees and participants. These were identified as: rising sea level and increased tidal waves, earthquakes, tsunamis, precipitation, dry period's coupled with unusual temperature variations and seasonal changes. A participant

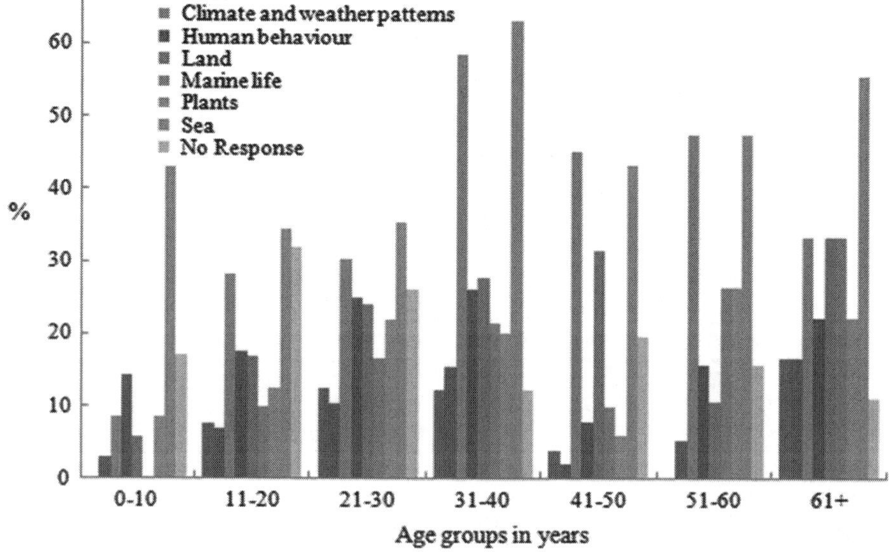

Fig. 4.6 Observations of changes to the environment by age group $N = 444$

said, "People have been moving away from the sea [they are concerned about] tsunamis and sea level rising."

As most villages are coastal, rising sea levels were a significant concern for participants. They described how communities had been forced to relocate, losing their cultural connections to the land, and severing traditional linkages of cultural heritage and knowledge. The importance of this was highlighted by a woman who said, "Culturally people depend on nature for survival. Way back in history we have passed knowledge from generation to generation... we have been nurtured by the land so it is like a mother to us." Another said, "Some of them have moved out for their original places of their ancestors. They had been living there for most of their lives."

One interviewee observed changes in sea level when he said: "At some places, when I was a student, the water stopped at a different place but now it keep [sic] coming up on the land and some places different to when I was a kid..." Another interviewee from the Big Bay area said, "[Sea level rising is a] Big problem. At one time I didn't see all of this but now we see it is becoming a great threat to the generations that are up and coming. Like in Sanma sometimes people live close to the river and now they have to move."

Participants described rising sea levels as eroding coastal areas, and often the most fertile lands were being lost, thus increasing the risk of food insecurity. A community leader from Luganville said, "The coastal and inland erosion by [the] sea and rivers is becoming more noticeable and frightening for low-lying areas where farms and communities are located. The implications for agricultural practices cannot be overlooked especially with the fast increase of the country's population." Another said, "All around parts of Santo they have experienced the sea level rise. They have the coconut plantations which are being destroyed by the sea. Also the infrastructure close to the sea is being destroyed. Coconut plantations would have been there for over 30–40 years."

The concern surrounding changes in weather patterns was heightened due to the impacts on agriculture resulting in decreased productivity. This concern for food security was expressed by a member of the Port Olry community who stated, "Plant it in the proper dry season, for when the plant should grow, and then it rains so that the plant doesn't grow, so now we don't know when to plant or what will grow when, so we don't have enough food." Another said, "People getting hungry as there is not enough food, as... production....is reduced, due to the changing seasons. [They] can't pay school fees as they cannot produce food to sell." Food security was a significant concern socially, economically, and culturally. The consequence of decreased productivity resulted in citizens needing cash to buy food for their families.

People noted less demarcation of the seasons. One interviewee noted, "...before there used to be distinct seasons like [the] windy season or rain season... but all of a sudden the seasons are changing and mixing, like it will rain for weeks in the dry season. I foresee a lot [of] flow [on] effects due to weather pattern changes." They described how higher levels of precipitation had multiple consequences including an increase in mosquitos and vector borne diseases such as malaria. An interviewee said: "When it rains a lot... there are more mosquitoes and malaria."

Kastom: Traditional Ways in a Contemporary World

In the qualitative results it was found that the majority of participants believed that Indigenous knowledge could assist in the future sustainable management of the community and environment. They described how this could occur, through strengthening and empowering chiefs to practice Kastom. Interviewees believed that the chiefs needed to continuously evolve their practices to accommodate the contemporary world including technological progress and systems of governance so they are "paddling with two hands." A community member said, "before we relied on Kastom. Now with mobile phones, ideologies within the community is becoming separated." Many interviewees, including chiefs, believed that the optimal pathway for the future was selecting the best from both worlds. They described the importance of 'Tabu' which is part of Kastom involving a chief placing a ban on activities such as fishing in a location, as an environmental protection method. This was explained by interviewees who said, "tabu, stops people from touching [plants], to keep it, to preserve, to make it important. Some of the plants and animals we have to keep." And "If you are always fishing in one place you destroy it, if you want to preserve something you must go some other place, need to move around to make sure there are enough fish. Make a farm for fish so that the fish can grow else they are always small. You must make some areas tabu so that the fish can grow so that you have got enough. Always money, focusing on money so they are fishing in one spot and not moving destroying the sea."

They viewed Kastom as a fundamental to ensuring the sustainable use of natural resources at a local level. However, Kastom is reliant on the community respecting the chiefs, a practice which has changed with chiefs having decreasing influence. This was occurring for a variety of reasons including internal migration where people, particularly young people, were leaving villages and relocating to larger centres such as Port Vila.

Participants stated that they believed education systems needed to have increased emphasis on Kastom, due to its important teachings about interactions with the environment. One interviewee raised this concern: "the way of our forefathers are time-tested and still applicable today but this is not being preserved through education curriculum." Another interviewee said that "everyone needs to be educated so that everyone can be involved and make a difference."

Discussion

This study identified that people in Espiritu Santo are highly concerned about the state of the environment. When compared with other sectors (e.g. social, cultural and economic) 'the environment' was found to be a significantly higher level of concern. Children to 50 year olds were found to have the highest levels of environmental concern. This result can be explained as people of all ages described their escalating levels of concern surrounding the direct and indirect impacts of GCC on the environment. Children were found to be the highest group who were 'extremely concerned', highlighting their fears for the future.

Participant's responses were based on their perceptions, observations and experiences that climate change was impacting their livelihoods and safety. They were concerned about sea level rise, changing climatic patterns and extreme weather events such as: tidal waves, tsunamis, cyclones and earthquakes. They described the indirect effects, such as less defined seasons that impacted on diminishing agricultural production, decreased income and food for their families due to prolonged droughts and/or increased rainfall. Other indirect impacts were an increase in vector borne diseases and erosion. According to the literature their concerns were justified as Vanuatu is a nation at the forefront of the impacts of climate change (BOM 2006; Brown Weiss 2008).

Most respondents believed that climate change was a problem 'now'. They prioritised 'the sea' and 'climate and weather patterns' as the main areas where they had observed changes. As Vanuatu's villages are mostly situated in coastal locations and rely on fishing as a food source (BOM 2006), it is logical they would have observed changes predominantly in the sea. More people in older age groups observed changes in 'the sea' and 'climate and weather patterns' when compared with responses from younger respondents. Older citizens had the advantage of time over young people, where they could observe environmental changes and draw from their memories.

Participants stressed the need to embrace the advancements of the contemporary world in areas such as technology, including the uptake of mobile phones. Equally they described the importance of continuing and strengthening traditional environmental management practices, particularly as communities adapt to changing landscapes due to the impacts of climate change. The need to plan for a sustainable future was articulated frequently and is supported by the foundation of sustainable development, intergenerational equity (Brundtland 1987). This was encapsulated by a young participant who said that people need to look after, "...their physical environment and making the most out of it for themselves as well as remembering to leave something intact for their future generations... Also learning to adapt intelligently to the modern changes that seem to be overtaking them in rapid succession."

People described the effectiveness of traditional practices of environmental protection managed by the chiefs through Kastom, a view that was supported through the literature (ESTA 2010; Cassidy 2004; Romulus and Lucas 2000). They raised their concerns surrounding the diminution of power of the chiefs which was compounded by trends of urbanisation as people, especially young people, are leaving the places of their ancestral heritage to find work in the larger towns and cities. This trend is a global phenomenon, not unique to Vanuatu (UNFPA 2007). Citizens explained how local chiefs monitor and govern environmental health 'on the ground' through Kastom which includes 'tabu' or a banning of detrimental activities. Such a localised approach to protecting life supporting ecosystems, and managing human interventions with nature, can assist the process of adapting to the impacts of climate change (Inda and Rosaldo 2002).

The literature highlighted how the Government of Vanuatu has formally recognised the importance of customary law (Kastom), as reflected in the nations Constitution and the governance roles of the Councils of Chiefs (RV 2006, Chap. 5). This model provides an exemplar for nations considering how the cultural heritage and environmental knowledge of Indigenous peoples can be incorporated in to future frameworks for Climate change adaptation.

This project provided the opportunity to test Intergenerational Democracy in an international context as previously it had only been applied previously in Australia. The model proved to be successful in capturing the views of people of all ages and was readily adopted in Vanuatu. As with Australian studies, it was found that surrounding some issues age specific trends were measured and in other responses influences such as gender, rural or urban areas of residence were the most predominant variables. The application of ID in Vanuatu affirmed the value of the age based approach as a method of capturing whole-of-community views. However the researcher cautions that secondary variables, such as, gender and urban versus rural place of residence, should be coupled with age in any future studies that apply Intergenerational Democracy. Age should not be the sole variant of analysis.

Conclusion

Through the voices of the people of Espiritu Santo, we have learnt about their direct experiences and concerns surrounding the impacts of climate change, as a developing nation, recognised as in the 'front line'. Their long-standing relationships with nature highlighted the dependencies of all people on the planets natural resources to survive. This was summarised by an interviewee who said, "Without those things [environment] we cannot exist. We grow crops, we use woods to cook, we use coconuts for oil, everything we need without those things [we have] nothing. We use palms for our shelter, local woods to build the houses, everything around we need." Citizens from Vanuatu described how climate change was impacting their livelihoods now. They listed they changes have observed regarding: the sea and marine ecosystems, less demarcation of seasons impacting on agricultural activities and subsistence farming, increased in severity of climatic events such as long periods of rain and drought. They were concerned about inland and coastal erosion impacting on farming land, villages and road infrastructure. People described an increase in vector borne diseases such as malaria which they attributed to increased mosquito populations due to longer wet periods. They also spoke of unusual plant diseases impacting on their agricultural productivity which they also attributed to changing climatic conditions.

The importance of traditional knowledge and law, or Kastom, was described as an effective way of managing, monitoring and governing

(continued)

ecosystem health. As the planet faces the increasing impacts of climate change, it has become critical to develop new ways for people to live and adapt within viable and changing natural systems. Learning from the past, through Indigenous knowledge systems and practices can assist in guiding contemporary and increasingly urbanised societies to adopt new approaches to environmental management.

Most importantly in this study, through the application of Intergenerational Democracy, we heard from the elderly who confirmed the environmental changes they had witnessed over their lifetimes, together with the children who fear for their futures. These children of Vanuatu represent children from across the planet, all of whom have the right to a viable future.

Acknowledgements Thank you to the people of Espiritu Santo for their investment in this project, with special mention to: High Chief of Sanma Council of Chiefs (Supenatavuitano Council of Chiefs) Chief Charlie Rani, Chief Martino Nev Paramount Chief of Port Olry and The Chief of Mantantas, Rex Thomas Tandak, Carol Grosskopf, Courtney Angus, Ketty Napwatt, Leo Charlington, Leisarur Wolem, Julia Callachor, Xiaoming Ma, Leo Charlington, Leonie Woods and Laura Hugh. Thank you also to the Australian Government for funding the project through the Australia Research Award Program.

References

Australian Bureau of Meteorology (BOM) (2006) Report on the stakeholders workshop on application of climate predictions in Vanuatu. AusAID Project: enhanced applications of climate predictions in Pacific Island Countries (PI-CCP). Accessed 7th August, 2010, http://www.bom.gov.au/climate/pi-cpp/vut/vanuatu_ws_200601.pdf

Brown Weiss E (2008) Climate change, Intergenerational equity and International Law. Vermont J Environ Law 9:616–627, Accessed 9th September 2012, http://vjel.org/journal/pdf/VJEL10071.pdf

Brundtland G (1987) Our common future: the World Commission on Environment and Development. Oxford University Press, Melbourne, Australia

Cassidy F (2004) Thoughts of local residents of Espiritu Santo on Australian residents. online, Available: http://eprints.usq.edu.au/6579/1/Cassidy_2004_SV.pdf, (Accessed 18th February, 2010)

Constitution of the Republic of Vanuatu, 30 July 1980

Costello A, Abbas M, Allen A, Ball S, Bell S, Bellamy R, Friel S, Groce N, Johnson A, Kett M, Lee M, Levy C, Maslin M, McCoy D, McGuire B, Montgomery H, Napier D, Pagel C, Patel J, Puppim de Oliveira J, Redclift N, Rees H, Rogger D, Scott J, Stephenson J, Twigg J, Wolff J, Patterson C (2009) Managing the health effects of climate change. The Lancet, London, United Kingdom, Accessed 16 May 2009, http://www.thelancet.com/climate-change

Cubasch U, Wuebbles D, Chen D, Facchini MC, Frame D, Mahowald N, Winther JG (2013) Introduction. In: Stocker TF, Qin D, Plattner GK, Tignor M, Allen SK, Boschung J, Nauels A, Xia Y, Bex V, Midgley PM (eds) Climate change 2013: The Physical Science Basis. Contribution of Working Group I to the Fifth Assessment Report of the Intergovernmental Panel on Climate change. Cambridge University Press, Cambridge, UK

Davies K (2012) Intergenerational democracy: rethinking sustainable development. Common Ground Publishing, USA

Espiritu Santo Tourism Association (ESTA) (2010) Language, culture and history. Online, Available: http://www.espiritusantotourism.com/ (Accessed 17th March, 2010)

Inda J, Rosaldo R (2002) The anthropology of globalization: a reader. Blackwell Publishers, Oxford, United Kingdom

IPCC (2007a) Summary for policymakers. In: Parry ML, Canziani OF, Palutikof JP, van der Linden PJ, Hanson CE (eds) Climate change 2007: impacts, adaptation and vulnerability. Contribution of working group II to the fourth assessment report of the intergovernmental panel on climate change. Cambridge University Press, Cambridge, UK, pp 7–2

IPCC (2007b) Climate change 2007: Working group II: impacts, adaptation and vulnerability. Site accessed 28th April 2014. http://www.ipcc.ch/publications_and_data/ar4/wg2/en/annexessglossary-a-d.html

IPCC (2012) Summary for policymakers. In: Field CB, Barros V, Stocker TF, Qin D, Dokken DJ, Ebi KL, Mastrandrea MD, Mach KJ, Plattner GK, Allen SK, Tignor M, Midgley PM (eds) Managing the risks of extreme events and disasters to advance climate change adaptation, A special report of working groups I and II of the intergovernmental panel on climate change. Cambridge University Press, Cambridge, UK, pp 1–19

IPCC (2013) *see* Cubasch et al

National Advisory Board on Climate change and Disaster Risk Reduction (NAB) (2012) website accessed 2nd April 2014, http://www.nab.vu/what-nab-1

Republic of Vanuatu (RV) (2007) Disaster Risk Reduction and Disaster Management National Action Plan 2006–2016. Accessed 13th July, 2010, http://www.sopac.org

Romulus G, Lucas PHC (2000) Landscape stewardship: New directions in conservation of nature and culture. From the Caribbean to the Pacific: community conservation in Small Island States. Paper for The George Wright Forum 17(1);47–54, The George Wright Society Inc., Accessed 18th February, 2010, www.georgewright.org/171romulus.pdf

Shiva V (2006) Earth democracy: justice, sustainability and peace. Zed Books, United Kingdom

Statute of the International Court of Justice art. 38(1)(b)

Stern N (2006) Stern review: the economics of climate change. Government of the United Kingdom, United Kingdom. Accessed 13 February 2007, http://www.hm-treasury.gov.uk/media/8A8/C1/Summary_of_Conclusions.pdf

United Nations (1992) United Nations Framework Convention on Climate change. Accessed 10 October 2012, http://unfccc.int/resource/docs/convkp/conveng.pdf

United Nations (2012) United Nations framework convention on climate change, essential background; Science; Kyoto Protocol; Cancun Agreements. Accessed 3 October 2012, http://unfccc.int/2860.php

United Nations Population Fund (UNFPA) (2007) State of World Population Report, the United Nations. Accessed 10 January 2009, http://www.unfpa.org/swp/2007/english/introduction.html

Vanuatu Cultural Centre (VCC) (2007) The traditional ecological knowledge (TEK) and traditional resource management (TRM) Database Project. Accessed 24th February, 2010, http://www.vanuatuculture.org/site-bm2/trm/20070801_TRMdatabase.shtml

Vanuatu National Advisory Committee on Climate Change (VNACCC) (2009) Vanuatu Meteorological Services, Preliminary Climate and Sea Level Changes for Vanuatu Through the Application of SimCLIM, 28 November 2009, Press Release. Accessed 8th August, 2010, http://www.prweb.com/releases/Climate_Changes/Sea_Level_Changes/prweb3267604.htm

Vanuatu National Statistics Office (VNSO) (2009) Vanuatu 2009 Census Preliminary Data Report. Accessed 12th March 2010, http://www.spc.int/prism/country/vu/stats/Projects/mics-07.htm

World Heritage Centre UNESCO (WHC) (2010) Vatthe conservation area. Accessed 18th March, 2010, http://whc.unesco.org/en/tentativelists/1973/

Chapter 5
Urban Poor Women and Climate Change in India: Enhancing Adaptive Capacity Through Communication for Development

Sakshi Saini, Savita Aggarwal, and Geeta Punhani

Abstract Women form a disproportionately large share of poor in countries all over the world. Climate change brings with it drought, floods, deforestation and scarcity of natural resources making the lives of poor women in developing countries much harder since they have to struggle much more to fulfill their roles and responsibilities. It is therefore imperative that efforts be made for the adjustment in systems to minimize the negative effects of climate change and enhance the adaptive capacity of the poor especially women, to climate change.

The present study has been conducted to raise the level of awareness and knowledge of urban poor women, change their attitude and behavior to deal with climate change and thereby enhance their adaptive capacity, through the use of systematic and strategic communication, on a statistically defined sample from five major regions of Delhi, India. A Communication for Development (C4D) module was administered to 150 women, each group comprising of 15–20 women, in selected slums across the five regions of Delhi. The impact assessment of the 2-day communication intervention revealed that there was maximum change in awareness and knowledge followed by change in attitude of women to various aspects of climate change. The change in behavioral intent of women was however limited and not statistically significant due to the lack of reinforcing and enabling factors. Overall, there were significant gains in the scores of awareness, knowledge, attitude and behavioral intent (AKAB) of women towards climate change adaptation and mitigation. Our study has conclusively shown that C4D approaches when effectively planned and delivered can be used to enhance significantly the climate literacy as well as the adaptive capacity of urban poor women to climate change.

S. Saini (✉) • S. Aggarwal • G. Punhani
Urban Improvement, Aga Khan Foundation, Hazrat Nizamuddin PO, Sunder Nursery, PO Box 3253, 110014 New Delhi, India
e-mail: sakshi.saini.02@gmail.com; savitaaggarwal@gmail.com; s.punhani@hotmail.com

W. Leal Filho (ed.), *Climate Change in the Asia-Pacific Region*, Climate Change Management, DOI 10.1007/978-3-319-14938-7_5

Introduction

Women have shared a very close relationship with water since it is at the core of their traditional household responsibilities. Studies conducted in different parts of the world show a clear linkage between gender and water. Women in developing countries bear on an average more than 64 % of the burden of water collection (WHO/UNICEF 2010). In India women across different states bear more than 87 % of the burden of water collection (MOSPI 1998–1999; Venkateswaran 1995). Rural women walk long distances everyday to collect water and spend almost 150 million working days every year on the activity. This amounts to an annual monetary loss of almost ten billion rupees (Suvira 2003). Women in urban poor localities may not walk that much, but have to wait in long queues for water. This leads to loss of schooling as well as time poverty for other productive or child-care activities (UNDP 2006; Koolwal and van de Walle 2013). Almost 50 million girls do not attend primary school since they share the burden of water and fuel collection with their mothers (UNICEF 1998). In a study conducted in Gujarat, has shown that women were already spending 3 4 h/day on water collection but in summers, they had to spend an additional 2 h on the activity. They did so by cutting down on microenterprise work leading to loss of income and independence (James et al. 2002).

Globally 780 million people in the world still do not get their drinking water from improved sources and about 2.5 billion people lack access to safe sanitation facilities (WHO/UNICEF, JMP 2012). Almost all of this population lives in rural areas or slums of developing countries. In India, cities in most of the 35 states and union territories have a high percentage of population living in slums and related settlements who do not have access to piped water supply on premises. Greater Mumbai in Maharashtra has 54 % of its urban population living in slums followed by Kolkata with 32 %, Chennai with 18.9 % and Delhi with 18.7 % (Census 2001).

While on the one hand, rapid urbanization, industrialization and physical growth of cities are pushing the demand for water and aggravating the already complex situation, on the other hand, availability of water faces a serious threat from human induced climate change. Globally, it has been projected that, by 2080, for a 3–4 °C rise in temperature, more than three billion people may face increased water stress out of which up to one billion people would be from Asia (Arnell 2004). In India, the per capita availability of water has reduced drastically from 5,177 m^3/capita/year in 1951 to a mere 1,654 m^3/capita/year in 2007 and is expected to further go down to below 1,140 m^3/year by 2050 (MoWR 2011). Besides, due to wide spatial and temporal variability in the availability of water, India is already facing a water crisis like situation, which may get aggravated by climate change (INCCA 2010; MoEF 2012).

The decreased availability of overall water supply in future will negatively impact the quality of life of people and more so the women who are hitherto responsible for procurement and management of water. The slum population already faces high level of vulnerability because of lack of adequate housing, poor access to basic services, low/no education and limited income, is very likely to bear the maximum brunt of climatic extremes and related environmental adversities. It is imperative that women and children will be affected much more because they spend greater amounts of time at home as compared to men. Collection of water and other resources for the families already takes a very large chunk of women's time. The condition may further deteriorate due to climate change induced shortfalls of resources leading to magnified traditional roles of women at the same time diluting their economic role. It is therefore important to enhance the adaptive capacity of women to enable them to face the challenges thrown by climate change. Adaptive capacity is the ability of a system to adjust to climate change in order to moderate the potential damages, to take advantage of opportunities, or to cope with the consequences.

It is well accepted that different sections of population may face differential vulnerability to climate change (UNDP 2009; Nellemann et al. 2011). Access to information, knowledge and skills along with institutions, infrastructure, technology, economic wealth and equity are some of the instruments of enhancing adaptive capacity of people to climate change (IPCC 2001). Given the information, knowledge and skills of appropriate practices, poor women who constitute one of the most at risk groups can find sustainable solutions to reduce the vulnerability of their families and lead climate resilient lives. This is supported by a lot of evidence such as in the case of Honduras in 1998, when Hurricane Mitch struck, a community by the name of *La Masica* reported no casualties. This was because 6 months earlier, a disaster agency had provided gender sensitive community education on early warning systems and hazard management (GGCA 2009). Also, an innovative Action Aid- supplied project in Nepal had seen women's empowerment make rapid progress through the use of video discussions about climate change (Khamis et al. 2009). These studies clearly indicate the positive role that awareness and knowledge enhancement of women can play in dealing with climate stresses, extremes and disasters.

The Nobel Peace prize to the IPCC raised awareness amongst the urban literate population about climate change and its impacts. However isolated studies conducted in different parts of the world show that there is still a considerable lack of awareness among people about climate change (Scott 2000; Dilling 2007). A study on the perception of level of threat of climate change across countries showed that in developing countries in Asia, did not perceive climate change as a threat whereas in European countries, the perception of threat was moderate. Even people who had heard of 'global warming' had a limited level of understanding and perceived it as changes in weather or something caused by ozone depletion (Bord et al. 1997; Mortsch et al. 2000). The coverage given to climate change in various media is often too scientific for the common citizen especially poor people with low or no literacy to understand. This acts as an obstacle for spreading awareness and information about the issue of climate change. Climate change communication is a challenging task because climate change is a complex issue involving several uncertainties and it lacks urgency since it is not a felt need of the people (Dilling 2007).

Since it is the poor who despite least contributing to climate change, suffer the most from its impacts due to their higher vulnerability but low adaptive capacity, it is important to examine their climate related literacy in order to enable them to take appropriate adaptive actions. There are hardly any studies conducted both globally and nationally, which study the level of awareness of the poor urban or rural women to climate change. It is therefore an area, which needs further exploration.

The present study therefore was planned to gauge the current levels of awareness and knowledge of urban poor women to various aspects of climate change based on which a need based communication intervention could be designed for them. The second objective of the study was to assess the impact of the climate change related communication intervention on the adaptive capacity of urban poor women in terms of change in awareness, knowledge, attitudes and behavioral intent of the women.

Methodology

The study has been designed in five phases. A communication based assessment (CBA) was undertaken in the first phase of the study, with the help of interviews, observations, focus group discussions and in-depth interviews to engage the stakeholders in identifying their needs and problems. The second phase involved designing the communication intervention using traditional, print and modern media to bring about changes in awareness, knowledge, attitude and behavioral intent of the women towards climate change adaptation and mitigation. The third phase comprised of scripting and designing of various media (print, audio, video, traditional), their pre-testing, modification and production followed by training of personnel to use the communication media and the module. Finally this phase involved the actual delivery of the message through administering of the communication module. This was done by collaborating with a local Non-Governmental organization working on environmental issues in slums in different regions of Delhi. This organization had already built rapport with the community, possessed office with rooms for seating about 20–25 women and electricity to facilitate the communication intervention. The fourth and the fifth phases comprised of evaluating the communication module and assessing its impact on the climate linked adaptive capacity of women.

The study was conducted in the slum areas of Delhi. Geographically all the regions of Delhi have different demographic and physical characteristics. To get a better insight into the problems and vulnerabilities faced by the urban poor women across the city, a statistically defined sample was drawn from all the five regions of National Capital Territory of Delhi namely North, South, East, West and Central (Fig. 5.1).

Development communication methodology used in the present study as indicated in Fig. 5.2:

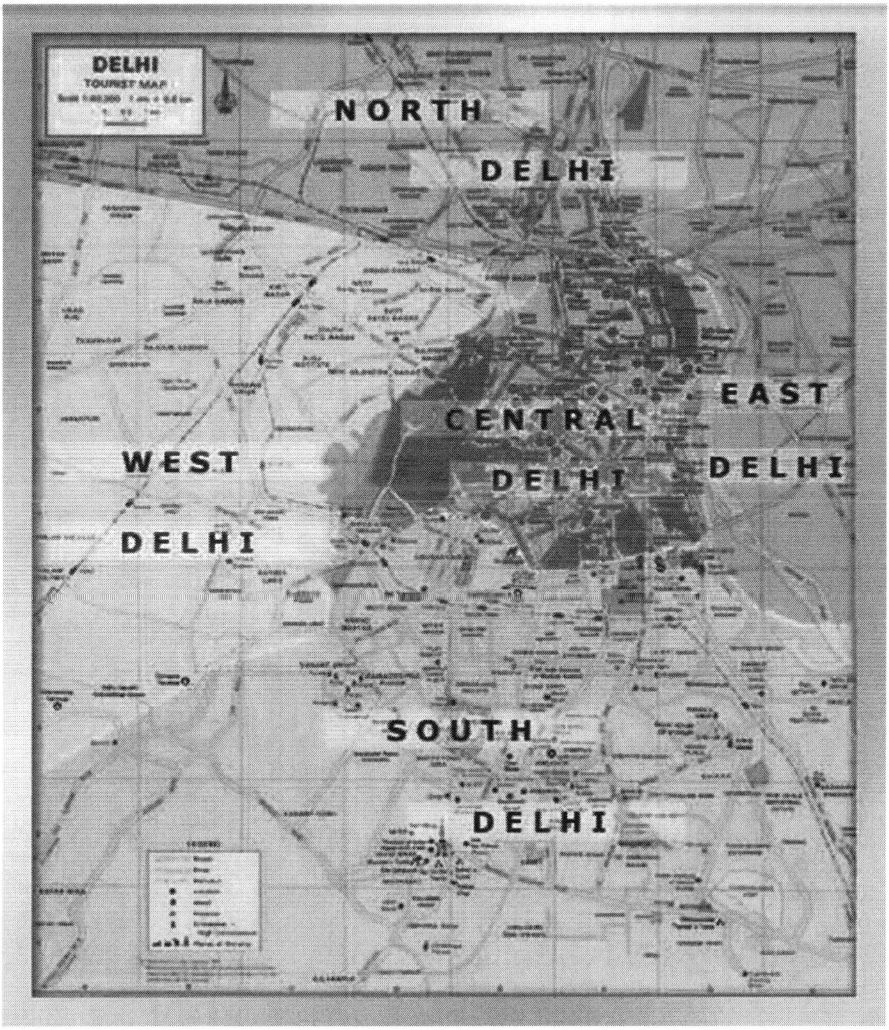

Fig. 5.1 Five regions of National Capital Territory of Delhi selected for the present study

A pilot study was conducted on 60 households from 4 slums of Delhi representing two regions to assess the variation in the most important components relating to awareness and knowledge level of urban poor women to climate change.

The sample size was calculated applying the formula:

$$\mathbf{n = t^2 x\ p(1-p)/m^2}$$

Where: **n** = required sample size

t = confidence level at 95 % (standard value of 1.96)

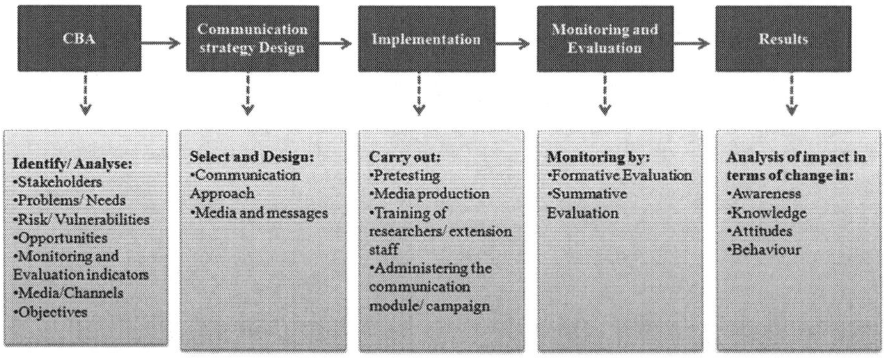

Fig. 5.2 Activities comprising each phase of the study (adapted from Paulo Mefalopulos)

p = estimated/observed percentage of the parameter
m = margin of error at 5 % (standard value of 0.05)
n = **288**

Accounting for some attrition, a sample size of 300 urban poor women was taken, divided equally amongst the five regions of Delhi thus, making a sample size of 60 women from each of the five regions. A two stage sampling design was used to select the sample for base line data collection. In the primary stage, four slums were randomly selected from each of the five regions of NCT of Delhi. The secondary stage comprised of selecting the households for the study using systematic circular sampling technique. Information regarding the total number of houses in the selected slum was obtained from the local leaders of the slum. The selection of the first house for data collection was randomized by draw of chits numbered one to ten. Thereafter every n^{th} house was selected depending upon the number of households in the slum so that the total number of selected households in each slum was 15 (Fig. 5.3). This process ensured that the selected households were geographically scattered in the selected slum and gave a fair representation of that area.

A combination of quantitative and qualitative data collection tools was used for the study. While a specially designed interview schedule was used to collect quantitative data, PLA tools such as focus group discussions and in-depth interviews were used to collect the qualitative data.

Development of the Communication Module

The communication based assessment of the urban poor women revealed almost negligible knowledge about various aspects of climate change. The women said that they did not know about the issue, were not educated and therefore could not say anything about it. On further probing, the women aged 40 years and above, if they were aware of any change in the weather in the last two to three decades or more

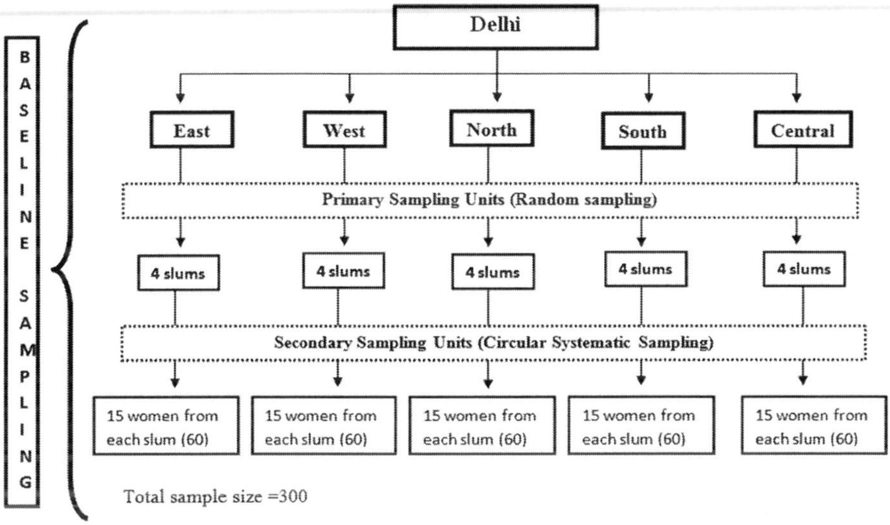

Fig. 5.3 Selection of primary and secondary sampling units for conducting the Communication Based Assessment (CBA)

especially in their respective villages in relation to agriculture (in which most of they had been deeply involved). They were also asked the possible causes of such changes (if any) in weather over a long-term period. The women said that changes in weather including rainfall and temperature were due to God's will and humans had no control over them. A small number of women said their villages were cooler as compared to the intense heat of the cities. Considering the existent knowledge level of women, an eight hours communication module spread over 2 days was designed. One of the objectives of the communication intervention was to create awareness among the urban poor women about climate change, its causes and impacts on their lives and livelihoods in current and future climates with special reference to water resources. Another objective was to impart knowledge and skills to the women about appropriate and contextually suitable mitigation strategies and water linked adaptation strategies to climate change at the household level. The final objective of the communication module was to bring about a change in the awareness, knowledge, attitudes and potential behavior of women towards climate change adaptation and mitigation strategies with special reference to water resources.

Some of the key components of the communication module were:

- Ice breaking games to increase the comfort level of the group and get them talking on the issue
- Causes and impacts of climate change on lives and livelihoods of people especially women
- Linkage between climate change, water and women

- Vulnerability of women to climatic stresses and extremes with special reference to water resources
- Contextual adaptation and mitigation strategies with special reference to water for domestic use
- Adaptation strategies to deteriorating quality of water-demonstration of water disinfection/purification techniques at the household level
- Recapitulation of causes, impacts, adaptation and mitigation strategies
- Question and answer sessions, competitions and quiz
- Administration of the AKAB assessment tool

Since the available communication materials were either too scientific or complicated besides not being in the local language, most of them could not be used for training the urban poor women. Therefore, several print (posters, flip charts, flash cards), traditional (puppet show, skit/drama) and modern media (public service announcements, documentaries, short films, camera mediated exercises) were scripted, designed and field tested to enable the delivery of the communication module to the urban poor women. This process was undertaken by a synergy of professionals from the field of climate change, gender, mass communication, development communication as well as trainers at the grassroots level.

The communication module was administered through NGOs in the five regions of Delhi. Participatory approach was followed throughout the implementation of the communication module. One of the problems encountered during administration of the communication intervention was to get the women participate in the intervention since they did not seem to be convinced about the issue of 'climate change' and were unsure what was going to be transacted in the 2 days. Others were willing to attend the program only on 1 day and had to be persuaded to attend for 2 days. This was due to the perception of 'climate change' as an unfelt need leading to a low level of motivation among the women.

Another constraint of the present study was the transaction of the communication intervention on two consecutive days rather than more days spread over a longer time period to enable better absorption of information and skills by the women. The lack of sufficient time gap for assessing the impacts of the intervention to be observable was another limitation of the present study. However, it has been noted by IPCC, Assessment Report 5, that measuring and attributing adaptation is a challenging task because adaptation outcomes take time to be identifiable. At the same time the outcomes are subject to constant influence by other developmental activities in the region.

Results

Profile of the Respondents

Most respondents had been living in Delhi for more than 3 years and were below the poverty line. The state of women's education was dismal as almost three-fourths of the women were illiterate. Among the rest, 18 % women had education up to primary level, 6 % up to secondary level and less than 1 % women had any form of higher education. The differences in level of education across the five zones of Delhi were very minor (Table 5.1). As highlighted by the Focus Group Discussions, most women who had been to school also reported that they were at best semi-literate, since they had dropped out of school in first, second or third grades and had never attempted to read anything after that.

Three fourths of the women were housewives and had not taken up paid employment because it was not easy to find suitable employment coinciding with their free time. The remaining slum women were gainfully employed, as domestic helpers, petty sellers or as daily wagers in construction or in factories.

Assessment of Change in Awareness, Knowledge, Attitude and Behavior After the Communication Intervention

Before and after the communication intervention, the AKAB assessment tool was administered to the women in order to assess the change in their awareness, knowledge, attitudes and behavioral intent of the urban poor women towards various aspects of climate change.

Changes in Awareness to Climate Change

Before the intervention, most of the women had not heard of the term 'climate change' either through the mass media (radio/television) or interpersonal networks and were also not aware of its causes and impacts. Only 1–2 % of the women had

Table 5.1 Educational level of respondents

	Zones of Delhi (percentages)					Average % (Delhi slums)
Educational level	East	West	North	South	Central	Delhi
Illiterate	73.33	80.00	76.67	71.67	78.33	76.00
Primary level	21.67	13.33	16.67	18.33	18.33	17.67
Secondary level	5.00	6.67	6.67	8.33	1.67	5.67
Higher education	0.00	0.00	0.00	1.67	1.67	0.67

Source: Primary survey

reported experiencing extreme heat, erratic rainfall patterns especially in their respective villages.

After administering the communication module all the respondents (100 %) said that climate change was indeed a reality to which they themselves could now relate. The women said they had been experiencing frequent floods and droughts in their villages in addition to very erratic rainfall but were ignorant about the reasons due to lack of education.

Change in Knowledge of Women About Climate Change and Related Issues

The knowledge of women to climate change was very dismal prior to the communication intervention. The women did not know about the phenomenon of climate change, its causes and impacts. However, after the communication intervention there was a substantial change in the knowledge of women about different aspects of climate change.

Causes of Climate Change

After the intervention almost half of the respondents said that a combination of factors such as burning coal for electricity generation, excessive use of petrol and diesel for vehicles were the major causes of climate change. More than one-third of the women said that deforestation was the major cause of climate change, whereas one-fourth of the women attributed climate change to increasing number of cement and steel industries because of increasing population. Only 12 % women thought that climate change was due to natural causes. A very small fraction of women said that farm animals and agriculture were responsible for climate change (Fig. 5.4).

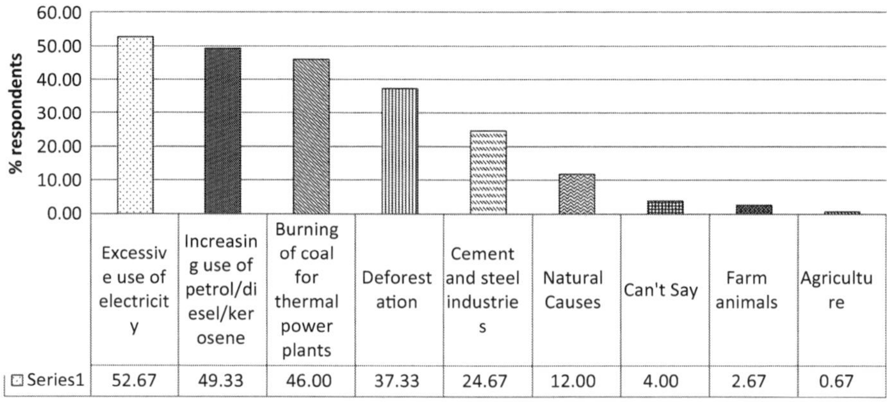

Fig. 5.4 Knowledge of respondents about causes of climate change (n = 150) (Source: Field work)

They were closely involved with agricultural activities in their villages and felt that agriculture was essential to feed everyone and therefore could not be considered as a contributor to climate change.

Impacts of Climate Change

The communication intervention resulted in considerable gain of knowledge about the impacts of climate change and most of the women said that climate change had several impacts. More than 46 % women reported at least three impacts of climate change, another 49 % knew about four to six impacts and a much smaller number (4.67 %) knew of seven or more impacts of climate change. Majority of the women (73 %) said that increase in the temperature would be the main impact of climate change followed by 67 % of the respondents who said that change in rainfall would be the major impact (Fig. 5.5).

Mitigation Strategies for Dealing with Climate Change

Before the communication intervention, the women could not relate to the measures to reduce the pace of climate change since they did not understand the phenomenon itself. After the conduct of the communication intervention there was a remarkable gain in knowledge of women with regard to the strategies for climate change mitigation. Almost three-fourths of the women reported that saving electricity and fuels such as petrol, diesel, kerosene etc. as well as planting of trees were important for reducing the pace of climate change (Fig. 5.6).

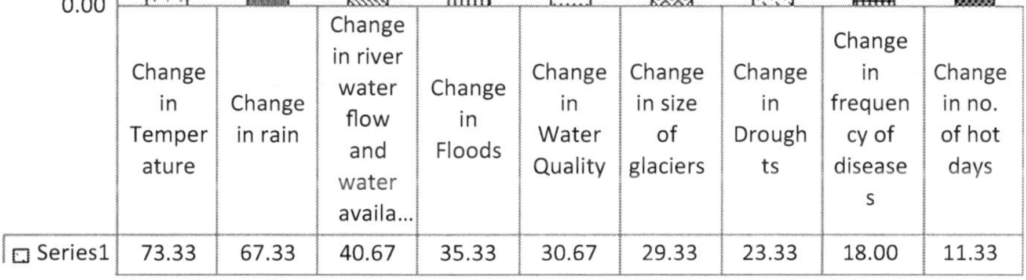

	Change in Temperature	Change in rain	Change in river water flow and water availa...	Change in Floods	Change in Water Quality	Change in size of glaciers	Change in Droughts	Change in frequency of diseases	Change in no. of hot days
Series1	73.33	67.33	40.67	35.33	30.67	29.33	23.33	18.00	11.33

Fig. 5.5 Knowledge of respondents about impacts of climate change-post intervention (n = 150) (Source: Field work)

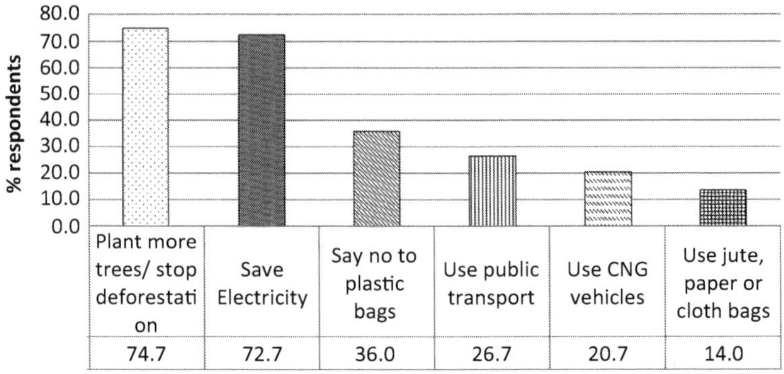

Fig. 5.6 Knowledge of climate change mitigation strategies-post intervention (n = 150) (Source: Field works)

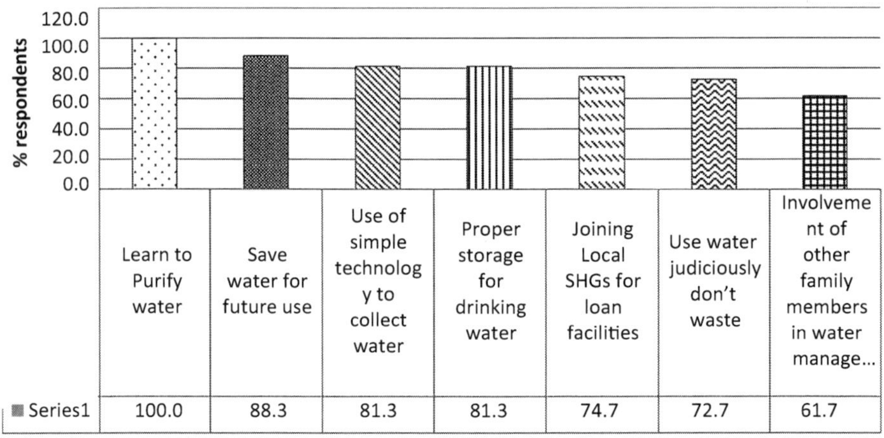

Fig. 5.7 Knowledge of women regarding water mediated adaptation strategies-post intervention (n = 150) (Source: Field work)

Adaptation Strategies to Deal with Climate Change with Special Emphasis on Water Management at the Household Level

Almost all the respondents felt that since the climate was changing there was an urgent need for everyone to adapt to it. All the women said that there was a need to purify or disinfect water especially for drinking purpose. More than 88 % women said that water was already in short supply and must be conserved for future use. A large number of women (72–81 %) felt that they should use simple technologies to collect and store water such as the use of plastic pipes, carts and cycles, proper storage containers and tanks. More than 60 % respondents felt that women were overburdened with water management responsibilities and a redistribution of water related roles within the household and involvement of other family members could provide relief to them (Fig. 5.7).

Changes in Attitude of Women

Climate Change

Before the intervention, majority of the respondents were either neutral or disagreed to the occurrence of climate change but this changed drastically after the communication intervention. A large majority (82–98 %) of the women agreed that climate was changing due to which people were facing problems. However, it was interesting to note that only half of them felt climate change was a serious problem. They felt that there were other more serious issues such as unemployment, poor income, inadequate housing, lack of medical and education facilities, which required more urgent attention (Table 5.2).

Causes of Climate Change

Before the intervention, since most of the women did not know about climate change, they were neutral to the causes of climate change. After the communication intervention more than 86 % women disagreed that climate change was due to God's will and that high consumption of electricity and deforestation have nothing to do with climate change. A large majority (84 %) of the women agreed that the primary cause of climate change is burning of fuels such as petrol/diesel and oil (Table 5.3).

Impacts of Climate Change

Even before the intervention, 78–82 % women agreed that frequency of rainfall at present had changed as compared to earlier times and drought and floods occurred more frequently at present. After the communication intervention, a larger majority of women (94–98 %) felt that frequency of rainfall, drought and floods has changed at present as compared to earlier times and agreed that climate change will have a negative impact on the quality of life of future generations. In addition 85–100 % women (as compared to 50 % women before intervention) felt that climate change will impact them negatively as it will enhance their work burden of water collection, purification and storage (Table 5.4).

Change in Behavioral Intent

Mitigation Strategies

Even before the communication intervention, a large majority of the women and their families were already practicing several mitigation strategies suggested to

Table 5.2 Attitude of women to climate change pre and post intervention

Statements of attitudinal scale	Pre Intervention			Post Intervention		
General statements- about climate change	Agree	Neutral	Disagree	Agree	Neutral	Disagree
Weather in the past 30 years has changed	12.00	84.00	4.00	98.00	1.67	0.33
Climate change is a serious problem faced by people at present	3.00	81.00	16.00	49.33	46.00	4.67
Climate change will be more of a problem in future	3.00	87.00	10.00	94.00	6.00	0.00
If the present rate of coal and oil use continues, climate change will continue to occur at a fast rate	9.00	79.00	12.00	84.00	6.00	10.00

Source: Field work

Table 5.3 Attitude of women to causes of climate change pre and post intervention

Statements of attitudinal scale	Pre intervention			Post intervention		
Causes of Climate change	Agree	Neutral	Disagree	Agree	Neutral	Disagree
Change in climate is God's will	8.00	82.00	10.00	8.00	6.00	86.00
The primary human activity that causes climate change is the burning of fossil fuels such as coal and oil	4.00	86.00	10.00	84.00	14.00	2.00
High consumption of electricity has nothing to do with climate change	10.00	80.00	10.00	24.00	0.00	76.00
Cutting of forests has no relationship with climate change	4.00	88.00	8.00	12.00	2.00	86.00

Source: Field work

Table 5.4 Attitude of women to Impacts of climate change pre and post intervention

Statements for attitudinal scale	Pre intervention			Post intervention		
Impacts of climate change	Agree	Neutral	Disagree	Agree	Neutral	Disagree
Frequency of rain has changed now as compared to that 50 years back.	82.00	18.00	0.00	94.00	4.00	2.00
One hears more about droughts and floods at present as compared to earlier.	78.00	22.00	0.00	98.00	2.00	0.00
Climate change will reduce the quality of life of our children & grandchildren in the future.	3.00	81.00	16.00	94.00	6.00	0.00
Climate change and women						
Climate change will more adversely affect women.	9.00	81.00	10.00	90.00	10.00	0.00
In future women will be more burdened by water collection.	50.00	50.00	0.00	84.00	12.00	4.00

Source: Field work

them in the present study such as minimizing the use of electricity by switching off electrical equipment when not in use, use of CFLs, coloring walls with light colors, using public transport and the use of CNG vehicles. This was because of very low levels of income of families and very poor availability of natural as well as other economic resources to families, which strictly called for minimizing their use. These families used cycles or public transport to travel long distances and walked to the nearby destinations. They could not afford to buy or hire petrol/diesel driven private vehicles.

After the communication intervention, the other respondents (who were small in number and were practicing the above mitigation strategies only to some extent) said they would use these strategies in future. 12 % women who were non-users of CFLs (Compact Fluorescent Lamps) wanted to switch over to them and another 14 % women thought they would like to try the use of CFLs. The constraint felt by most women in the use of CFLs was their high cost. As many as 26 % women said they would like to adopt better waste disposal techniques in the future while 24 % women said they would use less plastic bags in future.

Water Linked Adaptation Strategies

A large number of families (56–100 %) found certain adaptation strategies to be very easy and were practicing them to some extent even before the communication intervention. These were judicious use of water, use of covered and clean utensils for water storage, taking the help of other family members in water collection for the household. Even before the intervention, almost 60 % women were members of some women's group, which collected money from every member on a monthly basis and provided lump sum money to a member by rotation. The women said they used this money for meeting personal expenses (clothes, jewelry, and gifts). Some women said they could think of utilizing the money for purchase of water management related equipment, which could improve the quality of their day-to-day lives. After the intervention, more women (24 %) as compared to (4 %) earlier felt that they would like to install a water tank in their house to provide relief to them as well as use cycles/carts to bring water at least to some extent. For adopting these strategies the women said they could not decide on their own and would have to request/motivate the male members of the family. A large majority (88 %) also said they would install taps in water containers already possessed by them. With regard to purification/disinfection of water, even after the communication intervention, more than two-third women still said they would not like to invest in the purchase of water filter and would also not use boiling as a method of water purification. They preferred the use of solar disinfection technique and chlorine tablets for water purification/disinfection since these did not involve much cost. However, due to high inputs of time and effort, the women were not sure whether they would be able to adopt these methods.

Table 5.5 Change in scores of AKAB of the respondents after the communication intervention

Variable	Pre test mean	Post test mean	Mean difference	t score	Sig. (2-tailed)[a]
Awareness	0.32	2	1.68	31.734	0.001
Knowledge	27.42	44.78	17.36	31.467	0.001
Attitude	76.68	85.36	7.6735	17.041	0.001
Behavioural intent	59.49	60.88	1.3866	1.323	0.188
AKAB	163.91	193.02	29.11	20.21	0.001

[a]95 % level of significance

Change in Awareness, Knowledge, Attitude and Behavior of Women to Climate Change

A paired samples *t*-test was conducted to assess the change in awareness, knowledge, attitude and behavioral intent of the urban poor women with respect to various aspects of climate change before and after the communication intervention.

Overall, the AKAB scores before the intervention were ($M = 163.91$, $SD = 10.56$) which, after the communication intervention changed to ($M = 193.02$, $SD = 12.80$; $t\ (149) = 20.21$, $p < .001$). Hence, it can be concluded that the communication intervention had a very significant impact on the overall, awareness, knowledge, attitude and behavioral intent (AKAB) of the urban poor women to climate change (Table 5.5).

A holistic overview of the results of AKAB change after the communication intervention reveal that there was a 36.5 % change in the overall mean scores of awareness, knowledge, attitude and behavior of the respondents to climate change and related aspects. A breakup of the scores indicates the maximum change occurred in the average score of awareness component (74.66 %) followed by the knowledge level of women (59.66 %). The percent change in the attitude of women towards various issues pertaining to climate change was much lower, (13.4 %) followed by change in their behavioral intent towards climate change adaptation and mitigation strategies (4.1 %). It is important to note that despite a significant gain in the awareness and knowledge components, there was a limited change in the attitude and intent to behavior change of the women (Fig. 5.8).

This can be attributed to several reasons. One of the reasons for the differential change in the components of AKAB was that the women had almost no awareness and knowledge about various aspects of climate change prior to the communication intervention and had negligible scores in both these dimensions thus depicting high gain in scores after the intervention. By contrast, the women had scored substantially in the attitude and behavior components even before the communication intervention thus limiting the scope for gain after the communication intervention.

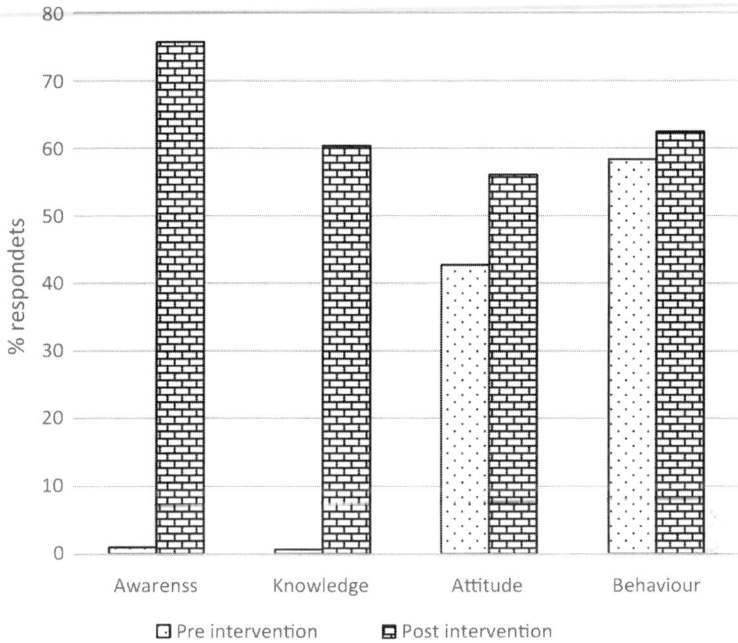

Fig. 5.8 Change in AKAB scores after the communication intervention (Source: Field work)

Several families despite not having knowledge of climate change and related aspects were already practicing several of the suggested mitigation (cutting down fuel/electricity consumption, use of public transport) and adaptation strategies (saving water) owing to their very low levels of income and a general lack of availability of resources. With respect to the other adaptation strategies suggested, the women had several apprehensions related to finance, lack of support of other family members including older women and men, lack of space, poor level of motivation due to temporary and rented housing which were under constant threat of demolition. The women were therefore not ready to adopt these adaptation strategies on their own without the consent or agreement with the male and other senior members of the household.

The retention of awareness, knowledge, change in the attitudes and behavior of urban poor women to various aspects of climate change needed to be assessed after a certain time period to study the impact of the communication intervention. A recall test on the women respondents comprising of 50 % of the original sample was therefore undertaken, after a gap of 4 weeks of conducting the communication intervention. The average AKAB retention of the various causes in the recall was 25.35 % as compared to 36.5 % immediately after the intervention.

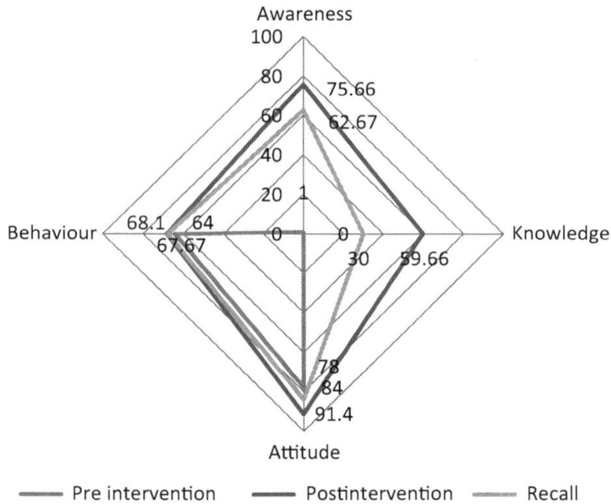

Fig. 5.9 Comparison in the scores of post intervention and recall scores

Figure 5.9 depicts the AKAB scores before, immediately after and four weeks after the communication intervention.

The results indicated a significant retention in the overall awareness, knowledge, attitude and behavioral intent of the urban poor women to various aspects of climate change. The knowledge can be enhanced or retained if put to use and alternatively can decline further if not used. It is therefore imperative to maintain a continuous sustained flow of information and knowledge to the vulnerable groups through various communication channels including interpersonal, group and mass media.

Conclusion

The present study has shown that a contextual and need based communication intervention was instrumental in enhancing the awareness, knowledge and attitude of the urban poor women to various aspects of climate change. However, it had a marginal impact on changing behavioral intent of the urban poor women to various strategies and practices for climate change adaptation and mitigation. This was because the families had very few resources including money to invest in more facilities, were not willing to invest because of temporary and rented housing, most women did not have decision making powers and faced several socio-cultural barriers in even suggesting to their family members about any purchase or lifestyle related changes. The women could undertake only small changes, which did not clash with the interests of other family members. The women also felt that it was the Government's responsibility to provide them with the basic services such as housing, water and sanitation since they were so poor.

(continued)

Though there exists a body of evidence to show that enhanced climate literacy of people has an intrinsic value and the more the members of a community understand the problem of climate change, the more will be the likelihood of supporting the changes brought about by various sectors including the Government and Non-Government organizations. It has also been seen that knowledge based initiatives may also work as a precursor to behavior based initiatives (Spehr and Curnow 2011). Communities are more likely to cope well with climate change if they have knowledge of potential threats and resources to adapt to them. However, knowledge and information are only one of the key determinants of adaptive capacity of communities at the local level. The other determinants being asset base (availability of key assets to allow people to adapt), appropriate institutional environment (to make capital and assets accessible to people and to innovate) and a flexible and forward looking governance which can anticipate, make changes and plan for the future (Jones et al. 2010).

Though the interaction between the above components remains a matter of further study, it is apparent that in order to make knowledge more meaningful to people, it is necessary to invest in all the above determinants of adaptive capacity. It is also important to make climate change personally relevant to people to motivate them to action. These efforts have the potential to enable the most vulnerable groups of society to take appropriate action towards climate change mitigation and adaptation and lead more empowered lives.

References

Arnell NW (2004) Climate change and global water resources: SRES emissions and socio economic scenarios. Glob Environ Change 14:31–52

Bord RJ, Fisher A, O'Connor RE (1997) Is accurate understanding of global warming necessary to promote willingness to sacrifice? Risk Health Saf Environ 8(4):339–354

Census of India (2001) Retrieved from Ministry of Statistics & Programme Implementation, 2007. Compendium of Environment Statistics India, 2007, accessed on July, 12, 2012.

Dilling L (2007) More communication of climate change science won't spur problem solving. News Center University of Colorado Boulder. Retrieved from http://www.colorado.edu/news/releases/2007/67.html (Accessed on March 4, 2013)

INCCA: Indian Network for Climate Change Assessment, MoEF (2010) Climate change and India: A 4x4 assessment, A sectoral and regional analysis for 2030s

IPCC (2001) Impacts, adaptation and vulnerability. In: McCarthy JJ, Canziani OF, Leary NA, Dokken DJ, White KS (eds) Contribution of Working Group II to the third assessment report of the intergovernmental panel on climate change. Cambridge University Press, Cambridge, p 1000

IUCN, UNDP, GGCA (2009) Training manual on gender and climate change. San José, Costa Rica. Marzo

James AJ et al (2002) Transforming time into money using water: a participatory study of economics and gender in rural India. Natural Resources Forum. Retrieved from: http://josiah.berkeley.edu/2007Fall/ER275/Readings/DP1-3/james%20etal.2002.pdf. Accessed on December 2, 2011

Jones L, Ludi E, Levine S (2010) Towards a characterisation of adaptive capacity: a framework for analysing adaptive capacity at the local level, Overseas Development Institute, 2010. retrieved from http://www.odi.org.uk/sites/odi.org.uk/files/odi-assets/publications-opinion-files/6353.pdf, accessed on February 1, 2012

Khamis M, Plush T, Zelaya I (2009) Women's rights in climate change: using video as a tool for empowerment in Nepal. Gender Develop 17(1):2009

Koolwal G, van de Walle D (2013) Access to water, women's work and child outcomes. Econ Dev Cult Change 61(2):369–405

Ministry of Environment and Forests, (MoEF) (2012) India second national communication to the United Nations Framework convention on climate change. Ministry of Environment and Forests, Government of India

Ministry of Statistics and Programme Implementation (MOSPI) (1998–1999) Time use survey: summary of findings. Retrieved from: http://www.mospi.gov.in/mospi_time_use_survey.htm. Accessed on August 12, 2010

Ministry of Water Resources (MoWR), Government of India (2011) Restructuring the Central Water Commission, Vol. 1, Retrieved from www.indiawaterportal.org. accessed on December 20, 2012.

Mortsch L, Bradley B, Andrey J, Warriner K, Fischer A (2000) A survey of climate change beliefs: a case study of the Canada Country Study participants. Proceedings of the international conference on climate change communication, Environment Canada, June 2000

Nellemann C, Verma R, Hislop L (eds) (2011) Women at the frontline of climate change: Gender risks and hopes. A rapid response assessment. United Nations Environment Programme, GRID-Adrendal. pp 119–125

Scott D (2000) Climate change communication, Proceeding of an International conference. Climate Change Action Fund and Government of Canada. Sage, Delhi

Spehr K, Curnow R (2011) Are your objectives really about behavior change? Or they are about knowledge and attitude. Community change behavior environment research education. Retrieved from http://www.communitychange.com.au/insights-and-tools/21-insights-and-tools/improving-your-program/are-your-objectives-really-about-behaviour-change/12.html accessed on January 20, 2012

Suvira S (2003). Going parched in the land of troubled water. Sahara Times, May 18 2003 New Delhi

UNDP (2006) Human Development Report. Beyond scarcity: power, poverty and the global water crisis. United Nations Development Programme. Macmillan, New York

UNICEF (1998) Accessed from Morna Lowe, C. (2000). Mainstreaming gender in water and sanitation, Literature review for the SA Department of Water and Sanitation, Johannesburg.

Venkateswaran S (1995) Environment, development and the gender gap. Sage, New Delhi

WHO/UNICEF (2010) Joint monitoring programme report. Progress on sanitation and drinking water for water supply and sanitation. UNICEF, New York and WHO, Geneva

WHO/UNICEF (2012) Joint monitoring programme report. Progress on sanitation and drinking water for water supply and sanitation. UNICEF, New York and WHO, Geneva

Chapter 6
Local Community Perception of Climate Change and Scientific Validation: A Review of Initiatives and Perspectives in the Indian Region

Prakash Rao and Walter Leal Filho

Abstract Recent studies have indicated that South Asian countries including India will particularly be vulnerable to climate change impacts. These include changes in temperature and rainfall regime, agriculture yields, sea level rise, extreme weather events, melting of glaciers, changes in ecosystem function, shifts in species distribution etc. Of late some of these impacts are being felt with greater frequency in the Indian subcontinent e.g. extreme weather events such as cyclones on India's eastern coast and droughts in the western states of Rajasthan and Gujarat. Serious health related problems are also being attributed to climate impacts include outbreaks of vector borne diseases like malaria.

The present study reviews the various efforts to document and articulate people's perceptions and knowledge of the impacts of climate change in diverse ecosystems in India. The study analyses the relationship between local community knowledge of climate change and the available scientific evidence and suggests that people's local knowledge of climate indicators conforms to existing scientific data. Effectiveness of citizen's initiatives to rapidly document climate change at pan India level have also been reviewed. The paper recommends adaptive strategies at local, national and regional level for limiting climate change and intervention across several sectors for developmental planning. The paper also suggests various stake holder based policy measures for integration with long term social and technological changes for mitigation and adaptation strategies for climate change.

P. Rao (✉)
Department of Energy and Environment, Symbiosis Institute of International Business (SIIB), Symbiosis International University (SIU), G.No.174/1, Hinjewadi 411 057, Pune, India
e-mail: prakash.rao@siib.ac.in

W.L. Filho
Hamburg University of Applied Sciences, Lohbruegger Kirchstraße 65, Sector S4, Hamburg, Germany
e-mail: walter.leal@haw-hamburg.de

W. Leal Filho (ed.), *Climate Change in the Asia-Pacific Region*, Climate Change Management, DOI 10.1007/978-3-319-14938-7_6

Introduction

Since the last industrial revolution, emissions resulting from anthropogenic activities have led to a substantial increase in the atmospheric concentration of greenhouse gases. Recent research suggests that rising emissions and resultant warming of the earth's atmosphere, may lead to a rise of about 4.8 °C above pre industrial levels.

The changes brought about by rapid climate variability are likely to impact not only the ecological and socio economic impacts dimensions of society but also could threaten the future growth and economic activities of several countries in the Asia Pacific region. Some of the critical ecosystem hot spots and landscapes in the Asia Pacific region (Myers et al. 2000) are also the most vulnerable to climate change. the Key indicator of climate change include extreme weather events (drought, frequent heatwaves, cyclones, deep depressions), increased agricultural losses, sea level rise, glacial retreat, coral bleaching, species declines etc. As a consequence ecosystem dependent communities across the world in both developed and developing countries are already facing the adverse impacts of climate change, it is also said that tropical countries are likely to be more vulnerable than developed countries.

Scenarios complied by the Intergovernmental Panel on Climate Change (Parry et al. 2007) suggest than unless humans dramatically reduce greenhouse gas emissions, we will see a doubling of pre-industrial carbon dioxide concentrations resulting in an increase of the earth's temperature from between 1.1 and 6.4 °C (depending on estimates for low and high scenarios), with recent modeling suggesting upwards of 11 °C by the end of the century (Stainforth et al. 2005). The last decade has been observed as the warmest with India and South-East Asia experiencing frequent extreme climatic events. While recent climate models predict an increase in rainfall patterns regional change may be different (Rupa Kumar et al. 2006).

The South Asian region is one of the most densely populated regions of the world has also some of the most diverse and critical ecosystems across a large land mass. The subcontinent has a primarily agriculture based economy, with a long coast line and a mountainous Himalayan range in the north. In this context the region is prone to any major climate variation affecting not only ecosystems about also people and communities. Using people's knowledge and local perception it is possible to develop strategic interventions to address the adaptation needs of local communities and ecosystems based on impact studies.

The knowledge base of local communities on climate related issues can greatly shape the climate science as well as help in taking policy decisions for mitigation and climate adaptation (Chaudhary and Bawa 2011; Reidlinger and Berkes 2001; Naess 2013). This unique approach of combining people's knowledge and validating it with science based facts is an interesting approach to studying climate change at local level (WWF 2005a, b).

Methodology

The study used the "Climate Witness" approach (WWF 2005a, b) as a methodology, which aimed at documenting the perception of the community at various stakeholders' level on the adverse impacts of the climate change. This methodology primarily was developed to record stories and grass roots details of climate impacts amongst communities where local knowledge about climate, weather, land, ecosystems etc. was available in the form of historical and anecdotal records (Chaudhary and Bawa 2011; Joshi and Joshi 2011). The approach was adopted to study and understand community knowledge about the adverse impacts of climate change in areas like traditional knowledge systems and community responses, involvement of different levels of stakeholders in developing a response mechanism, including local concerns about climate change in overall development planning. In doing so, the study attempted to assess various community responses in climate impacted ecosystems in the Himalaya, coastal regions and the deltaic regions of the Sundarbans.

Linking Ecosystems, Community Knowledge and Climate Change in India

Understanding climate impacts in vulnerable and ecosensitive regions of the Indian subcontinent like the Himalayas, Sundarbans and the coastal regions requires a knowledge depth of how local communities perceive these environmental issues and their response mechanisms. This study reviews some of the available facts on climate change to local citizens knowledge as well as related initiatives at a pan India level. In a vast country with diverse ecosystems it was appropriate therefore to consider 2–3 important and significant locations which also have a key relationship with climate change in terms of their impacts in the region. While the entire Indian regions is seen as being impacted for the purpose of this study we considered the coastal fisheries of southern India, the deltaic regions of the Sundarbans and the western Himalaya.

Coastal Ecosystems

The coastal region of India is perhaps one of the most productive and ecologically diverse landscapes covering over 7,500 km of coastline. The importance of promoting regional fisheries in enhancing local livelihoods through sustainable measures is an important aspect of maintaining local ecological balance (MoES 2010). A vast majority of coastal communities are currently facing stress form various pressures like large scale development along costal sites which threaten to affect their survival. Climate impacts are likely to further add to the growing changes and decline in the productivity of marine ecosystems in India's coastal sites. Several institutions have been involved in research and advocacy by studying some of the impacts caused by climate change in selected fishing communities in southern India including finding ways and mean of building resilience of local communities against the adverse impacts of climate change.

Sundarbans

Sundarbans is the world's largest mangrove ecosystem, a UN heritage site, a WWF designated Global 200 eco region and now under severe stress due to sea level rise and associated problems. The approximately four million human population residing in the Indian Sundarbans are severely stressed due to sustained variability in local weather coupled with occurrence of frequent cyclonic storms and depressions impacting their overall livelihood. The threat of decline of endangered species like Tiger, Turtles and some of the rare mangroves is also a major issue in this unique ecosystem (Loucks et al. 2010). This is also likely to adversely affect the overall ecological balance and increase the vulnerability of the region. Immediate attention is required from the world community as well as stakeholders at the local level, to develop an effective coping mechanism to reduce the vulnerability of the region.

Himalaya

The vast number and range of glaciers and perennial river systems originating from the Himalaya mountain range in the north are the major source of freshwater supplies to the subcontinent. Due to rising temperature and changing precipitation pattern, the glaciers are expected to show higher rates of retreat, (Xu et al. 2009) and with possible implications on fresh water availability and agricultural productivity in the South Asian region (WWF 2005a, b). The direct evidence of real climate change impacts in the Himalaya requires immediate attention of the international community as well as regional and national policy makers for consideration in any future development planning.

Community Response and Interventions

Recognizing the value of information available from the broader community and also the changing framework of traditional knowledge and local perceptions of coastal communities, the study reviews various initiatives across key ecosystems in

India and the coping capacity of these communities from potential climate related impacts.

*Coastal Ecosystems (Gulf of Mannar, India and the False Trevally (*Lactarius lactarius**)**

In the past few years changes in rainfall, currents, and sea level associated with global warming, are already affecting the world's coastal ecosystems and fisheries. The recent IPCC report has also provided ample evidence of the implication of climate change on our biodiversity and the increasing vulnerability of some of our critical ecosystems and consequences for livelihoods of people (TERI 1996). Tropical Asia would probably experience the highest impact of present day climate variability and therefore is more prone to global climate change (Nicholls and Leatherman 1995).

Erratic weather and monsoon patterns, along the Indian coastline along with frequent extreme climatic events like cyclones are major threats to the ecosystem including in some cases low-lying islands some of which are already facing partial submergence resulting in shoreline changes (Patwardhan et al. 2003). A recent study (McGranahan et al. 2007) indicates that one tenth of the global population live in coastal areas that lie within just ten metres above sea level. The study also brings home the fact that nearly two-thirds of urban settlements with more than 5 million inhabitants are at least partially in the 0–10 m zone while on an average, 14 % of people in the least developed countries live in the zone (compared to 10 % in OECD countries).

Coastal ecosystems are particularly sensitive to physical and biochemical changes with reference to:

- Increased level of flooding, loss of wetlands and mangroves and saline water intrusion into freshwater habitats.
- Severity and increase of cyclonic events leading to coastal erosion, loss of ecological diversity along shorelines.

The Fourth Assessment Report of the IPCC (Parry et al. 2007) has suggested that climate change is likely to have significant impacts on the coastal regions of India. Some of these include:

- Increased frequency of hotter days and multiple-day heat wave in the past century with increase in deaths due to heat stress in recent years.
- Sea-level rise has led to intrusion of saline water into the groundwater in coastal aquifers and thus adversely affecting local freshwater resources. e.g. for two small and flat coral islands at the coast of India, the thickness of freshwater lens was computed to decrease from 25 to 10 m and from 36 to 28 m respectively, for a sea level rise of only 0.1 m.

- Warmer climate, precipitation decline and droughts in most delta regions of India have resulted in drying up of wetlands and severe degradation of ecosystems
- Ganges-Brahmaputra delta: More than 1 million people are likely to be directly affected by 2050 from risk through coastal erosion and land loss, primarily as a result of the decreased sediment delivery by the rivers, but also through the accentuated rates of sea-level rise.

In coastal regions like the Sundarbans delta in West Bengal, the recent drastic changes in weather conditions and monsoon patterns, along with frequent extreme climatic events like cyclones are major threats to the ecosystem of the region. Climate change induced by anthropogenic activities is thought to be behind the observed rise in sea level, lengthier summers, and a dramatic increase in rainfall over the past 15–20 years. The already marginal economy of human populations dependent on single crop agriculture, fishing and harvesting of forest resources is also adversely affected by changes such as sea level rise, increase in salination, changing patterns of rainfall, and increase in moisture content in the atmosphere leading to increasing incidences of vector-borne diseases (WWF 2010). This has increased their vulnerability and possibly their dependence on the forest resources.

Similarly fluctuations in the sea surface temperature along the coasts of Bay of Bengal and Arabian Seas have also resulted in changes and decline in the availability of fish species some of which are of good commercial value. Impact of climate change on regional fisheries can be ranked in terms of likelihood (for either warming or cooling) of impacts. Most of this knowledge comes from empirical studies over the recent 50 years, (WWF 2005a, b) when weather and environmental records became fundamental to explaining individual species' behavior and population responses to changes in local conditions (Patterson and Samuel 2004) or through peoples' knowledge base.

Most of our knowledge on regional fisheries and related climate impacts originates from scientific studies over the recent 50 years, when weather, local climate and environmental changes became fundamental to explaining individual species distribution and behaviour patterns (Patterson and Samuel 2004).

The fisheries are seen as highly sensitive to climate change are also often the most affected by other human disturbances like infrastructure, local migration along rivers, encroachment of wetlands, habitat loss and other issues of human population growth.

A study coordinated by the author, along with Dr. J. K. Patterson, in 2004 in the Gulf of Mannar region analyzed the effect of climate change on the fishery of False Trevally (*Lactarius lactarius*) and the reduction in the income of the dependent small scale fisherfolk community. The study helped in understanding the present status of this fishery as also identify shift in breeding grounds or migratory patterns of the fish species. False Trevally is an economically and culturally important fish in India and found near the Rameshwaram coast of south east India. The species is generally seen at depths ranging from 15 to 90 m. Over the past few years there has been a steady decline in the catch of this fish both as a consequence of human disturbance and changes in the ocean temperatures. Human activities involving use of destructive fishing practices and regular violation of coastal laws have also been responsible for the decline of the species. This has resulted in the movement of the species to other regions along the coast including the east coast of Sri Lanka (Patterson and Samuel 2004).

Sundarbans

The study in the Sundarbans primarily focused on documenting local community knowledge and perception about the adverse impacts of climate change through the "Climate Witness" Initiative (WWF 2005a, b). The idea of studying people's perception and knowledge of extreme weather events originated from the strong indicators of climate change that were available in the region through the various scientific studies that were being carried out by various academic institutions and universities. Climate change was seen as of one of the factors responsible for sea level rise, flooding, salt water intrusion, increased frequency of extreme weather events apart from loss of land and property of local communities. The east coast of India itself is highly prone to cyclones and storms generated in the Bay of Bengal. The records of natural calamities in Sagar Island itself show several such incidents, which claimed several human lives and caused damage to property. Historical records also suggest that storm surges in 1668 killed almost 60,000 people in Sagar Island suggesting the presence of a large population on the island.

The Sundarbans is part of the world's largest delta (80,000 sq.km) formed from sediments deposited by three great rivers, the Ganges, Brahmaputra and Meghna, which converge on the Bengal Basin. It consists of 102 low lying islands in the Bay of Bengal and forms one of the world's richest mangrove ecosystems with 34 true mangrove species (MoEF 1989). The faunal diversity of this area is also significant with a significant population of tiger in the region (Loucks et al. 2010). The Islands selected for the study were located in the South Western corner of Sundarbans except Chhoto Mollakhali and Bali islands situated at the North eastern part of the delta. There are more than 100 islands spread over the entire Sundarban region of which the sea facing ones are influenced by both the tidal action and delta forming process. This coastline is remarkable for its highly productive mangrove forest and nutrient rich backwaters nourishing the aquatic diversity of the Indian east coast.

A field survey was carried with the community in those islands, which experienced loss of landmass over the past few decades. These included direct interactions with local communities several impacts were categorized on people's lives and livelihood. These included soil erosion, loss of landmass, damage of embankment, siltation, unsustainable livelihood practices, population pressure, storms & cyclones, effect of tidal waves, river flows, policy and planning process.

The study revealed that ecosystem dependent communities like farmers and fishermen formed a major part of the work force of these islands. In absence of industries and other developmental activities livelihood options were limited here.

Nearly 61.85 % of the respondents surveyed were involved with farming and fishing, while fishing was an exclusive occupation for nearly 10 % of respondents. The general livelihood pattern indicated that majority of the inhabitants were vulnerable to climate related adversities.

Local community perception of perceived changes in weather were mostly linked to erratic weather patterns rather than temperature. The delayed onset of monsoon over the past few years had implications for local agricultural productivity. Average maximum temperature in Sundarban delta ranged between 35 and 39°C while the minimum level was between 12 and 15 °C. Delayed monsoon and untimely rain often hampered the agricultural productivity leading to crop loss. An extended period of heat was observed during last few years. Majority of the farmers reported pest attacks on account of irregular climate pattern.

The interventions response of the community in developing adaptation responses has been primarily a localized effort with village communities implementing short term actions as a reactive response to the threat of climate change impacts. Some of these responses included:

- Shifting of farming time in anticipation of shifting of monsoon season
- Diversification into different weather resistant crops
- Construction and renovation of ponds and canals for rain water harvesting and use in winter cultivation
- Constructing of mud-barrages around the island to protect it from incursion of saline water
- Reforestation activity (mangroves) on the mud barrage
- Alternative livelihood options for proper substitution of certain livelihood activities like baby prawn/ shrimp catches, timber use etc.

Rise in sea surface temperature in the monsoon season had a correlation with the frequency and intensity of tropical cyclones. Tidal amplitude ranged between 4.5 and 5.5 m during April to September which had resulted in inundation. Local residents reported very high frequency of thunder and lightning during storms in last 10–15 years. In their opinion depression and cyclonic storms occurred more frequently than earlier.

The study also highlighted the local community knowledge and perception of climate related impacts through development of case studies, testimonials by local witnesses and by producing a documentary film. A film Sundarbans—Future Imperfect was released during a side event organized at the United Nations Framework Convention on Climate Change, Conference of Parties—10 at Buenos Aires, Argentina in December 2004. The film records witness case studies on perception of climate change impacts in the Sundarbans as well as scientific evidence on ecosystem changes.

Western Himalaya

Impacts on high mountain systems including glacial retreat are amongst the most directly visible signals of global warming in many parts of the World. On a time scale recent glaciations occurred around 20,000 years ago as part of the earth's paleoclimatic history. Although the recession of glaciers has been suggested by some scientists as a natural phenomenon, in the later half of twentieth century, an increase in the rate of retreat has been observed in most glaciers around the world including the Himalayas.

The Himalayan region has the largest concentration of glaciers outside the polar caps. With glacier coverage of 33,000 km^2, the region is aptly called the "Water Tower of Asia" as it provides around 86,000,000 cubic meters of water annually (Immerzeel et al. 2010). These Himalayan glaciers feed seven of Asia's great rivers: the Ganga, Indus, Brahmaputra, Salween, Mekong, Yangtze and Huang Ho and ensure a year round water supply to about one billion people.

The changes in climate variability have led to a rapid retreat of mountain glacier systems which are considered the lifeline of river basins and ecosystems. Scientific studies have shown that a vast majority of glaciers are retreating at a startling rate in the Himalayas as a result of various factors including climate change (Kumar et al. 2009). The threat of glacial melt is likely to have serious implications on several sectors of society and livelihoods exacerbating the stress from existing developmental activities. Sectors which are likely to be affected include, water, agriculture, power, and biodiversity. Past work by glaciologists and climatologists have found that the accelerated rate of glacial melt in the some of the important glaciers in Uttarakhand in Western Himalaya will have serious consequences for the freshwater ecosystems of the Ganga basin, with long term impacts for biodiversity, people and livelihoods as well as regional food security (Rees and Collins 2004; Rao et al. 2008). This will not only mean repercussions on the region's agricultural productivity and industrial activity, but also on the Terai ecosystems and species like the Ganga river dolphin. Some of the important glaciers in the region include the Gangotri, Yamunotri, Dokriani, Pindari and Milam glaciers which form important components of the watershed.

Community knowledge at local level and responses have suggested that some of the major concerns of the Western Himalaya relate to the availability and requirements of water, agriculture (Vedwan and Rhoades 2001) biodiversity, and power.

Water

The supply of freshwater resources is likely to increase as perennial snow and ice volume decrease. Glacial melt runoff in Himalayan glaciers has been highest in the summer months (May–August). Communities have noticed changes in agriculture environments and related water resources through reduced crop yields of important crops like wheat and rice in the fertile belt of the Indo Gangetic Plains. The increased threat of extreme events—floods and droughts leading to Glacial Lake Outburst Flood (GLOF) events could threaten mountain communities as well as infrastructure projects downstream.

Agriculture

The reduced availability of freshwater in terms of quantity, timing and flow in the western Himalaya has also led short fall in supply in the perennial rivers which in turn has affected agricultural productivity. Increases in temperature and water stress are expected to lead to decline in crop yields up to 30 % according to some studies. Changes in phenology and timing of growth due to increased surface temperatures may have a deleterious effect on local vegetation composition. A case in point is the recent production loss in apple crop in many parts of the Himalaya due to climatic variability.

Biodiversity

Community knowledge system and anecdotal information amongst people in the Uttarakhand Himalayas also reveal that riparian ecosystems in the Himalayan belt comprising deciduous temperate forests are likely to be affected (including shifts in forest boundaries) due to changes in water flows under changing glacial melt (Rao et al. 2008). Many of them report that changes in local surface temperature have resulted in upward movement of tree line which in turn has affected species composition and vegetation types. Although we do not have information on how this may impact faunal diversity, it is expected that certain ecosystem dependent species may decline over time and also move shift their distribution range as a consequence.

Power

Although community knowledge systems are very local in nature and perceptions are often seen only in terms of climate impacts on lands, species, vegetation etc. An important aspect to consider is the availability of power in the Himalayan terrain. The perennial rivers of the Himalayan region are an important source for hydro power generation in India. The Ganga basin has nearly 11,000 MW of hydropower potential which suggests that there has been renewed emphasis on exploiting the hydro power potential of such mountain river systems through development of various types of dams for hydro projects. These include run-of-the river hydro power plants which are considered as environmentally friendly by many. However changes in melt water flows could impact energy security through lack of enough water to run a hydropower plant at full potential. The rapid development of such dams could also cause other indirect effects like environmental impacts of fragile forest ecosystems as well as a livelihood issues like rehabilitation (Thakkar 2009). Local communities have often cited the serious ecological impacts of power plants in a fragile mountain ecosystem which not only has implications for biodiversity but also affects natural river flow, submergence of entire communities and erosion of riverine landscapes. The recent Uttarakhand floods in June 2013 is a case in point where large scale flooding in the upper reaches due to an extreme weather coupled with unplanned hydropower dam construction resulted in serious damage to property and life.

Conclusions

Climate change as an issue is still relatively less understood and strengthening capacity through awareness generation is an important part of building resilience. The role of climate adaptation efforts in the context of natural ecosystems has gained considerable momentum in the past few years through National and International efforts under the UNFCCC as well as through the intervention of civil society institutions. Other activities include promoting the role of grass roots level civil societies through development of resource centers and creating local knowledge networks to raise the level of local development planning.

The use of people knowledge in building adaptive strategies is an important element of climate adaptation for ecosystems and landscapes. While the science is fairly clear understanding peoples priorities and their capacity to absorb climate impacts must be an integral part of climate adaptation (Byg and Salick 2009). Interventions through adaptation efforts are possible to build the resilience level across ecosystems like Himalayas, coastal region and deltas like the Sundarbans. These could be achieved through different approaches where people, livelihood and ecosystems are considered as an integral part of the biosphere. Adaptation to climate change and reducing vulnerability in these l regions may be dependent on several factors including gender, access to information, technology, infrastructure etc. A case in point is gender based vulnerability where women in coastal regions often tend to face higher mortalities during natural disasters like cyclones, and extreme weather events. For example, in the 1991 cyclone in Bangladesh, women constituted nearly 90 % of the total deaths out of about 140,000.

Initiatives in the three key landscapes for building the resilience and adaptive capacity of the region of these critical and fragile ecosystems should be considered not only from an ecosystem perspective but also from the view point of understanding the requirements of people and their livelihood. In all the three cases, local community knowledge network documentation is important as they have a strong livelihood dependency factor on ecosystem thereby increasing their vulnerability to climate related stress.

(continued)

The involvement and role of multiple levels of stakeholders is therefore a challenge in developing suitable adaptation frameworks in these impacted regions. These could be summarized in terms of:

(a) Effective intergovernmental coordination between various coordination departments particularly those which a have a direct relevance to issues related to the affected ecosystem. Using the case of Sundarbans, this could involve Ministries of Forests, Irrigation, Agriculture, Fisheries, Revenue, Inland waterways, rural development, etc.
(b) The role played by academicians and research scientists is very vital in building a common understanding of the scientific validation of the problem. In the case of the Gulf of Mannar, False trevally case, it is important to bring together fishery scientists, social and developmental scientists, oceanographers, marine biologists as well as climatologists for understanding the factors leading to decline in the density of False Trevally as an economically viable fish for the communities living in the Gulf of Mannar region.
(c) Communications and media workers often play a major role in dissemination and documentation local and traditional knowledge systems and key issues of the region to local communities, policy makers at state and, national and international level effectively. One of the challenges in this area is the lack of such information being made available across all sections of society particularly at the level of policy makers for effective implementation of policies.

References

Byg A, Salick J (2009) Local perspectives on a global phenomenon-climate change in eastern Tibetan villages. Glob Environ Change 19:156–166. doi:10.1016/j.gloenvcha.2009.01.010

Chaudhary P, Bawa K (2011) Local perceptions of climate change validated by scientific evidence in the Himalayas. Biol Lett. doi:10.1098/rsbl.2011.0269

Immerzeel WW, van BeekL PH, BierkensM FP (2010) Climate change will affect the Asian Towers. Science 328:1382–1385

Joshi A, Joshi PK (2011) A rapid inventory of indicators of climate change in the middle Himalaya. Curr Sci 100(6):831–833

Kumar R, Areendran G, Rao P (2009) Witnessing change: Glaciers in the Indian Himalayas. WWF India and Birla Institute of Technology Publication, p 38

Loucks C, Barber-Meyer S, Hossain MAA, Barlow A, Chowdhury RM (2010) Sea level rise and tigers: predicted impacts to Bangladesh's Sundarbans mangroves. Clim Change 98:291–298

McGranahan G, Balk D, Anderson B (2007) The rising tide: assessing the risks of climate change and human settlements in low elevation coastal zones. Environ Urban 19(1):17–37

MoEF, Sunderban Biosphere Reserve (1989) Project Document No. 10. Ministry of Environment and Forests, Government of India, New Delhi. (http://www.sadepartmentwb.org/Sundarban_Biosphere.htm)

MoES Annual Report (2010) Ministry of Earth Sciences, Government of India, New Delhi, India. (http://www.dod.nic.in/AnnualReport/AR2009-2010.pdf). pp 98

Myers N, Mittermeier RA, Mittermeier CG, da-Fonseca GAB, Kent J (2000) Biodiversity hotspots for conservation priorities. Nature 40:853–858. doi:10.1038/35002501

Naess LO (2013) The role of local knowledge in adaptation to climate change. WIREs Clim Change 4:99–106. doi:10.1002/wcc.204

Nicholls RJ, Leatherman SP (1995) Potential impacts of accelerated sea level rise on developing countries. J Coast Res 14:324

Parry ML, Canziani OF, Palutikof JP, Van der Linden PJ, Hanson CE (Eds) (2007) Climate change 2007, Impacts, adaptation and vulnerability. Contribution of Working Group II to the Fourth Assessment Report of the Intergovernmental Panel on Climate Change. Cambridge University Press, Cambridge United Kingdom and New York, NY, pp 976

Patterson JK, Samuel D (2004) Effect of climate change on the loss of *Lactarius Lactarius* fishery and dependent fishermen in the southeast coast of India. Internal Technical Report submitted to WWF India. New Delhi, pp 32

Patwardhan A, Narayanan K, Parthasarathy D, Sharma U (2003) Impacts of climate change on coastal zones. In: Shukla PR, Sharma SK, Ravindranath NH, Garg A, Bhattacharya S (eds) Climate change and India-vulnerability assessment and adaptation. Universities Press (India), Hyderabad, India, pp 326–359

Rao P, Areendran G, Sareen R (2008) Potential impacts of climate change in the Uttarakhand Himalayas. Mountain Forum Bull 8(1):28–30

Rees G, Collins D (2004) An assessment of the potential impacts of deglaciation on water resources of the Himalaya—Sagarmatha Project Summary Report

Reidlinger D, Berkes F (2001) Contributions of traditional knowledge to understanding climate change in the Canadian Arctic. Polar Rec 37:315–329. doi:10.1017/S0032247400017058

Rupa Kumar K, Sahai AK, Kumar KK, Patwardhan SK, Mishra PK, Revadekar JV, Kamala K, Pant GB (2006) High Resolution climate change scenarios for India for the 21st century. Curr Sci 90(3):334–345

Stainforth DA, Aina T, Christensen C, Collins M, Faull N, Frame DJ, Kettleborough JA (2005) Uncertainty in predictions of the climate response to rising levels of greenhouse gases. Nature 433:403–406

TERI (1996) The economic impact of a one metre sea level rise on the Indian coastline: methods and case studies. Tata Energy Research Institute. [Report No 93/GW/52, submitted to the Ford Foundation. New Delhi

Thakkar H (2009) India's national action plan on climate change: A civil society view- "There is little hope here". South Asia Network on Dams, Rivers and People Publication, New Delhi, India, p 56

Vedwan N, Rhoades RE (2001) Climate change in the Western Himalayas of India: a study of local perception and response. Clim Res 19:109–117

WWF (2005a) An overview of glaciers, glacier retreat and its subsequent impacts in the Nepal, India and China. WWF Nepal Program. Kathmandu, p 67

WWF (2005b) Are we putting our fish in hot water? WWF Climate Change programme, WWF International Report, Gland, Switzerland, p 15

WWF (2010) Sundarbans: future imperfect-climate In: Danda A (ed) Adaptation report. WWF India Publication, New Delhi, p 30

Xu J, Grumbine RE, Shreshtha A, Eriksson M, Yang X, Wang Y, Wilkes A (2009) The melting Himalayas: cascading effects of climate change on water, biodiversity, and livelihoods. Conserv Biol 23:520–530

Chapter 7
Managing Environmental Migration to Improve Economic and Social Outcomes in Developing Asia and Pacific

Bart W. Édes and François Gemenne

Abstract This paper assesses that millions of persons are displaced annually in the Asia and Pacific region due to large concentrated populations residing in areas exposed to environmental risks. Some of those displaced end up becoming migrants, establishing their home in a new location. Migration influenced by environmental factors is contributing to the urbanization trend in the region. Displacement and migration are expected to increase in the coming years due to the impacts of climate change, and the increasing number of people residing in areas at risk of extreme environmental events or slow-onset phenomenon, such as rising sea levels. In addressing environmentally driven migration, a key policy aim should be to curtail the likelihood of forced displacement. "Climate proofing" of physical infrastructure and government support to vulnerable and poor populations can strengthen resilience. Although environmentally driven migration generally unfolds within a country's borders, it can also take on an international character. For this reason, governments need to collaborate in tackling this challenge. If the threat of environmentally driven migration is properly managed and planned, benefits can accrue to migrants, host communities, and communities of origin.

B.W. Édes (✉)
Governance, Poverty Reduction and Social Development Division, Regional and Sustainable Development Department, Asian Development Bank, 6 ADB Avenue, 1550 Mandaluyong, Metro Manila, Philippines
e-mail: bedes@adb.org

F. Gemenne
CEARC, OVSQ, University of Versailles Saint-Quentin (UVSQ), 11 Boulevard d'Alembert, 78280 Guyancourt Cedex, France

CEDEM, University of Liège, Boulevard du Rectorat, 7/45 Building B31, 4000 Liège, Belgium
e-mail: francois.gemenne@sciencespo.fr; Francois.Gemenne@uvsq.fr; F.Gemenne@ulg.ac.be

W. Leal Filho (ed.), *Climate Change in the Asia-Pacific Region*, Climate Change Management, DOI 10.1007/978-3-319-14938-7_7

Introduction

Asia and the Pacific is the most disaster-prone region in the world. Millions are displaced annually in the region by extreme environmental events, including monsoons (cyclones), floods, storm surges, and drought. Over the coming decades, climate change is expected to increase the number and severity of storms, and contribute to coastal erosion and rising seas. Combined with large and growing numbers of people living in environmentally precarious areas, these trends, combined with economic inequalities and other societal transformations, will stimulate migration.

Discourse on climate change and migration tends to be dominated by humanitarian catastrophes and doom-and-gloom scenarios in which migration is presented as a threat to security, or as a crisis to be avoided at all costs. This chapter posits that environmental migration can actually be managed and anticipated. Although perhaps counterintuitive, migration need not signal helplessness and a failure of adaptation to climate change.

On the contrary, if governments help to building resilience[1] of environmentally threatened communities, improve the capacity of destination areas to accommodate more inhabitants, and begin to facilitate proactive, voluntary movement away from areas facing the most likely environmental disruption, migration can be harnessed as a practical and powerful tool to enable adaptation, promote sustainable development, and avoid humanitarian crises.

An Urbanizing Region at Risk

Asia and the Pacific will be among the global regions most affected by the impacts of climate change, be they creeping at a gradual pace or sudden catastrophic events. Such impacts include significant temperature increases, changing rainfall patterns, greater monsoon variability, sea-level rise, floods, and more intense tropical cyclones (Cruz et al. 2007). Asia and the Pacific is particularly vulnerable because its already elevated exposure to environmental risks, high population density—particularly along coastlines, and vulnerability of particular social and economic groups.

Asia and the Pacific accounted for around half of the global frequency of intense natural disasters between 1971 and 2010. Intense hydrometeorological disasters accounted for 72 % of intense natural disasters recorded in the region over the same period (ADB 2013a). Future increases in monsoon-related precipitation extremes are very likely in East, South, and Southeast Asia (IPCC 2013).

Within the broad and diverse region, climate change is expected to take the heaviest toll on the Pacific, South Asia, and Southeast Asia. Globally, eight of the ten countries with the largest forecasted population at risk from sea level rise in

[1] "Resilience" is the ability of countries, communities, businesses, and individual households to resist, absorb, recover from, and reorganize in response to natural hazard events, without jeopardizing their sustained socioeconomic advancement and development (ADB 2013b).

2050 are located in South Asia and Southeast Asia (Wheeler 2011). In addition, most large Asian delta areas are sinking due to groundwater withdrawal, floodplain engineering, and trapping of sediments by dams (IPCC 2013). Sea-level rise threatens critical rice production areas along coasts and in delta areas, e.g., Bangladesh and the Mekong Delta area (IPCC 2013).

There is also a serious risk of greater food insecurity and the breakdown of food systems linked to warming, drought, flooding, and precipitation variability and extremes in Asia and the Pacific, particularly for poorer populations. Climate change is also projected to reduce raw water quality and pose risks to drinking water quality even with conventional treatment, due to interacting factors of higher temperatures, increased sediment, nutrient, and pollutant loadings from heavy rainfall; heightened concentration of pollutants during droughts; and disruption of treatment facilities during floods (IPCC 2014). Climate change is also expected to exacerbate poverty and health problems, which will interplay with other environmental disruptions.

Combined, these developments could contribute to population displacements of unprecedented scale in the coming decades. According to the International Displacement Monitoring Centre, more than 90 million[2] people—equivalent to Viet Nam's entire population—were displaced by storms, floods, mudslides, and typhoons (cyclones) in Asia and the Pacific in the 5 years 2009–2013. The accompanying table shows that the countries experiencing the largest displacement due to weather in 2013 are all in developing Asia.

While most displaced persons go back home when circumstances permit, many cannot return to their communities, or choose to settle elsewhere for reasons that often relate to safety, security, or livelihood opportunities. In addition, untold numbers leave their homes partly because of creeping environmental threats, such as farmers fleeing rural areas after a string of drought years, or fisherfolk gradually pushed away from coastlines by erosion and storm surges.

Between May and August 2010 alone, some 10 million people were displaced in Pakistan by unusually heavy monsoon rains, and another 15 million were forced to depart their homes in the People's Republic of China due to flooding. In November 2013, one of the strongest tropical storms ever recorded—anywhere—pummelled the Visayas region of the Philippines, dislocating four million people, or about one-third of the population in this central section of the country.

Small island developing countries often have a higher relative disaster risk than larger countries because most of their population and assets are exposed to hazards such as tropical cyclones, while their economies may be concentrated in one or a few vulnerable sectors, such as tourism (UNISDR 2011). The Pacific includes lightly populated states facing the long-term prospects of inundation due to sea rise, especially Tuvalu and Kiribati. In fact, in June 2014, a family from Tuvalu was granted residency on humanitarian grounds by the Immigration and Protection Tribunal in New Zealand after claiming to be threatened by climate change in its home country.

---

[2] The figure includes weather-related events which displaced at least 100,000 people. Source: IDMC annual estimates of displacement due to disasters: http://www.internal-displacement.org/publications.

In March 2014, the Marshall Islands experienced the worst king tides in more than three decades, forcing evacuation of people in various areas including Majuro Atoll, which in some places is just 30 cm above sea level. In Carteret Islands, situated 53 km northeast of Bougainville in Papua New Guinea, storm surges and invasive high tides over the years have led to start-stop efforts to relocate the population to Bougainville to escape persistent flooding. Eight thousand kilometres to the north, on the Bering Sea coast, hundreds of residents of Newtok, Alaska, are in the process of relocating to higher ground as their village is expected to be fully submerged within a handful of years. The Ninglick River, which empties into the sea, is fast eroding the land around the community. Thawing permafrost and diminishing sea ice are reducing natural barriers to wave erosion.

Asian countries with the largest number of persons displaced by weather-related natural disasters in 2013

Country	Displaced event	Month	Total displaced (per event)	Total displaced (per country)
Philippines	Typhoon Haiyan (local name Yolanda)	November	4,095,000	6,721,000
	Typhoon Trami (local name: Maring)	August	1,744,000	
	Typhoon Nari (local name: Santi)	October	406,000	
	Tropical Storm Shanshan (local name Crising)	February	223,000	
	Typhoon Utor (local name Labuyo)	August	129,000	
	Floods (intertropical convergence zone effects)	October	124,000	
People's Republic of China	Floods	June–July	1,577,000	4,878,000
	Typhoon Fitow	October	826,000	
	Typhoon Usagi	September	587,000	
	Typhoon Utor	August	513,000	
	Typhoon Soulik	July	500,000	
	Floods	July–August	354,000	
	Typhoon Trami	August	190,000	
	Typhoon Haiyan	November	181,000	
	Hainan Floods	December	150,000	
India	Floods	June–October	1,042,000	2,042,000
	Tropical Cyclone Phailin	October	1,000,000	
Bangladesh	Tropical Cyclone Mahasen	May	1,100,000	1,100,000
Viet Nam	Typhoon Haiyan	November	800,000	1,015,000
	Typhoon Nari	Mid-October	109,000	
	Typhoon Wutip	End September–October	106,000	

Source: International Displacement Monitoring Centre (2014), Global Estimates (2014): People Internally Displaced by Disasters, Geneva

In addition to being on the frontline of the impacts of climate change, Asia and the Pacific has experienced massive and rapid socioeconomic transformation and industrialization. It boasts a population of about 4.4 billion people, about three-fifths of the world's total. The region is also home to the most important source of migrants worldwide (about 80 million out of 200,000,000). Mobility has increased substantially in recent years, due to rising incomes, regional integration, and greater access to communications and transportation.

Indeed, one of the most striking demographic trends in the region in recent decades has been an increase in the level and complexity of population mobility. Migration within countries, especially from rural to urban and peri-urban areas, has had major societal impacts. Environmentally induced migration has contributed to increasing urbanization. Existing migration corridors and channels, both internal and international, are expected to be travelled by future migrants.

Patterns of migration within Asia and the Pacific—and between this vast region and others—are complex. Most migration takes place within countries, although there is substantial international movement. This is influenced by a range of factors, including geography; cultural, historical, and political links; employment opportunities; and government policies. Climate change can be expected to both accentuate and alter these trends as it mingles with other motivations for people to relocate.

A major current running through the socioeconomic development of Asia and the Pacific is urbanization. By the middle of this century, Asia's urban population is expected to increase by 1.4 billion, and will account for over 50 % of the global population (UNDESA 2012).

Environmental migration is intertwined with urbanization, and at the same time strengthens it. Many of those displaced by Typhoon Haiyan (2013) in the Visayas region of the Philippines relocated to larger urban centers in the country, including Cebu City and Metro Manila.

Several large Asian cities have grown swiftly over the past two decades and now struggle to support their large populations. Environmentally driven migration will add to the strain. Further, the urban destinations of many migrants are themselves at great risk due to climate change. The movement of people to cities puts pressure on governments to improve urban planning. With an increasing inflow of people, needed actions include zoning laws and incentives that encourage people to settle in less vulnerable areas, and greater investment in basic infrastructure. Many Asian cities lack the carrying capacity to accommodate continued growth without major changes. Urban development patterns will therefore need to be rethought so that cities can grow in a more sustainable way and provide adequate services to their populations, including new arrivals driven by environmental and other reasons.

A particular feature of Asian urbanization has been the emergence of "mega-cities"—large, complex urban areas of 10 million or more. The massive growth of these urban agglomerationsin coastal areas has enlarged the number of people exposed to the risks posed by climate change. It is ironic that many of the vulnerable urban areas being fed by environmental migration are actually at great risk themselves due to short- or long-term hazards such as intense storms and rising seas. Climate change is expected to increase the frequency of extreme events—especially floods—in coastal megacities. Due to population growth, the number of people exposed to these flooding events, and at risk of displacement, will also grow.

If cities are planned and managed in anticipation of further large in-migration, they can act as transformational agents that reduce urban poverty and energize an entire country. Unplanned cities contribute to poverty and create vulnerability as evidenced by living circumstances in slums and informal settlements. Involving stakeholders and taking a holistic approach to urban development cannot only reinforce the role of cities as engines of growth, but also promote improved positive interactions with rural areas (Yuen 2014).

Using Migration as a Tool of Adaptation

Long-term environmental change, rising seas, coastal erosion, and reduced agricultural productivity will stimulate future migration (Lilleor and Van den Broeck 2011). Where environmental conditions slowly worsen, those who have sufficient resources, information and contacts, and are sufficiently motivated or entrepreneurial, will migrate in search of a better life elsewhere. Those with few assets will move later, or not at all, unless a sudden catastrophic event presents them with no immediate alternative. The most vulnerable households are able to use migration to cope with environmental stress, but their migration is an emergency response that creates conditions of debt and increased vulnerability, rather than reducing them (Warner and Afif 2013). Migration decisions depend on many factors, including availability of insurance before a major storm, the effectiveness of emergency relief, the presence of income-producing opportunities, and whether or not homes, businesses, and equipment remain intact and usable (Newland 2011).

Displacement risk increases when populations that lack the resources for planned migration experience higher exposure to extreme weather events, in both rural and urban areas, particularly in developing countries with low income. Expanding opportunities for mobility can reduce vulnerability for such populations. Changes in migration patterns can be responses to both extreme weather events and longer-term climate variability and change, and migration can also be an effective adaptation strategy (IPCC 2014).

Migration can contribute to improved livelihoods, and provide resources to environmentally threatened communities to strengthen their capacity to withstand natural hazards. Out-migration can serve as a way of coping with climate change, and as a mechanism to reduce poverty and increase resilience in affected areas. For example, migrants from rural areas can migrate to take urban jobs and remit funds to their home communities. Seasonal and temporary migration can generate large development impacts, at least on the households that supply workers (Gibson et al. 2013).

In fragile environments, migration is a common response to extreme vulnerability and is essential in satisfying basic needs. It can lead to improved livelihoods or act as a sort of insurance strategy for households through diversification of income sources (Foresight 2011).

Migrant networks can contribute to improved resilience to disasters by implementing humanitarian, adaptation, and development projects, facilitating access to information, lobbying in the political sphere, and channelling donations and remittances (Barnett and Webber 2010; ADB 2012).

Migration also can benefit communities of destination, most notably by addressing labor and skill shortages in growing urban areas of destination, stimulating economies through entrepreneurial activities, and filling demographic gaps created by aging populations (Foresight 2011).

While there is fear in many developed countries that immigration from less developed states will result in the loss of jobs, particularly for less skilled workers, the evidence does not bear this out. On the contrary, migration tends to stimulate investment, induce task specialization of host country workers, and possibly even raise demand for all workers. Evidence shows that migrants respond to demand for their labour, so enhancing legal migration opportunities for low-skilled workers does not necessarily increase total immigration as long as there is enforcement of labour regulations applicable to employers (Dadush 2013).

International cooperation is critical to addressing the cross-border aspects of migration. There are existing guidelines, principles, and agreements that can and should be utilized by sovereign states. For example, ASEAN member states have negotiated a framework to facilitate movement of skilled labour and professionals.

Patterns of migration within Asia and the Pacific—and between this vast region and others—are complex. Most migration takes place within national borders, although there is substantial international movement. This is influenced by a range of factors, including geography; cultural, historical, and political links; employment opportunities; and government policies. Climate change can be expected to accentuate and to alter these trends as it mingles with other motivations for people to relocate to other places.

Adapting and Building Resilience

In addressing environmentally driven migration, a key aim should be to curtail the likelihood that slow onset climatic changes or extreme environmental events lead to *forced* displacement of a short- or longer term nature. The obvious priority is climate change mitigation, i.e., a reduction in greenhouse gas emissions. Yet changes in weather patterns over recent years and decades are extremely unlikely to be altered by any realistic scenario of reduced emissions. Thus, complementary efforts of climate change adaptation are required.

Adaptation efforts that reduce migration compelled by environmental events include flood protection; land reclamation; engineering for more productive land use (such as terraces, wind-breaks, and irrigation); and the development of crop varieties that are more resistant to flooding, drought, higher temperatures, and increased salinity (Newland 2011). Adaptation is also being advanced through the mainstreaming of climate adaptation action into subnational development planning, early warning systems, integrated water resources management, agroforestry, and coastal reforestation of mangroves (IPCC 2014).

Many people living in the above mentioned Carteret Islands have resisted relocation and have instead taken active adaptation measures with the aim of remaining in their communities. Residents have built seawalls and planted mangroves in an effort to stave off the encroaching sea. While their efforts may in the longer term prove unsustainable, they have strengthened their resilience. The attitude of these island dwellers is similar to that of others living in low-lying parts of Asia and the Pacific. They do not wish to abandon their land and culture.

Unfortunately for small islands, the efficacy of traditional community coping strategies is expected to be substantially reduced in the future. One can expect, over time, a loss of livelihoods, coastal settlements, infrastructure, ecosystem services, and economic stability in such islands (IPCC 2014).

In Cuttack, a flood-prone city in India that lies between two cities, a federation of women's savings groups have used hand-held Global Position System devices to map all 331 informal slum settlements in the city. The resulting accurate digital maps enable the government to identify dwellings a risk of flooding, and to contact local leaders before rising waters pose a danger. The maps have also been used in determining use of public funds for slum upgrading (Scherr 2014).

There are three basic core needs that must be satisfied to strengthen resilience and provide a sturdy foundation for inclusive, sustainable development in the face of disaster risks. These needs include conducting a risk assessment (understanding who and what is at risk), risk reduction (reducing impact of specific natural hazards), and residual risk management (minimizing further indirect and secondary consequences following a major event) (ADB 2013b).

Risk assessment at the community level involves determining vulnerability to natural hazards while taking into account exposure, sensitivity, and adaptive capacity. Such an assessment may involve the formulation of a detailed plan of exposed areas and vulnerable spots that factor in the physical environment and demographic data (such as where people are living and their socio-economic status). Planning and zoning measures are increasingly being adopted to limit settlement in high-risk areas. Most of those displaced in December 2011 in the southern Philippines by Tropical Storm Washi were living in informal settlements. The settlements were officially recognized as high-risk areas, but poor people were forced to build there and to use techniques and lightweight materials unable to withstand floods (Internal Displacement Monitoring Centre 2013).

After the destructive Indian Ocean tsumami of December 2004, buffer zones were declared in various coastal areas (e.g., in Sri Lanka, Indonesia's Aceh Province). Construction of houses, buildings, markets, etc. was curtailed within 200 or more meters from the sea. Although a prudent measure, establishing a buffer zone can also be controversial in affected coastal communities as people may already live in restricted areas (or demand to return to such areas to rebuild destroyed homes). Further, varying topography along a coast may warrant adjusting the depth of a buffer zone from place to place. Finally the impact of buffer zones on livelihoods and community cohesion also should also be taken into account.

Resilience can also be enhanced by "climate proofing" the construction of physical infrastructure to ensure that design standards take into account disaster risks. Such standards can make a major difference in human well-being, as structural damage is a leading cause of initial displacement after a severe storm. Considerations including choice of site, the integrity of the building foundation, elevation above the ground, and impact-resistant walls. Efforts can also be made to protect and reinforce natural tools, such as drainage channels, wetlands, natural reservoirs, and shoreline buffers.

Other engineered options to reduce exposure and vulnerability include flood levees, water storage facilities, flood and typhoon (cyclone) shelters, storm and wastewater management, transport and road infrastructure improvements, elevated and floating houses, and reinforcement of power plants and electricity transmission grids.

When Severe Tropical Storm Washi struck the east coast of Mindanao island in the southern Philippines, it displaced 430,900 people. A case study of the storm response made a number of recommendations to limit the impact of future severe storms. One recommendation was to provide government and civil society actors the skills and information to act on hazard and risk maps, and improve the provision of relief assistance. The report noted that, "the risk of disaster-related displacement will continue to rise unless local officials more effectively address the exposure of vulnerable communities to natural hazards and displacement" (IDMC 2013).

Disaster risk management bolsters the resilience of vulnerable communities. Climate-proofing of urban infrastructure can be a key part of disaster-risk management in the context of climate change. The 2005 Hyogo Framework for Action (HFA) provides a comprehensive 10-year plan for disaster risk management that has been adopted by 168 member states of the United Nations. The HFA is built upon five priority actions and provides resources for improving disaster resilience.

In 2005, the Association of Southeast Asian Nations (ASEAN) countries reaffirmed and complemented these priorities in the ASEAN Agreement on

Disaster Management and Emergency Response. Yet regional cooperation should be strengthened in the field of disaster risk management, especially with regard to technology and knowledge transfer. Overall, disaster risk management should be more consistently mainstreamed into adaptation policies, with a view to preventing forced displacement.

In the case of strengthening the capacity of cities to withstand environmental assault, pursuit of risk-resilient urban development requires enabling measures related to urban governance and both human and financial resources. These measures include participatory urban development strategies, adoption of a systems approach for urban planning and management, and improved coordination among urban institutions (ADB 2013c).

Affordable, risk-relevant insurance also builds resilience and improves livelihoods and security. In rural areas, farmers can plant crops with the confidence that if disaster strikes, they will be compensated for their losses. Yet very few people living in vulnerable settings in developing Asia and the Pacific have insurance. In March 2014, ADB and the Government of Bangladesh agreed to develop innovative crop insurance products that give small-holder farmers in Bangladesh income protection from extreme storms. The demonstration project, financed by the Governments of Bangladesh and Japan, will design and pilot crop insurance products over a 3-year period in selected parts of the country.

Public works and cash transfer programs are also being utilized to strengthen resilience of vulnerable communities. Other practical forms of social assistance for use before, during or after disasters include food banks, subsidized healthcare and basic services for those who cannot afford them, and disaster and humanitarian relief. A study carried out to support Ho Chi Minh City, Viet Nam to adapt and cope with climate change and variability identified two main measures to reduce poverty and vulnerability of poor people: (1) livelihood protection and interventions to promote diversification for households and protect food security and incomes, and (2) social protection and health insurance schemes (ADB 2010). These recommendations are also highly relevant to many rural and urban communities elsewhere in Asia and the Pacific.

While government must play the leading role in guiding climate change adaptation within individual countries, findings from a micro-scale analysis in Can Tho, Viet Nam warrant reflection. The analysis found that "actual adaptation measures are often based on action and agency at household level, rather than on formal broad-scale strategies driven and implemented by state agencies or local government." It further "illustrated that the actual financial costs and human resources have to by the households themselves" (Birkmann et al. 2010).

Facilitating Movement

A key policy challenge for decision makers in Asia and the Pacific will be how to harness migration's potential as an adaptation strategy to climate change. Policy and program interventions will be required to enable people to use existing migration channels, as well as new ones. A particular group that requires attention are poorer residents of large cities vulnerable to climate change impacts. They often live in areas most at risk of experiencing both gradual and sudden environmental hazards. It is important that mechanisms and resources be made available to the poorest populations to encourage their participation in migration as a way of adapting to climate change. Too often, these vulnerable populations are given inadequate attention when policies are made. Any initiative policymakers design to aim at better assisting those who leave should not forget those who are forced to stay; one should work toward the establishment of a genuine right to mobility.

While most climate change-related migration will occur within countries, there is likely to be an increase in international migration as well. Although most migration in Asia and the Pacific is within countries, there is considerable movement of people across borders within the region, and between Asia and the Pacific and other parts of the world. Decisions about what persons to legally admit into a country are of course the domain of governments. Thus, more dialogue and cooperation among governments is needed to constructively address cross-border population movements.

A promising area for international cooperation is the negotiation of agreements between migrant-sending countries and countries that are experiencing labour shortages. Advanced industrialized economies with declining fertility rates and aging populations could benefit by welcoming more migrant workers from disaster-prone developing countries with people interested in earning higher wages abroad. Under the Recognized Seasonal Employer partnership between New Zealand and several Pacific island countries, a few thousand low-skilled workers are employed in New Zealand's horticultural and viticulture sectors. The migrants can work from seven to nine months each year, and may return the following year. The Pacific countries[3] involved include several that are very vulnerable to climate change. Separately, some developed countries in other world regions (e.g., United States, Finland) have adopted policies allowing those who have fled natural disasters to seek at least temporary asylum.

The "Colombo Process" has since 2003 provided a forum for Asian (and other) states to discuss effective management of overseas employment programs. These and other examples provide a base for discussions on how to cooperate in the management of future migration, including migration driven by climate-related factors. In addition, the Nansen Initiative, launched in October 2012, has organized subregional discussions among government officials on human mobility in the context of disasters and climate change in Asia and the Pacific.

Rather than establishing a new category of climate-induced migrants—which is very difficult for a number of methodological and political reasons, countries will probably prefer to work within currently recognized migration categories when addressing climate-related migration. Few, if any, destination countries are willing

[3] Kiribati, Samoa, Tonga, Tuvalu, and Vanuatu.

to accept a new category of "climate change migrants." Many migration channels are already in place, and the demographic reality in those low fertility and aging countries is that there will be an increased need for migration to sustain their workforces in the future.

Social protection, provided through mechanisms like cash transfers and active labour market programs, can facilitate migration by lowering costs for migrants. It can help migrants connect with labour markets, develop job-relevant skills, subsidize acquisition of administrative and legal documentation, and direct costs and opportunity costs of relocation, such as compensation for lost assets and earnings. Social protection can also support social networks that facilitate migration and, as noted earlier, reduce risk through insurance (Johnson and Krishnamurthy 2010).

While internal migration to safer areas within most developing Asian countries is feasible, the situation of small Pacific islands is very much different. Residents of island states may have few practical domestic options. Significant barriers exist to migration, including the expense of traveling, visa requirements in other countries, and problems associated with accessing housing, services, and work in new destinations" (Barnett and Chamberlain 2010).

Emigration from the Pacific tends to closely follow certain paths, such as from Polynesia and islands governed in free association with New Zealand and the United States, to destinations in Australia, New Zealand, and the United States (Bedford 2000). "Once migrants from the Pacific Islands have established themselves in their new destinations, they help others in their social networks to overcome the barriers to migration. Thus, migration from places where the legal barriers are minimal—for example from Niue to New Zealand or from rural to urban areas within island states—tends to follow distinct patterns as people move to places where they have family and friends who can help them settle" (Barnett and Chamberlain 2010).

Migrants are typically often among the most marginalized groups in society. Governments should take action to reduce their vulnerability and risk, and protect them from abuses." In particular, governments should "provide migrants with access to the same basic services as current residents (education, health, water, and sanitation); actively promote a positive image of migrants to facilitate their integration into receiving communities; apply internationally recognized standards and principles on human mobility, as well as good practice on involuntary resettlement; and codify and enforce land ownership" (ADB 2012).

Under various possible scenarios, many low-lying countries in the Pacific (and elsewhere—like the Maldives) may be at least partially submerged within this century unless very substantial sums are invested in physical infrastructure that can keep the rising sea at bay. Along with low elevation coastal areas in developing Asia, remaining in place may not be an option for the children, grandchildren or great grandchildren of people presently residing in such places. Yet preparatory action undertaken now can ease the adjustment over time as carefully managed resettlement is organized and implemented.

Conclusion

By acting now, governments can reduce the prospects of humanitarian crises and manage the increased migration flows that are anticipated. If properly managed, facilitated relocation of environmentally threatened populations can be beneficial for the migrants, the host communities and the communities of origin. This poses a specific challenge in Asia and the Pacific because of the high environmental exposure of the region, and because of its demographic growth and growing urbanization.

This chapter has sought to highlight some of the challenges that need to be addressed for better management of environmental migration in the region. Policy choices by Asian and Pacific governments will determine if the future patterns of mobility related to climate impacts in the region are characterized by forced displacement or by voluntary, planned migration. Timely action and international cooperation can turn climate-induced migration from a threat into an opportunity to improve livelihoods, reduce poverty, meet labour force needs, boost economies, and strengthen links between communities and countries.

I must be kept in mind that the poorest and most vulnerable are often unable to move to more secure locations even as climatic conditions worsen. These populations will need particular attention in the elaboration of disaster risk management and adaptation policies, and in the extension of social protection coverage. In many cases, migration will need to be facilitated, not avoided.

At the same time, those who wish to stay should, within practical constraints, have the possibility to do so. Among other things, this means that livelihoods will need to be protected, so that forced displacement does not occur. The most cost-effective way for a country to strengthen resilience and reduce climate-related risks over the medium-term is by incorporating risk reduction into development planning, land use policies, building standards, and environmental management. Such an approach can, over time, lessen more intensive risks by directing development to less exposed areas, mitigating hazards, reducing vulnerability, and curtailing forced migration (UNISDR 2011).

Climate-induced migration can be addressed through a variety of policy interventions but needs to be firmly established within a development agenda. Too often, the displacement of people by climate change has been presented as a threat and addressed from a security standpoint, prompting maladaptive policy responses.

Governments should provide the most vulnerable communities with migration options, including where feasible the right to stay where they live. This will not come without cost. Yet there are a variety of means by action on environmental migration can be financed, including risk-based

(continued)

insurance mechanisms, and incorporating negative externalities into fiscal measures.

The way we will address environmental migration will be a key element of the way we will face climate change. Future policies and actions on climate change adaptation should systematically integrate considerations of possible displacement of populations. Doing so will reduce the incidence of humanitarian crises, and better highlight the role of migration as a tool for the promotion of sustainable development.

References

ADB (2012) Addressing climate change and migration in Asia and the Pacific. Manila

ADB (2013a) Climate-related disasters in Asia and the Pacific. ADB Economics Working Paper Series, No. 358. Manila

ADB (2013b) Investing in resilience: ensuring a disaster-resistant future. Manila

ADB (2013c) Moving from risk to resilience: sustainable urban development in the Pacific. Manila

Asian Development Bank (ADB) (2010) Ho Chi Minh City—adaptation to climate change summary report. Manila

Barnett J, Chamberlain N (2010) Migration as climate change adaptation: implications for the Pacific. In: Burson B (ed) Climate change and migration—South Pacific perspectives. Institute of Policy Studies, Wellington

Barnett J, Webber M (2010) Accommodating migration to promote adaptation to climate change. The World Bank, Washington, DC

Bedford R (2000) Meta-societies, remittance economies and Internet addresses of contemporary human security in Polynesia. In: Graham D, Poku N (eds) Migration, globalization and human security. Routledge, London and New York

Birkmann J et al (2010) Adaptive urban governance: new challenges for the second generation of urban adaptation strategies to climate change. Sustain Sci 5:201

Cruz RV, Harasawa H, Lal M, Wu S, Anokhin Y, Punsalmaa B, Huu Ninh N (2007) Chapter 10: Asia. In: Parry ML, Canziani OF, Palutikof JP, van der Linden PJ, Hanson CE (eds) Climate change 2007: impacts, adaptation and vulnerability: contribution of working group II to the Fourth Assessment Report of the Intergovernmental Panel on Climate Change. Cambridge University Press, Cambridge, UK

Dadush U (2013) The effect of low-skilled labor migration on the host economy. Washington, DC. Global Knowledge Partnership on Migration and Development, World Bank

Foresight (2011) Migration and global environmental change. Final project report. The Government Office for Science, London, United Kingdom

Gibson J, McKenzie D, Halahingano R (2013) Development impacts of seasonal and temporary migration: a review of evidence from the Pacific and Southeast Asia. Asia Pacific Pol Stud 1 (1):18–32

Intergovernmental Panel on Climate Change (IPCC) Working Group II (2013) Fifth assessment report, Chapter 24—Asia (final draft). Stanford, CA

Intergovernmental Panel on Climate Change (IPCC) Working Group II (2014) Fifth assessment report, summary for policy makers: Climate change 2013: impact, adaptation, and vulnerability. Stanford, CA

Internal Displacement Monitoring Centre (IDMC) (2013) Disaster-induced internal displacement in the Philippines—the case of tropical storm Washi/Sendong. Geneva.

Internal Displacement Monitoring Centre (IDMC) (2014) Global estimates 2014: people internally displaced by disasters. Geneva

Johnson CA, Krishnamurthy K (2010) Dealing with displacement: can "Social Protection" facilitate long-term adaptation to climate change? Glob Environ Change 20(4):648–653

Lilleor HB, Van den Broeck K (2011) Drivers of migration and climate change in less development countries. Glob Environ Change 21(Suppl):S70–S81

Newland K (2011) Climate change and migration dynamics. Migration Policy Institute, Washington, DC

Scherr J (2014) Slum dwellers face climatic disasters. Natural Resources Defense Council Staff Blog (http://switchboard.nrdc.org/blogs/jscherr/slum_dwellers_face_climatic_di.html)

United Nations Department of Economic and Social Affairs (UNDESA) (2012) World urbanization prospects the 2011 revision. United Nations, New York

United Nations International Strategy for Disaster Reduction (2011) Strengthening climate change adaptation through effective disaster risk reduction. Briefing Note 03. UNISDR, Geneva, Switzerland.

Warner K, Afif T (2013) Where the rain falls: evidence from 8 countries: how households use content and erosive migration to manage the risk of rainfall variability and food insecurity. Climate and development. Taylor and Francis, London

Wheeler D (2011) Quantifying vulnerability to climate change: implications for adaptation assistance. Center for Global Development, Washington, DC

Yuen B (2014) Migration and slums in urban Asia. In: Chatterjee S (ed) Ending Asian deprivations: compulsions for a fair, prosperous and equitable Asia. Routledge, New York

Chapter 8
Urban Environmental Governance: Bangladesh as a Case in Point

Nidhi Mittal, Luca Petrarulo, and Nipunika Perera

Abstract Given the rapid urbanisation trends and the strategic and economic significance of cities in Asia, sustainably managing urban environments needs to be an urgent priority. Taking Bangladesh as a case in point, the authors have sought to validate their core premise that effective urban environmental governance at all levels is crucial for building sustainable and climate resilient cities. Bangladesh prides itself with exemplar macro-level environmental policy and governance instruments, but these have failed to translate into practice, due to weak policy enforcement and implementation on account of a number of systemic barriers and constraints. This paper maps the role and dimensions of urban environmental governance and the systemic barriers, opportunities and pathways for strengthening this at all levels in the Bangladesh context.

The paper draws on the concept of multi-level governance to rationalise and define some enabling structures, mechanisms and processes to strengthen urban environmental governance, as a basis to build sustainable urban futures. Drawing on a range of primary and secondary evidence, the authors argue that addressing the systemic barriers especially at the municipal governance or 'meso' level is instrumental for the effective implementation and enforcement of national and sub-national 'macro' policies to deliver intended environmental outcomes and services for urban communities at the 'micro' level.

A harmonized, vertically and horizontally coordinated and collaborative effort by donors, civil society, and private sector, to address the persistent and entrenched systemic governance challenges and barriers, and replace these with systemic enablers especially at the local level, can stem the tide of this environmental deterioration and catalyse transformative changes in urban contexts. The authors conclude with a discussion on how the reflections and analysis drawn from the Bangladesh context may apply more widely and transfer across other South Asian contexts which share similar urbanization and environmental governance challenges.

N. Mittal (✉) • L. Petrarulo • N. Perera
Ricardo-AEA Ltd, Marble Arch Tower, 55 Bryanston Street, London W1H 7AA, UK
e-mail: nidhi.mittal@ricardo-aea.com

W. Leal Filho (ed.), *Climate Change in the Asia-Pacific Region*, Climate Change Management, DOI 10.1007/978-3-319-14938-7_8

Introduction

With half of the world's population living in urban areas today (OECD 2012), urbanisation is one of the foremost trends shaping the future of both developing and developed nations. It is said that the battles for sustainable development globally will increasingly be won and lost in cities (United Nations 2013). Asia is a microcosm of this global urbanisation trend and has been experiencing a steady urban growth and sprawl during the past decades (see Fig. 8.1). According to the IPCC's Fifth Assessment Report (IPCC 2014, Chap. 24), Asia's urban population is expected to increase by 1.4 billion, accounting for nearly half of the global population by the middle of the century (ibid). Nearly half of the Asian urban dwellers will be living in small cities, while 20 % are expected to reside in large urban agglomerations by 2050 (ibid).

City economies account for nearly 75 % of the world's economically productive activities and assets (UNGI 2013; DPU 2012), and the statistics in Asia are similar. However, this rapid urbanisation trend has also brought forth some acute environmental challenges. A number of cities and towns especially in low- and middle-income countries are experiencing severe environmental deterioration, and are especially vulnerable to climate change, with nearly half to two-thirds of cities with one million or more inhabitants, exposed to one or multiple hazards such as floods and cyclones (ibid). Asian urban environments have become a nexus where pressures such as climate change and population growth may collide and be exacerbated in the absence of effective local governance (Hardoy et al. 2001; Bicknell et al. 2009).

According to Satterthwaite (2009), "good governance" in urban contexts, especially in developing country contexts, "needs to combine economic policies that support city prosperity with good social policies". Bringing in a central thrust on

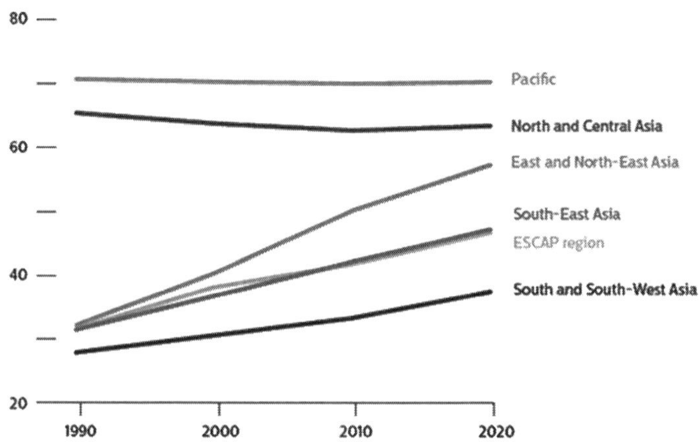

Fig. 8.1 Urbanization in Asia and the Pacific across sub-regions, 1990–2020 (%). *Source*: UN-ESCAP (2013), original data from United Nations (2012)

'environment', Puppim de Oliveira et al. (2013) argue that urban environmental governance is key to reconciling and harmonizing the opportunities and challenges resulting from urban growth, balancing socio-economic-political pressures and resolving the development versus the environmental degradation conundrum. Multi-level governance has been postulated as essential to ensure responsibilities are distributed across levels and actors (Bache and Flinders 2004). In relation to climate change, multi-level governance is necessary for joint actions by public, private and non-governmental actors at various levels and to address mitigation and adaptation. This calls for the narrowing or closing of the policy "gaps" between the various levels of government via vertical and horizontal cooperation (Corfee-Morlot et al. 2009). UNEP (2009) and Godfrey and Savage (2012) further postulate that a cooperation across the public, private and civil society is critical to move towards sustainable urban futures.

Given the strategic importance of sustainable urban environments in Asian contexts, multi-level governance provides a framework for cooperation across actors in environment and climate change issues, and has been drawn on as an analytical framework for this paper. The authors are guided by a core premise that effective urban environmental governance at all levels is crucial for building sustainable and climate resilient cities. The aim of the paper is to answer the following question: *what is the role of urban environmental governance in delivering sustainable urban environments and how might this role be enhanced at the local, sub-national, and national levels?* This paper takes Bangladesh as a case in point.

Unplanned, ad hoc and often uncontrolled development in Bangladesh has resulted in an escalating rise in population without a commensurate expansion in services and facilities needed for sustainable urban environments (Hardoy et al. 2001). The rising incidence of urban poverty also means that the most marginalised poor living in settlements in and around cities such as Dhaka, Rajshahi and Chittagong, are exposed to a number of risks due to where they live, often lack reliable basic services such as water and sanitation, and suffer from high degrees of socio-economic vulnerability (Satterthwaite and Mitlin 2013). Further compounding the problem is that a number of cities are situated in low-elevation coastal zones and are facing increasing threats of climate change (Bicknell et al. 2009). In the absence of appropriate effective urban governance that can stem the tide of the various environmental drivers and pressures, and manage the pace of urban development, the accelerating pace of environmental degradation in cities and towns in Bangladesh is a growing concern.

This paper probes the factors that constrain or enable environmental governance in urban contexts, and reflects on the systemic constraints or barriers that need to be replaced with systemic enablers or facilitators if the notion of sustainable cities is to become a reality in Bangladesh. The specific research questions that the paper attempts to address are:

- What is the role of environmental governance in addressing environmental deterioration in urban areas?

- What are the key dimensions of urban environmental governance and how do these play out at the national, sub-national and local levels?
- What are the systemic barriers to effective urban environmental governance?
- What are the entry points for transforming environmental governance in urban areas?

Methodology

The analysis presented here is the outcome from the cumulative experience of the authors' professional and research engagements in Bangladesh and South Asia over a period of 5 years. The findings on the state of climate and environmental change and related governance dimensions in the study were based on a multi-method approach to ensure the robustness of evidence. A comprehensive review of grey and policy literature alongside more than 30 academic publications constituted the backbone of the research. The literature review was complemented by consultations with a diverse range of stakeholders—donors, academia, think tanks, NGOs, local and national government institutions, communities and international experts undertaken during missions in Bangladesh between 2011 and 2014. These involved semi-structured interviews, workshops and group discussions. Direct observations and consultations with communities, municipal government institutions, academia and civil society were also held in the cities of Dhaka, Rajshahi and Chittagong to unpack the different drivers and impacts of environmental and climate change of urban communities, their evolving needs, attitudes and perceived challenges and weaknesses in environmental governance.

A brief review of theoretical constructs on urban environmental governance is presented in section "Urban Environmental Governance Discourses: A Brief Review". The aim here is not to validate or criticise the various frameworks, rather, to provide an academic context for the country-focused case study that follows. Section "Case Study: Bangladesh" presents the state of play of urban environmental governance in Bangladesh, followed by some discursive and analytical reflections on addressing systemic barriers and catalysing systemic enablers and pathways for change in Chap. 4. The paper ends with some concluding remarks, limitations and reflections on opportunities for further research.

Urban Environmental Governance Discourses: A Brief Review

The urban governance and environmental governance discourses are complementary and are framed across the academic and grey literature in varied ways, depending on whether the central thrust is placed on 'environmental protection or enhancement' or 'governance'. The authors have opted to frame urban governance as a sub-dimension within the wider environmental governance paradigm, as this provides a scope to investigate city dynamics in detail and their interplay with the wider national governance context. The academic literature review below follows this line of thought.

Environmental governance has been defined by UNEP (2009) as the set of "rules, practices, policies and institutions that shape how humans interact with the environment." Focusing specifically on cities, UN-HABITAT (2010) defines urban governance as the sum of the many ways individuals and institutions, public and private (formal and informal), plan and manage the common affairs of the city, and reconcile diverse interests to enable cooperative action. Sustainability is observed as an integral element which balances the social, economic and environmental needs of present and future generations (ibid). Narrowing the focus on the constituents of 'good governance' for low-income city dwellers, Satterthwaite (2009) postulates five dimensions—elected city governments; city governments with the capacity, resources and decentralised powers to act; formal and informal avenues to allow civil society to hold governments to account; organized urban poor groups that can engage legitimately with city governments to influence policies; and a rule of law that protects the interests of low-income groups and is not biased against informal housing or enterprise.

Linking the notions of environmental and urban governance, an operational definition of urban environmental governance can be deduced: *rules, practices, public, private and civil society networks that shape how environmental sustainability in urban contexts is formally and informally planned and collectively pursued.*

So, what does sound urban environmental governance look like in practice? Collective strategic action (CSA) as a pragmatic means for the civil society to challenge the dominant power status quo, organize itself, engage with public and private sectors and lead planned actions towards an equitable distribution of environmental resources (e.g., water and sanitation) has been observed as an effective approach in Bhopal, India (Mittal 2010). Mol (2009) has made similar observations in relation to citizen involvement in the Chinese context. Practical Action (2011) has further observed in Latin America that "integrated, long-term and locally-focused development plans involving multiple stakeholders; and participatory decision-making reinforces the social and equity dimension". This view is corroborated by Tyler and Moench (2012) who theorise that "resilience in urban contexts is focused on the interactions between people, their existing systems (such as electricity, water and ecosystems) and the rules and institutions that connect systems to people, such as land tenure systems, legal frameworks and planning processes".

Vedeld et al. (2014) expands on the concept of multi-level governance and the vertical and horizontal dimensions of action and influence within this: the vertical dimension spans across scales or levels of governance e.g., between different levels of administrative units (within which the state tends to have a dominant position); and the horizontal dimension explores the link between public and private actors across levels of social organization or between local jurisdictions. This links well with recent discourses by Puppim de Oliveira et al. (2013) and others on multi-stakeholder dialogues and a collective engagement and coordination between environmental and development-oriented agencies for improving decision-making effectiveness to create green economies in cities. In another study on urban low carbon transitions, Khan (2013) has observed "a shift from traditional hierarchical governance forms where the state is the regulator, to looser forms of governance where private actors such as business and NGOs increasingly participate in policy making" (ibid, p. 134), a phenomenon he terms as urban network governance. He postulates that this bears the risks of lack of control and accountability of governments, and potentially strengthening vested interests who are potentially substituting elected decision-making forums in cities, further impeding urban environmental governance.

Bontenbal (2009) presents an analysis on the enablers for city-to-city (C2C) cooperation and exchange to strengthen municipal knowledge and expertise, service delivery and civil society capacity on complex environmental issues. Peer-to-peer city networks have been instrumental in facilitating learning and exchange of experiences as well as catalysing inter-city cooperation in the 'Adaptation Strategies for European Cities' programme undertaken across 21 European Cities (Ricardo-AEA et al. 2013).

Against this theoretical backdrop, the next section uses Bangladesh as a case study for reflections on the drivers and pressures, systemic barriers and enablers of urban environmental governance.

Case Study: Bangladesh

Environmental and Demographic Context

Bangladesh is a low-lying country situated on the delta of the Ganges (Padma), Brahmaputra (Jamuna), and Meghna Himalayan river systems, which create a network of over 230 distributaries. Two-thirds of the country is flood plains and wetlands. Bangladesh prides itself with rich and diverse ecosystems such as the Sundarbans mangroves in the Southwest, the "Haors", flood-prone natural water basins in the Northeast, dry lands in the Northwest, the Chittagong Hill Tracts (CHT) in the Southeast, coastal zones and marine ecosystems in the South and flat marshy areas in the middle-lower parts of the country.

In addition, Bangladesh is one the most densely populated countries globally, with its current 155 million people expected to reach 177 million in 2021, 228 million in 2051 and a daunting population density of 1,015 people per km^2 (Planning Commission 2013). Both population growth and its density are recognised as serious environmental stressors (ibid).

Significantly, given its location, socio-economic vulnerabilities and various other bio-physical factors, Bangladesh is one of the world's most vulnerable countries to climate change. Indeed, it is estimated that in the period 1980–2010, 234 disasters occurred in Bangladesh, disrupting the lives of over 300 million people, causing 191,836 deaths, and resulting in over US$17 billion of economic damages (Prevention Web 2013). The combination of rising sea level and more frequent and intense storms continues to exacerbate these risks. The recently published IPCC fifth assessment report has emphasized that cities in Asian megadeltas like Bangladesh (e.g., Dhaka) are at high "risk of severe harm and loss due to climate change-related hazards and various vulnerabilities" (IPCC 2014, Chap. 19). The IPCC (2014, Chap. 5) forewarns that Bangladesh is expected to suffer "very high impacts and associated annual damage and adaptation costs of several percentage points of GDP".

Urbanisation: A Driver of Environmental Change

A key trend shaping Bangladesh today is the unprecedented and rapid urbanization it has experienced over the past decades. The percentage of the population of Bangladesh living in urban areas has increased rapidly over the last four decades, from 8 % in 1974 to nearly 30 % in 2014 (see Fig. 8.2). Current projections suggest that nearly a third of the Bangladeshi population may live in cities by 2021 (Planning Commission 2013) and more than half by the middle of this century (Brown and Dodman 2014) (see Fig. 8.2).

Burgeoning population growth, on-going migration of populations to urban areas due to a multitude of economic, education and trade factors are the key drivers for urbanisation. The rising threat to food security and livelihoods in parts of rural Bangladesh given slow-paced climate-induced changes such as sea level rise and consequent salinization of riverine and groundwater sources, inundation, desertification, and river erosion have also led to an increased incidence of seasonal migration and skewed the migration inflows to urban and peri-urban areas.

While tier 1 metropolitan cities like Dhaka and Chittagong that are the engine of the Bangladesh economy confront some grave environmental challenges, emerging urban centres are also experiencing increasing environmental concerns. Tier 2 and 3 towns and cities face evolving political and socio-economic contexts that limit the ability of municipal authorities to provide healthy living environments for a growing base of city population. Population density, poor urban planning, economic growth and governance barriers are all factors that underpin the massive environmental degradation in cities.

The pressures of climate change only serve to exacerbate the pre-existing environmental deterioration challenges (Alam and Rabbani 2007). Dhaka, the capital city which receives the highest influx of new migrants a day from coastal, peri-urban and rural zones amongst Bangladeshi cities is particularly at risk. According to the IPCC (2014, Chap. 24), Dhaka has been noted as one of the

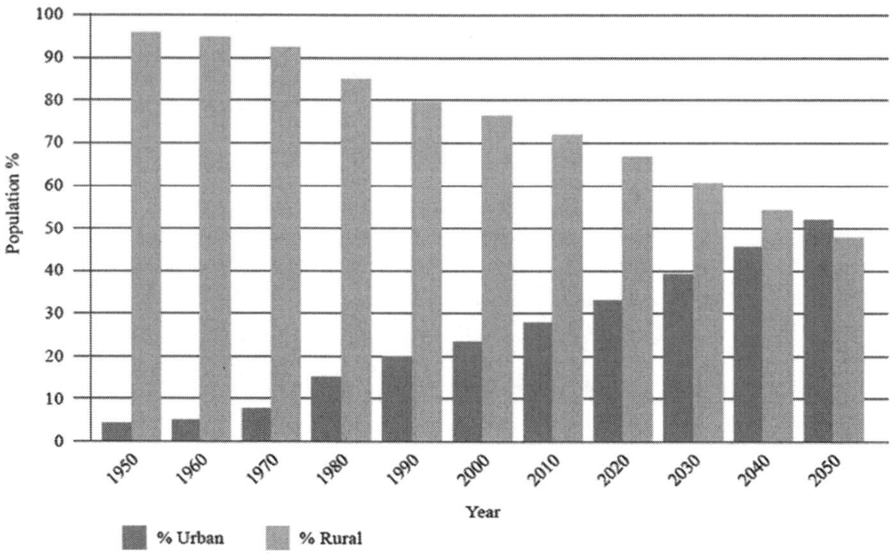

Fig. 8.2 Rapid urbanisation in Bangladesh, 1950–2050. *Source*: Brown and Dodman (2014), original data from UNDESA (2012)

most vulnerable Asian cities in terms of population exposure to coastal flooding by the 2070s. This city of nearly 15 million inhabitants faced devastation and heavy economic losses from floods in 1988, 1998 and 2004 (IIED 2007). Declining fresh water resources, falling water tables, flooding due to unpredictable and often heavy monsoons and health epidemics are other significant climate-induced concerns increasingly prevalent in towns and cities.

According to Banks (2012), there is a rising incidence of pervasive urban and peri-urban poverty and vulnerability across major cities like Dhaka, Chittagong, Rajshahi and Barisal, with nearly a third of the population lives in overcrowded slums (Table 8.1). A substantial portion of economically marginalised groups live in extremely poor and environmentally degraded conditions, often in informal or 'illegal' settlements with no entitlements to land, housing, basic infrastructure or services (ibid). A plethora of urban environmental issues linked to water, land and air merit urgent consideration, including water and sanitation access, groundwater scarcity and quality (including arsenic contamination), surface water pollution, affordable housing and land tenure, solid waste management (SWM) and exposure to outdoor and indoor air pollution.

The widespread urban environmental degradation in Bangladesh is attributed in large part to the inadequate planning processes and overwhelmed local governance capacities to cope with increasing population demands. As confirmed by numerous consultations with local stakeholders, city administrations in Bangladesh are facing substantial challenges often exacerbated by deficient governance mechanisms such as lack of coordination, competition between different government levels, polarised and often litigious party politics, informal power relations, and inadequate financial

Table 8.1 Basic characteristics of low-income settlements in Bangladesh (2005)

City corporation	City population (estimate)	Slum population (people)	Slum population (% total population)	Number of slums	Slum population density (person per km^2)	Non-slum density (persons per km^2)
Dhaka	9,136,182	3,420,521	37.4	4,966	220,246	19,677
Chittagong	4,133,014	1,465,028	35.4	1,814	255,100	15,543
Khulna	966,837	188,442	20	520	132,988	16,884
Rajshahi	489,514	156,793	32	641	67,236	6,796
Barisal	365,059	109,705	30	351	133,730	5,084
Sylhet	356,440	97,676	27	756	154,741	9,630

Source: Banks (2012, p. 7)

autonomy. These challenges are further aggravated by the lack of effective regulatory and institutional frameworks to coordinate, harmonize and integrate the work of the participating actors (del Valle 2011). Notwithstanding the political, strategic and economic significance of cities and towns in Bangladesh, the degradation in urban environments is recognised by many—Government included (MoEF 2012)—as a key obstacle in the country's path towards sustainable development.

Environmental Governance in Bangladesh

Institutional Governance and Political Context

Bangladesh has a highly complex and devolved national governance context involving a web of political and administrative institutional structures with differentiated responsibilities and legitimacies. The official count of Ministries in the Government of Bangladesh (GoB) is 43 and rises to 60 if national Divisions and Commissions are considered (GoB 2014). Since the governance of environmental issues is a cross-cutting theme, its jurisdiction spans a number of Ministries, devolved into Departments, with relevant divisions at the regional and local level. City and town administrative structures are governed by sub-national and national governance structures and mechanisms that are often inadequately ill-equipped in dealing with the escalating and evolving urban environmental challenges.

Bangladesh is divided into seven administrative divisions (see Fig. 8.3). These are then subdivided into a multi-tier governance system comprising 64 *zilas* (districts) and about 490 *upazila* (sub-districts) or *thanas*.

At the local level, metropolitan, municipal and rural areas follow different governance structures. In rural areas, the sub-districts are divided further into unions (or *wards*), comprising of several villages which are led by elected chairmen and councils. The ten Metropolitan Areas (cities) on the other hand, which represent the largest urban conglomerates, are governed by City Corporations. In contrast, the

Fig. 8.3 Administrative divisions and districts of Bangladesh. *Source*: http://bangladeshmap.facts.co/bangladeshmapof/bangladeshmap.php

smaller urban areas are headed by Municipal Corporations (or *Pourashavas*). City Corporations and *Pourashavas* are centrally supervised by the Local Governance Department (LGD).

The delivery of specific services is under the responsibility of decentralised boards and authorities which are coordinated by the Corporations, but respond to different ministries. In addition, the largest metropolitan areas, i.e., Dhaka, Chittagong, Khulna, and Rajshahi, are controlled by City Development Agencies (CDAs) which are responsible for their urban planning. The Ministry of Housing and Public Works controls the CDAs which are non-elected bodies.

As highlighted by local stakeholders and academic sources, in addition to overlap and complexity, the mechanism of political power devolution appears to be often functionally incomplete (Sohini and Pooja 2010). In fact, the scope of both political power and the main streams of financial resources of local authorities are centrally governed. For instance, government bodies at the division, *zila* and *upazila* level are run by officials directly appointed by the central government. Furthermore, fiscal decisions are mostly controlled at the national level and funds primarily come from the Annual Development Programme (ADP) through block grants which are vertically controlled by the ministries. Consequently, authorities at the lower governance level closest to local environmental problems and challenges have often the least control over which priority issues the funds are spent on.

National Environmental Legislations and Policy

Bangladeshi Governments over the years have been very active in the publication of environmental legislation, policies, strategies and plans, in order to address the numerous environmental issues that the country faces, and keep the country on a path of sustainable development. Some key national environmental and climate strategies, including those that have embedded an urban resilience and sustainability focus, are summarised in Table 8.2.

However, despite this prolific portfolio of environmental legislations, the deteriorating state of the environment demonstrates that the existence of laws and policies alone are insufficient to tackle the consequences of environmental change in Bangladesh.

Systemic Barriers Constraining Urban Environmental Governance

Despite the multitude of national-level policies, the recognition of the 'sustainable cities' imperative, and the implementation of local-level projects and initiatives, urban environments in Bangladesh continue to deteriorate. This can be attributed to the prevalence of persistent systemic challenges that prevent environmental policies from being translated into widespread practice. Lack of effective urban environmental governance represents one of the main impediments to enabling sustainable and transformative environmental improvements at the local level. Specific barriers constraining urban environmental governance are included below:

Table 8.2 Brief overview of Bangladesh's key environmental legislations and policies

Constitution
Art. 18A: "[t]he State shall endeavour to protect and improve the environment and to preserve and safeguard the natural resources, bio-diversity, wetlands, forests and wild lives for the present and future citizens"
National plans with an environmental focus
Perspective plan of Bangladesh 2010–2021 ('Making Vision 2021 a Reality')
Sixth five year plan (2011–2016)
National sustainable development strategy (NSDS)
Delta plan, 2100
Bangladesh country investment plan (CIP)
National plans and strategies with a climate change focus
Bangladesh climate change strategy and action plan (BCCSAP)
National adaptation programme of action (NAPA)
National plan for disaster management (NPDM) 2008–2015

- **Incomplete devolution and subsidiarity:** As postulated by Satterthwaite (2009, 2013), decentralization of political and financial power is one of the key elements of healthy urban governance. However, as a range of local authority and community representatives consulted indicated, the political and fiscal devolution in Bangladesh appears to be incomplete. For instance, the allocation of financial resources is mostly taken at the national level as funds for local development primarily come from the Annual Development Programme (ADP), channelled down to sectorial programmes directly controlled by the relevant ministries. Therefore, as Sohini and Pooja (2010) corroborate, despite the articulated formal decentralised governance described in section "Environmental and Demographic Context", municipal authorities experience very limited fiscal autonomy which results in a substantial barrier to effective and efficient governance on the ground. This tends to disempower and reduce the political accountability of local officials, encouraging a blame shifting culture between city authorities and central government which tends to be unconstructive.
- **Overlapping institutional mandates with unclear division of responsibilities:** As previously explained, in Bangladesh (environmental) governance spans a multitude of ministries, departments, and other authorities that exercise jurisdiction over a plethora of policies, plans and legislations which are often disjointed and not harmonised. This often chaotic governance system cascades down from the national to the city governance level. Moreover, there are a daunting number of public institutional departments and boards operating in metropolitan areas. For example, a key informant reported that in Dhaka there are more than 50 authorities providing municipal services. Separately in the Rajshahi metropolitan area, water services are managed by at least six different authorities (if not more) reporting to a few different Ministries including Ministry of Agriculture, Local Government, Water resources, Housing and Public works resulting in the lack of coordination on how the various interlinked services (e.g., water and sanitation, water infrastructure, urban water supply, irrigation, water resource management and water resources planning) are planned, managed and delivered. The consequence of such fragmented governance and decision-making structures is that roles and responsibilities often become unclear and overlapping, resulting in inefficient and ineffective general urban environmental governance, including the provision of basic environmental services to local communities.
- **Poor horizontal and vertical coordination and harmonization:** Unclear political and administrative mandates of authorities at all levels as well as the competitive nature of budget allocation due to a highly centralized governance structure is often detrimental to the effective coordination and alignment of policies and actions among ministries and authorities at the same level (horizontally) and across multiple levels (vertically). A polarised and litigious political context also contributes to a biased allocation of resources and an uncollaborative attitude among institutions. The outcome is a competing, inefficient and ineffective administrative environment which becomes a barrier to

virtuous multi-level governance, postulated by many (Vedeld et al. 2014; Puppim de Oliveira et al. 2013; Khan 2013) as a crucial element of urban environmental governance.

- **Lack of accountability and continuity:** Although City Governments are directly elected, the political decision-making power is still fragmented across numerous non-elected bureaucrats in city authorities and decentralised ministerial and sub-national institutional bodies. These are not only incentivised to follow partisan (either personal or institutional) agendas, but they also rotate, often annually, across different chairs and departments. Furthermore, the lack of financial incentives and meritocracy often makes the retention of public officials with pivotal roles particularly challenging. This systemic lack of accountability and continuity in public governance has been presented by key informants in national and international NGOs, academic and donor groups, as a major barrier to good urban environmental governance as it encourages short-term-oriented policy making and implementation over a common long-term vision.
- **Informal power structures and relations:** From a number of stakeholder consultations over the years both with communities, as well as academia and civil society in Bangladesh and other South Asia contexts, it has emerged that law implementation and enforcement is often compromised by informal power structures or vested interests. This can be described as the *ability of powerful individuals or groups to pursue their personal and specific interests over and above the greater good of the society often through treating law as a marketable commodity. In other words they inappropriately influence law development or implementation to gain access to power or financial resources.* This may or may not involve corruption as exchange of money may not always occur. It appears to be a prevent systemic issue in a number of developing (and developed) contexts which exacerbates the incidence of social inequity, thereby constraining healthy urban environmental governance (Khan 2013). Land tenure, management and conflicts in peri-urban areas are a good example to illustrate the prevalence and implications of informal power structures. Here, rapid socio-economic development opportunities have often been translated into land encroachment, price pressures, and poorly transparent allocation of land which are conducive of unsustainable management of natural resources and environmental law breaches. Academic literature sources (Feldman and Geisler 2011; LADac 2012; Alam 2014) further support this analysis.
- **Lack of resources and capacity:** Consultations with academic, civil society and government stakeholders, highlighted that local governments and municipal public authorities in some cases lack the technical capacity (requisite skills and knowledge) and the capability (required human and financial resources) to adequately implement and enforce that impressive portfolio of national policies and legislation described above or support the scale up of high-impact community-based environmental actions and initiatives. This does not mean that all local government officials and administrators are insensitive to the degradation of natural resources and livelihoods. On the contrary, many local government officials met in Bangladesh were observed as inspiring, motivated and keen to

address local challenges constructively. Nevertheless, lack of technical capacities, human resources, and financial capabilities pose serious constraints to the ability of local officials to make a sustainable impact on environmental issues in urban areas. For instance, Alam (2014) in his study on Dhaka's increasing urban flooding reports that:

> *Severe underdevelopment of institutional capacities and lack of legal, human and material resources*, coupled with intricate social, political, and economic interactions between various agents and structures have created major obstacles that have prompted flood vulnerability in Greater Dhaka due to en masse flood zones and retention ponds encroaching by land and housing development projects.

- **Donors' priorities skewing urban agendas:** The authors' discussions with governments and international donors have revealed that the lack of available financial resources or devolution or policy enforcement mechanisms in the government ministries and departments, often cause political agendas and project development to be shaped to fit donors' preferences at the expense of neglecting imminent environmental problems. For instance, climate change adaptation and mitigation in cities have been accorded priority in the recent years in Bangladesh in megacities such as Dhaka over the brown agendas of air quality, water and sanitation and waste management. As a result an important river in the city, *'Buriganga, a significant river in the city has turned from a fresh water source to a waste dump'* in the words of a civil society practitioner.

The above systemic barriers have resulted in a **lack of weak environmental policy enforcement and implementation in Bangladesh.** This is particularly magnified in urban settings as local governments are obligated to provide a multitude of public services across complex and often convoluted institutional arrangements. Ineffective municipal governments often become **'local bottlenecks as opposed to local facilitators',** unable to deliver the requirements of national government or meet the needs of communities. Experts observe that "most local governments are weak and ineffective—and often refuse to allow any public services or investments in informal settlements, even when these house more than one-third of the city's population and workforce" (IIED 2007, p. 4).

In effect, municipalities in Bangladesh are unable to function effectively in their roles as service deliverers to the communities. This also leads to **good practice remaining isolated**, rarely being able to scale up or scale out. **Local environmental governance thus** needs to be strengthened to provide conducive conditions for national policies to be implemented to carve sustainable pathways and resilient solutions for city populations.

Discussion

Bangladesh has a strong base of innovative and exemplar pilot cases and an array of policies, plans, and legislations that provide a platform to build sustainable development and climate resilience across its cities and towns. However, the authors argue that the systemic barriers described above constrain the ability of the country to adequately progress towards that end. **The figure below illustrates the challenges that Bangladesh faces at the macro (policy)—meso (local governance)—and micro (community) levels and the potential entry points for transformative environmental change in urban contexts** (Fig. 8.4).

Addressing the systemic barriers at various levels is instrumental for the effective implementation and enforcement of national and sub-national 'macro' policies to deliver intended environmental outcomes and services for urban communities at the 'micro' level.

Figure 8.4 illustrates the horizontal and vertical multi-level governance flows of information, power and decision-making authority from national—sub-national—municipal—local levels. A number of systemic barriers act at various levels with varying implications, resulting in poor urban environmental outcomes attributable to—inadequate influence of policies on practice or research, and inadequate mechanisms to ensure that local evidence adequately informs policies, lack of horizontal

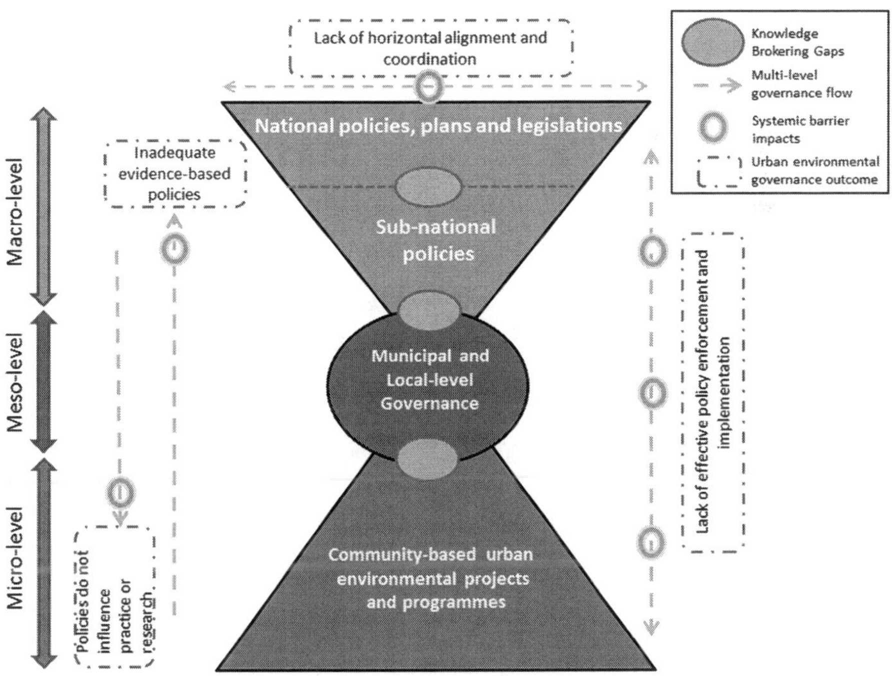

Fig. 8.4 Systemic barriers in multi-level urban environmental governance

alignment and coordination and lack of effective policy enforcement and implementation at all levels. A significant challenge worth highlighting is in terms of knowledge brokering gaps at various levels in terms of translating policies into a format accessible to policy enforcers and practitioners at local levels; and translating and communicating local-level evidence in a form that it can be coupled with scientific evidence to inform and shape higher-level policies.

The systemic barriers highlighted above can be potentially replaced with appropriate systemic enablers or facilitators to catalyse a virtuous cycle of multi-level governance. Addressing the bottlenecks especially at the meso-level merits focused consideration due to the prime importance of local governments in delivering environmental services to city populations.

The quality of governance at the local (district or municipal) level is a key factor in mitigating the risk faced by the most vulnerable urban poor with limited incomes or assets in Bangladesh (Tacoli 2010). Local governments influence "the quality of provision of *infrastructure* (for instance, drainage that limits risks of flooding), *disaster preparedness* (including warnings, measures taken to limit damage and, if needed, provision to help people move to safer areas quickly) and *disaster response* (for example, rescue services and appropriate emergency and health care services and programmes, to help those who lose their homes and livelihoods to rebuild these)" (IIED 2007). Thus, urban growth and environments need to be managed effectively not only to meet poverty and development needs of low-income groups but also to address growing climate change risks. This cannot be done by the markets but "only by governments working with, and accountable to, those who are most at risk" (ibid). Thus, local-level governance bottlenecks need to be adequately to enable sustainable urban environments. The next section "Pathways for Transforming Urban Environmental Governance" builds on the above analysis of systemic barriers to elaborate on the entry points and pathways to transform urban environmental governance.

Pathways for Transforming Urban Environmental Governance

In order to improve urban environmental governance in Bangladesh, multiple issues will need to be addressed across various levels. Shifting the environmental paradigm in cities would involve closer engagement with municipal governments to strengthen their capacity, but also working at higher echelons of governance to improve the processes of devolution and subsidiarity, to support the decentralisation of decision-making to local governments.

The academic evidence on the role and effectiveness of multi-level governance (Corfee-Morlot et al. 2009; Vedeld et al. 2014), backed by further direct consultations with key Bangladeshi informants and other secondary sources, has provided a rich basis for identifying enabling structures, mechanisms and processes (or systemic enablers) for strengthening urban environmental governance. These are:

- **Empowering and supporting community-based organisations to enforce local-level accountability at the highest levels.** This would involve developing their "technical", "organizational", "social" and "strategic" capabilities and institutional capacities to expand their room for manoeuvre and 'action spaces' at micro and macro levels, gain access to political decision-making processes to challenge the power differentials and status quo gradually and systematically (Mittal 2010). **Enhancing the government's capacity to address internal systemic barriers and resourcing gaps** that constrain delivery, and ensure devolved decision making (Bicknell et al. 2009). This may necessitate:
  - **Engaging with the relevant institutions within government** (ministries, agencies, departments, etc.) to strengthen or improve their functions over time in an effort to streamline urban policies, plans and service delivery on environment and climate change issues; and
  - **Addressing informal power structures** at the community and municipal governance level in cities and towns but also at the sub-national and national levels;
  - **Fostering the institutionalisation of knowledge** in municipal divisions and departments to address the lack of continuity and staff turnover that constrains institutional memory.
- **Facilitating horizontal and vertical alignment** through supporting the implementation of national environmental policies and legislation (top-down action) and scale up of local urban and peri-urban examples effectively by mainstreaming them into policies (bottom-up action) (Satterthwaite 2009). This may require:
  - **Empowering knowledge-brokering institutions and initiatives** that can bridge the gap among science, research, practice and policy action in cities;
  - **Engaging with diverse actors** working in urban areas (e.g., NGOs, consultants, research organisations). In the words of a policy expert: "civil society organisations could be judged by donors on how well they have capacitating the government agencies in the longer run, to incentivise them to engage effectively with the government" (Vedeld et al. 2014); and
  - **Supporting joint accountability and transparency frameworks and mechanisms** for improved monitoring, evaluation and learning for policy enforcement at sub-national and local levels, and integration of city-level planning and priorities at higher levels.
- **Strengthening the adaptive capacity of institutions** in terms of extending their decision-making horizon to be more long-term and able to plan for risks and uncertainty. This could be achieved through:
  - **Adopting a programme-based and impact-focused approach as opposed to a project-based or output and outcome-centred approach.** This may

mean giving precedence to less tangible activities like knowledge management over more 'visible' and 'tangible' large-scale urban infrastructure projects.
  - **Supporting the integration of up-to-date environmental and climate data into city level policy-making** and capacity building and recruitment of municipal authorities on state of the art information systems; and
  - **Scoping and framing pan-city or regional plans** and initiatives to catalyse peer-to-peer city-level collaboration and hubs in Bangladesh (Bontenbal 2009).

- **Donors, civil society, private sector and national and international institutions should seek to work collaboratively with the national government in a partnership of equals, to replace systemic barriers with enablers** at all levels, by supporting and strengthening existing policy and planning mechanisms, platforms and initiatives that support the vision of building sustainable cities (Godfrey and Savage 2012). This would help to deliver to the unmet needs of city citizens, support the growth and investments into local economies, whilst enhancing the quality of life for the current and future generations of urban populations.
  - **Key actors including the private sector or donors may need to see themselves as catalysts for governance reform and change as opposed to solution deliverers.** This will help to strengthen and empower governments, driving resource-efficient environmental and climate change strategies and plans ideally with a focus on initiatives that deliver long-term development co-benefits.

If such **systemic enablers (e.g., improved adaptive capacity, transparency, accountability and subsidiarity) are strengthened, Bangladesh can protect its urban environments from further degradation and accelerate progress along an environmental trajectory. This will help Bangladesh to perpetuate a virtuous cycle of effective urban governance, environmental sustainability and transformative resilience**.

The insights and recommendations here may have parallels beyond Bangladesh, in India and Nepal (Mittal 2010, 2011, 2012; Steele and Mittal 2012), and some wider learnings for other South Asian economies can be inferred. The Campaign for an Urban Sustainable Development Goal (SDG) (UNGI 2013) supported by a large number of regional governments and international organisations further reinforces this message and advocates that the "*dynamism of cities represents a major sustainable development opportunity, and it is essential to mobilize stakeholders, promoted integrated, city-level approaches, and accelerate progress towards sustainable development, including the end of extreme poverty*" (Satterthwaite 2014).

Conclusions

Concluding Remarks

Through a specific focus on Bangladesh, we have examined the role of urban environmental governance in enabling sustainable urban environments in the face of escalating environmental drivers and pressures. Specifically, the paper has attempted to unpack the dimensions of environmental governance in sustainably managing urban environments; the systemic barriers that constrain effective urban environmental governance; and the entry points for transforming urban environmental governance i.e., systemic enablers.

Despite the growing strategic significance of cities and towns in Bangladesh, urban environments continue to deteriorate mainly due to ineffective local environmental governance, which is symptomatic of the challenging institutional and political environmental governance context. Whilst there is a plethora of national policies and plans, as well as a number of noteworthy bottom-up local level project initiatives, these have been inadequate in stemming the deterioration in urban environments. This can be attributed to inadequate and often overwhelmed local governance planning processes and capacities which impede the effective implementation of national policies at the local level.

The prevalence of systemic barriers such as—incomplete devolution and subsidiarity; overlapping institutional mandates with unclear division of responsibilities; poor horizontal and vertical alignment coordination and harmonization; lack of accountability and continuity; informal power structures and relations; lack of resources and capacities; donor donor-driven urban agendas amongst other factors, lie at the heart of the problem. As a result, municipalities in Bangladesh are unable to function effectively in their roles as service deliverers to the communities. It is proposed that the above systemic barriers must be systematically addressed and substituted by enabling structures, mechanisms and processes or systemic enablers to support effective local governance.

A number of such 'enablers' that can strengthen urban environmental governance at the micro-meso-macro levels have been highlighted. These include:

- empowering and supporting community-based organisations to enforce local level accountability at higher levels;
- enhancing the government's capacity to address internal systemic barriers and resourcing gaps through addressing informal power structures, better streamlining policies and support institutionalisation of knowledge; and significantly empowering local governments through political and financial decentralisation and subsidiarity.

(continued)

Facilitating horizontal and vertical alignment of policies and practice through integration of environment and climate change in GOB and city administrations; joint accountability and transparency frameworks to ensure policy enforcement; and promoting knowledge brokering to support evidence-based policies and practice;

- improving the adaptive capacity of city institutions to embed long-term planning and accounting for uncertainty through programme-focused approaches, integration of climate and environment data into policies and peer to peer city networks and hubs.

While the reflections and analysis in this paper are rooted specifically in the Bangladesh urban environmental and governance context, there may be wider learnings for other Asian urban contexts to be inferred. South Asian economies, in particular, due to their shared history, present a number of parallels in terms of the widespread environmental degradation in urban areas, often on account of inadequate municipal capacities to cope with increasing demands posed by unplanned and ad hoc development in the face of rising population, economic growth and climate change. A number of countries in this region are characterised by a similar national governance framework involving a web of political and administrative institutional structures with different accountabilities and mandates. They also experience a divide between rural and urban administrative structures coupled with functionally incomplete mechanisms for political power devolution and financial autonomy at the local government level. These are further aggravated by the lack of effective regulatory and institutional frameworks to coordinate, harmonize and integrate the work of participating actors towards over-arching sustainable city imperatives.

The rising prominence of the environmental sustainability, climate resilience and low-carbon agenda in cities especially in fast growing South Asian economies presents an opportunity to horizontally and vertically align approaches across the private, public and civil society sectors to drive long-term transformative change in cities. The systemic barriers and enablers discussed in the context of Bangladesh may provide some pertinent entry points and pathways to transformative urban environmental governance in other contexts, and could be more widely applicable and transferable, albeit with nuanced context-specific considerations.

While systemic barriers need to be addressed at all levels, addressing them especially at the municipal governance or 'meso' level is a crucial underpinning for the effective implementation and enforcement of national and sub-national 'macro' policies to deliver intended environmental outcomes and services for urban communities at the 'micro' level. Since environment and climate change are cross-cutting issues, a harmonized, collaborative and

(continued)

coordinated vision and action by various actors through effective multi-level governance would go a long way in addressing the endemic and persistent urban environmental governance challenges to ultimately deliver to the vision of sustainable urban futures.

Such systemic enablers will be instrumental in helping fast growing South Asian economies, like Bangladesh, accelerate progress on their sustainable development trajectory and perpetuate a virtuous cycle of effective environmental governance and transformative resilience in urban environments.

Limitations of the Study

This paper provides a brief overview of the underpinnings of the environmental governance challenges in cities to provide some strategic reflections and recommendations moving forward on how these can be overcome.

However, governance, climate change and environment governance in the context of cities are multi-layered and cross-cutting issues. Furthermore, towns and cities are not a homogeneous group, and in Bangladesh they differ widely in their environmental contexts and governance dimensions both from each other and from other counterparts in Asia. This study is only a starting point in terms of probing deeper into one country's context i.e., Bangladesh and drawing some insights, reflections and analysis that may be broadly applicable especially across other South Asian urban contexts.

Some variations in terms of the degrees of reliability, consistency and accuracy of both literature sources and information obtained through formal or informal consultation of informants may be present as not all sources could be rigorously verified. Nevertheless, all possible efforts have been made to corroborate and triangulate the data and information used.

Further Research Opportunities

In order to ensure that some of the analysis and recommendations of this article are well-founded and can be actually operationalised, some areas merit further research and investigation.

Firstly, the climate change-environment nexus could be further unpacked and the convergences and divergences in the policy discourse and practice domains examined with a city lens. Further analysis may also be needed to integrate the urban governance challenges with the broader political economy context in Bangladesh and other South Asian countries to compare and contrast the barriers and opportunities to urban environmental governance, separating elements that may reflect the specific national or local Bangladeshi context and others that may apply to the wider region.

Furthermore, it would be beneficial to undertake a targeted engagement of stakeholders across the national, sub-national, and local levels to disaggregate the needs and capacities of institutions and investigate specific opportunities

(continued)

for strengthening environmental governance at each level. A review of the donor or government funded capacity building, agency coordination, collaboration mechanisms, as well as, private sector and civil society (autonomous or funded) initiatives in urban contexts, and to what extent these are effective or not in enabling a synergised multi-actor effort on urban environmental governance will add value. Focused research as highlighted above would help to focus and prioritise on-going and future efforts and carve sustainable pathways for change within cities.

References

Alam MJ (2014) The organized encroachment of land developers—effects on urban flood management in Greater Dhaka, Bangladesh. Sustain Cities Soc 10:49–58

Alam MJ, Rabbani MDG (2007) Vulnerabilities and responses to climate change for Dhaka. Environ Urban 19(1):81–97. Retrieved from http://eau.sagepub.com/content/19/1/81.full.pdf+html

Bache I, Flinders M (2004) Themes and issues in multi-level governance. In: Bache I, Flinders M (eds) Multi-level governance. Oxford University Press, Oxford

Banks N (2012) Urban poverty in Bangladesh: causes, consequences, and coping strategies. BWPI working paper 178. Brooks World Poverty Institute, University of Manchester, Manchester. Retrieved from http://www.bwpi.manchester.ac.uk/medialibrary/publications/working_papers/bwpi-wp-17812.pdf

Bicknell J, Dodman D, Satterthwaite D (eds) (2009) Adapting cities to climate change: understanding and addressing the development challenges. Earthscan, London

Bontenbal MC (2009) Strengthening urban governance in the south through city-to-city cooperation: towards an analytical framework. In Habitat Int 33(2):181–189

Brown D, Dodman D (2014) Understanding children's risk and agency in urban areas and their implications for childcentred urban disaster risk reduction in Asia: insights from Dhaka, Kathmandu, Manila and Jakarta. Working paper 6:2014. International Institute for Environment and Development, London. Retrieved from http://pubs.iied.org/pdfs/10652IIED.pdf

Corfee-Morlot J, Kamal-Chaoui L, Donovan M, Cochran I, Robert A, Teasdale P (2009) Cities, climate change and multilevel governance. OECD environmental working papers N° 14:2009. Organisation for Economic Co-operation and Development (OECD), Paris. Retrieved from http://www.oecd.org/regional/regional-policy/44232263.pdf

Puppim de Oliveira JA, Doll CNH, Balaban O, Jiang P, Dreyfus M, Suwa A, Moreno-Penaranda R, Dirgahayani P (2013) Green economy and governance in cities: assessing good governance in key urban economic processes. J Clean Prod 58:138–152, Special volume: Climate co-benefits in urban Asia

Del Valle J (2011) ELLA guide: addressing the urban challenge in integral and inclusive ways. Practical Action Consulting, Lima

Feldman S, Geisler C (2011) Land grabbing in Bangladesh: in-situ displacement of peasant holdings. Presented at the international conference of global land grabbing. Land Deals Politics Initiative (LDPI), University of Sussex, Brighton, 6–8 April 2011. Retrieved from http://www.iss.nl/fileadmin/ASSETS/iss/Documents/Conference_papers/LDPI/5_Shelley_Feldman_and_Charles_Geisler.pdf

GoB (Government of Bangladesh) (2014) Website of the government of Bangladesh. Online resource. Retrieved from http://www.bangladesh.gov.bd/index.php?option=com_weblinks&task=ministry&Itemid=152

Godfrey N, Savage R (2012) Future proofing cities: risks and opportunities for inclusive urban growth in developing countries. DFID, The Bartlett Development Planning Unit and WS Atkins, Epsom. Retrieved from http://r4d.dfid.gov.uk/pdf/outputs/urbanisation/FPC_Report_FINAL.pdf
Hardoy JE, Mitlin D, Satterthwaite D (2001) Environmental problems in an urbanizing world: finding solutions in cities in Africa, Asia and Latin America, 2nd edn. Earthscan, London
IIED (2007) Environment and urbanization brief 15: reducing risks to cities from climate change; an environmental or a development agenda? Environ Urban 19(1):1–6, Retrieved from http://pubs.iied.org/pdfs/10548IIED.pdf
IPCC (2014) Working Group II—climate change 2014: impacts, adaptation, and vulnerability. Cambridge University Press, Cambridge, Retrieved from https://www.ipcc.ch/report/ar5/wg2/
Khan J (2013) What role for network governance in urban low carbon transitions? J Clean Prod 50:133–139, Special issue: Advancing sustainable urban transformation
LADac (2012) Bangladesh—food security and land governance factsheet. IS Academy on Land Governance for Equitable and Sustainable Development, Utrecht
Mittal N (2010) Collective strategic action in access to water and sanitation services for environmental justice: the case of Madhya Pradesh urban services for the poor in Bhopal, India. Development Planning Unit, University College London, London
Mittal N (2011) The importance and role of community-based organisations in building urban resilience. In: Session 12 on institutional capacity dimensions of scaling up CBA. Presented at the fifth international conference on community-based adaptation, Dhaka
Mittal N (2012) An integrated framework for disaster risk reduction, climate change adaptation and resilience for save the children in Nepal: a brief review. Save the children (Unpublished internal workshop report)
MoEF (Ministry of Environment and Forestry) (2012) Rio +20: national report on sustainable development. Government of Bangladesh, Dhaka
Mol AP (2009) Urban environmental governance innovations in China. Curr Opin Environ Sustain 1(1):96–100
OECD (2012) OECD environmental outlook to 2050. Organisation for Economic Co-operation and Development, OECD Publishing, Paris. Retrieved from http://www.oecd-ilibrary.org/content/book/9789264122246-en
Planning Commission (2013) National sustainable development strategy 2010–21 (NSDS). Government of Bangladesh, Dhaka
Ricardo-AEA, ICLEI, Adelphi, Arcadis, Alexander Ballard, University of Manchester and Tecnalia (2013) Adaptation strategies for European cities (No. CLIMA.C.3/SER/2011/0030). Report for the EC—Directorate general for climate action. Ricardo-AEA, Didcot
Satterthwaite D (2009) Environment and urbanization brief 18: what role for mayors in good city governance? Environ Urban 21(1):1–6, Retrieved from http://pubs.iied.org/pdfs/10579IIED.pdf
Satterthwaite D (2013) The political underpinnings of cities' accumulated resilience to climate change. Environ Urban 25(2):381–391, Retrieved from http://eau.sagepub.com/content/25/2/381.full.pdf
Satterthwaite D (2014) Environment and urbanization brief 29: getting local governments, residents and enterprises to respond to the new IPCC assessment. Environ Urban 26(1):1–6, Retrieved from http://pubs.iied.org/pdfs/10703IIED.pdf
Satterthwaite D, Mitlin D (2013) Urban poverty in the global south: scale and nature. Routledge, London
Sohini P, Pooja RG (2010) Decentralisation in Bangladesh. National Council of Applied Research, New Delhi
Steele W, Mittal N (2012) Building equitable urban resilience: the challenge for cities. In: Otto-Zimmermann K (ed) Resilient cities 2: cities and adaptation to climate change—global forum 2011. Springer Science and Business, Dordrecht

Tacoli C (2010) Environment and urbanisation brief 20: governance, migration and local development. Environ Urban 22(1):1–6, Retrieved from http://pubs.iied.org/pdfs/10588IIED.pdf
The Bartlett Development Planning Unit (2012) Future proofing cities: risks and responses to inclusive urban growth in developing countries. The Bartlett Development Planning Unit, UCL, London. Retrieved from http://www.bartlett.ucl.ac.uk/dpu/news/future-proofing-cities
Tyler S, Moench M (2012) A framework for urban climate resilience. Clim Dev 4(4):311–326
UNDESA (2012) World population prospects: the 2012 revision. United Nations Department of Economic and Social Affairs, New York. Retrieved from http://esa.un.org/unpd/wpp/index.htm
UNEP (2009) Environmental governance. United Nations environment programme, Nairobi. Retrieved from http://www.unep.org/pdf/brochures/EnvironmentalGovernance.pdf
UN-ESCAP (2013) Urbanization trends in Asia and the Pacific. UN economic and social commission for Asia and the Pacific, Bangkok. Retrieved from http://www.unescapsdd.org/files/documents/SPPS-Factsheet-urbanization-v5.pdf
UNGI (2013) Sustainable development solutions network, 'Campaign for an Urban Sustainable Development Goal'. United Nations Global Initiative. Online resource. Retrieved from http://urbansdg.org/
UN-HABITAT (2010) UN-Habitat state of the world's cities 2010/2011: bridging the urban divide, Nairobi. Retrieved from http://sustainabledevelopment.un.org/index.php?page=view&type=400&nr=1114&menu=35
United Nations (2012) World urbanization prospects: the 2011 revision. United Nations, Department of Economic and Social Affairs, Population Division. CD-ROM edition – data in digital form. United Nations Publications, New York. Retrieved from http://esa.un.org/unup/CD-ROM/Urban-Rural-Population.htm
United Nations (2013) A new global partnership: eradicate poverty and transform economies through sustainable development. The report of the high level panel of eminent persons on the post-2015 development agenda. United Nations Publications, New York. Retrieved from http://www.post2015hlp.org/wp-content/uploads/2013/05/UN-Report.pdf
Vedeld T, Kombe W, Kweka-Msale C, Mareme Ndour N, Hellevik S (2014) Multi-level governance, climate change and floods in urban Africa. In: Pauleit S, Jorgensen G, Kabisch S, Gasparini P, Fohlmeister S, Simonis I, Yeshitela K, Coly A, Lindley S, Kombe WJ (eds) Urban vulnerability and climate change in Africa. A multidisciplinary approach, vol 4. Springer, London. Retrieved from http://www.nibr.no/filer/tekstfiler/Vedeld_et_al.pdf

Chapter 9
The Economic Value of Cyclonic Storm-Surge Risks: A Hedonic Case Study of Residential Property in Exmouth, Western Australia

Roberts Rebecca, Lynnath E. Beckley, and Malcolm Tull

Abstract Recent major coastal disasters in the Asia-Pacific region have resulted in a massive loss of life and high societal costs. Unbridled development, growing coastal populations and injudicious land planning amplifies the predicted disaster risk due to climate change and extreme weather events. As it typifies expanding coastal development in areas prone to extreme weather events, the town of Exmouth (NW Australia) was used to investigate economic strategies for coastal risk mitigation. Recent marina development, with a loss of disaster mitigating ecosystem functions, has increased risk to previously unaffected areas. The extent to which risk perceptions of cyclonic storm-surge inundation and flooding influenced the price buyers paid for residential property in Exmouth over the period 1988–2013 was examined using a Hedonic Price Model. The analysis indicated that prices did not reflect the real societal cost of risk. Due to the absence of a monetary signal, such as higher insurance premiums, buyers tended to be risk insensitive and give greater weight to coastal amenity. To internalize these costs, a mandatory private insurance scheme for high-risk properties, penalties for local councils undertaking unsustainable developments, and a hybrid economic instrument aimed at correcting the market failure in coastal land, is proposed.

R. Rebecca (✉)
Department of Lands, Government of Western Australia, 140, William Street, Perth, WA 6000, Australia
e-mail: rebeccaroberts007@gmail.com

L.E. Beckley
School of Veterinary and Life Sciences, Murdoch University, 90 South Street, Murdoch, WA 6150, Australia
e-mail: L.Beckley@murdoch.edu.au

M. Tull
School of Management and Governance, Murdoch University, 90 South Street, Murdoch, WA 6150, Australia
e-mail: M.Tull@murdoch.edu.au

W. Leal Filho (ed.), *Climate Change in the Asia-Pacific Region*, Climate Change Management, DOI 10.1007/978-3-319-14938-7_9

Introduction

Climate change and associated impacts on weather patterns could result in a substantial increase in more intense extreme events such as storms, cyclone and flooding over the next century (Knutson 2011). With predictions that the global coastal population will increase by 34 % by 2025, vulnerability to coastal disasters will grow, exacerbated by sea level rise (Intergovernmental Panel on Climate Change [IPCC] 2007). Coastal population growth in areas prone to periodic cyclonic storm-surge inundation, wind damage and erosion is of major concern.

Financially constrained local government agencies in remote northern parts can be driven to undertake injudivious coastal developments which can revitalise the economy of the town and provide additional rate revenue to the shire. Government policy failures can also result in market failures on the part of coastal property buyers in risk-prone areas. Especially so when potential buyers drive the demand for residential property and the premium associated with living on the coast provides a strong incentive for developers to build as close to the coastline as possible, despite the risks of living near the dynamic land-ocean interface.

Property buyers often fail to factor in the costs of living close to the coastline. Therefore, costs of disaster recovery and rehabilitation are passed onto the rest of society. For example, a major flood in Queensland, Australia in 2011, resulted in a temporary nationwide flood levy to fund recovery costs (Australian Government and Prime Minister's Office 2011).

This study is innovative in its examination of coastal disaster risk from an economic policy perspective, with its focus on government and market failure in coastal property development as the central facet of investigation. This study is particularly relevant in that the effect of coastal risk on residential property prices, particularly in cyclone-prone areas of Australia is not well known. This study demonstrated how coastal development in high risk areas, concomitant with decisions to purchase residential property close to the coastline can increase vulnerability.

The aim was to determine if coastal property buyers factored potential risk of cyclonic storm surge inundation and flood risk into their property purchase decisions. The town of Exmouth, located in northwestern Australia, was used as a case study, as it typifies expanding coastal and industrial development, in areas prone to extreme weather events. Cyclones are experienced in Exmouth every second or third year, on average. These cause strong winds, heavy rain and storm-surge, leading to inundation of the low-lying coastline. Further, flooding during cyclone events associated with run-off from Cape Range, about 314 m high and located east of the townsite, is of particular concern. In 1999, the town was affected by category five cyclone Vance, which damaged buildings and other major public infrastructure in the town.

Previously, town planning limited development to the northeastern part of Exmouth, in areas of higher elevation, away from the coastline and creeks. A policy shift has resulted in the establishment of residential development in low-lying areas

Fig. 9.1 Annotated aerial image depicting the geographical layout of the Exmouth townsite including the recently constructed marina

(Western Australian Government et al. 2004). Figure 9.1 depicts an aerial view of the town, including areas of recent development, older parts where previous development was carried out and location of the marina, relative to its proximity to the coastline.

The marina was constructed to provide residential real-estate that is high in coastal amenity value (Bowman Bishaw Gorham 1997). Properties in the marina are not more that 1–2 m above the mean seal level and are very close to the coastline. There has also been a loss of the 1–3 m foredunes and 8–10 m secondary dunes in this area (Western Australian Government et al. 2007). The vegetation covering the floodplain, serving to protect the area against erosion has also been lost.

Research has demonstrated the importance of disaster mitigating ecosystems in reducing effects of wind and waves associated with coastal disasters. Construction of the marina has also resulted in the loss of disaster-mitigating natural capital functions in the area. This is as a result of filling in, and compaction of, land immediately north of the marina, behind the coastal dune system. Originally this area functioned as a natural flood detention area, capturing excess water during heavy storms and cyclone events, and gradually allowing it to percolate into the ocean (Western Australian Government et al. 2007). Channels and other protective engineering structures have been built to compensate for the loss of these functions, but a recent study suggests that these might be insufficient to withstand large flooding or cyclonic storm-surge inundation events (Western Australian Government et al. 2007). Erosion, as a result of storm-surge, will exacerbate the situation, and there is additional concern in the context of an almost 1 m predicted sea level rise along the coast of WA by 2100 (Western Australian and Planning Commission 2013).

Method

The Hedonic Price Method (HPM), theoretically specified by Rosen (1974), was used to disaggregate the impact of risk factors (location in the flood zone or areas prone to cyclonic storm-surge) on residential property prices from other attributes of residential property in Exmouth. This method has been extensively used to determine the effect of various property characteristics, such as, the structure (e.g., lot size), the neighbourhood (e.g., average commute time), and the environment (e.g., degree of air pollution).

Over the last decades, the HPM has been increasingly applied to natural disaster risk, ranging from floods, earthquakes, fire, and volcanic eruptions to wind and erosion (Nakagawa et al. 2007, 2009; Keskin 2008). The underlying theory is that, just like positive environmental features, such as a scenic view can create an increase in property prices, greater exposure to a natural hazard can be reflected through diminished prices. Location of a property in a cyclone-prone area can influence an individual's perception of possible future threat to life and property, which may be reflected in the price paid at the time of purchase.

Property sales data over 1988–2013 for Exmouth were used in a Hedonic Price Model (HPM) to assess the perception of risk through location in areas prone to cyclonic storm-surge inundation, 100-year Average Return Interval Flooding and areas damaged by the 1999 Cyclone Vance.

Table 9.1 presents a summary of the data used in the analysis, including the source, and form into which they were transformed for this analysis. Most of the variables, except those depicting distance and lot size, were converted to dichotomous variables.

Proxies used in the HPM equation were in two forms: firstly, those representing risk exposure, based on location of property, and secondly, those representing risk awareness (Table 9.2). Variables that denote degree of exposure to risk are based on previous studies using a similar concept, including Bin et al. (2008) and MacDonald et al. (1987, 1990). These studies used dichotomous variables for risk, where the value of one denotes that a property is located in a risk-prone area and zero if it is not.

This study used information from a scenario analysis carried out for Exmouth, where the centroids of the transacted properties were overlaid with the storm-surge risk areas, to derive variables denoting areas at risk of storm-surge heights of 4 m. The coefficients of this variable measured the price difference between properties in the cyclonic storm-surge-prone area and other areas. The working hypothesis was that if a property was located in an area of risk of cyclonic storm-surge inundation, it would have a lower price than those located outside the risk zone,

Table 9.1 Summary of the variables used in the HPM analyses carried out for Exmouth properties (1988–2013)

Variables	Data source	Data type in original dataset	Converted form used in the HPM
Dwelling-specific variables			
Number of bedrooms, bathrooms, family rooms, dining rooms, games rooms, lounges, studies, kitchens, pools, carport or garage	WA Valuer General (Landgate)	Numeric	Dichotomous
Lot size		Numeric (m^2)	No change
Year built		Date (Year)	Dichotomous
Wall and roof material		Alphanumeric	Dichotomous
Property classification		Alphanumeric	Dichotomous
Property classes		Alphanumeric	Dichotomous
Land-use type		Alphanumeric	Dichotomous
Distance from urban amenities	Walk Score website (Walk Score 2013) to determine the straight-line distances of each of the sold properties from the nearest restaurant, coffee shop, grocery store, bookshop, pub and entertainment venue	Numeric (m)	No change
Risk variables			
Located in100-year ARI floodplain		Dichotomous	No change
Post-cyclone Vance		Dichotomous	No change
6 months		Dichotomous	No change
1 year		Dichotomous	No change
2 years		Dichotomous	No change
Inundated by a 4 m storm surge	Data derived from spatial analysis study by Roberts (2012)	Dichotomous	No change

ceteris paribus. The price differential would therefore provide the discount value on the price of risk in this area.

Correction for Auto-Correlation

As temporal autocorrelation results from property sales over multiple years, especially when annual sales in a market are sparse, the additional effect of time must be captured as part of the equation (Taylor 2003). Using several years of data may affect the stability of the HPM, where the method of ordinary least squares

Table 9.2 Variables used to denote risk exposure and risk awareness in the Hedonic Pricing Model, and the number of transacted properties for Exmouth (1988–2013)

Variables	Number of transactions
Inundated by a 4 m storm surge	44
Located in100-year ARI floodplain	180
Post-cyclone Vance	
6 months	52
1 year	93
2 years	150
Interaction variables	
100-year ARI * sold 6 months after Vance	4
100-year ARI * sold 1 year after Vance	6
100-year ARI * sold 2 years after Vance	11

regression violates the Gauss-Markov condition of being the best linear unbiased estimator (Gujarati 2006).

The HPM study for Exmouth, using the approach specified by Taylor (2003), used dichotomous variables for each year of the data over the study period, to correct for temporal auto-correlation. Other studies on flood risk, such as Morgan (2007) and Bin et al. (2008), have followed this approach, where dichotomous variables were specified for each year of the data. The data were also tested for possible effects of month and quarter of sale, but were found to be insignificant (at 0.01 and 0.05 levels).

Following Field (2009), bivariate correlation between sales prices and each of the variables in the dataset was carried out to determine those most suitable to be included in the final HPM. Those found to be significantly correlated ($p < 0.01$ and <0.05), were selected.

Limitations

A limitation of HPm studies it that, by the use indirect means to investigate economic decision-making, other factors such as risk tolerance are not considered. Another source of complexity pertains to cyclonic risk in HPM. Unlike earthquake risk for instance, where risk proxies can be derived from the location in relation to the fault line, the actual pathways of cyclones are less predictable. Further, a test on spatial autocorrelation where a result can exist between variables that are located in close proximity to one another (Anselin 1988) was not carried out. This was owing to the town covering a relatively small geographic area.

Results

Applying the HPM specification of Rosen (1974) and the results of the bivariate correlation to property sales prices, the following equation was specified.

Property Sale Price $= \propto_0 + \beta_I S_I$ (structural characteristics) $+ \beta_J N_J$ (neighbourhood characteristics) $+ \beta_K R_K$ (risk characteristics) $+ \epsilon$

where:

Structural characteristics $(S_I) =$ lot size (m^2), age of house (years), presence of a family room, property classification (flat, villa, house, single unit, duplex unit), wall material (iron, steel frame, brick, fibro, asbestos);
Neighbourhood characteristics $(N_J) =$ distance to the nearest grocery shop (km);
Risk characteristics $(R_K) =$ 4 m cyclonic storm-surge inundation risk, 100-year ARI flooding, sale of property within 6 months, 1 year and 2 years after cyclone Vance, and interaction effects between location in the 100-year ARI areas and sale after cyclone Vance;
$\epsilon =$ residual error;
$\propto_0 =$ intercept; and
β_I, β_J, $\beta_K =$ coefficients of the structural characteristics, neighbourhood and risk characteristics, respectively.

Risk Variables

Eight risk variables were tested representing cyclonic storm-surge inundation and flood risk in the townsite (Table 9.3). It was expected that they would have a negative impact on property price. Findings indicated, however, that this was not

Table 9.3 Summary of the coefficients and their level of significance for the risk variables under the HPM analyses carried out for Exmouth (1988–2013)

Variables	Number of transactions	Coefficients values and level of significance		
		Linear	Semi-log	Log-log
Inundated by a 4 m storm surge	44	83,241**	0.42**	0.35**
Located in a 100-year ARI floodplain	180	40,305**	−0.05	−0.04
Post-cyclone Vance				
6 months	52	13,887	0.06	0.06
1 year	93	15,181	0.08	0.08
2 years	150	8,103	0.08	0.05
Interaction variables				
100-year ARI * sold 6 months after Vance	4	78,660	0.51**	0.45**
100-year ARI * sold 1 year after Vance	6	61,900	0.36**	0.34**
100-year ARI * sold 2 years after Vance	11	48,454	0.22*	0.22*

**Correlation is significant at 0.01 level (two-tailed)
*Correlation is significant at 0.05 level (two-tailed)

the case. Prices were, in fact, much higher in areas prone to cyclonic storm-surge (4 m) and 100-year ARI flood risk. The devastation caused by category five cyclone Vance, which damaged buildings and major public infrastructure in the town, had no effect on price. Even the prices of properties located in 100-year ARI areas, that experienced flooding due to cyclone Vance, were not discounted for risk.

The first HPM analysis used the variables specified for the structural characteristics (S_I) and neighbourhood characteristics (N_J) as described in the above equation. The risk characteristic (R_K), tested was 4 m cyclonic storm-surge risk during the study period. There were 44 property sales transactions within this risk area. The summary statistics for this HPM are presented in Table 9.3. Linear, semi-log and log-log functional forms were significant at a 0.01 level.

The coefficient values, and their level of significance for the HPM, showed that the variable representing risk of 4 m cyclonic storm-surge (Table 9.3), was positive and significant at a 0.01 level. The finding of this model indicates that buyers paid higher prices for properties located in risk-prone locations as compared to other areas, and buyers did not discount for risk as expected.

An examination of the data found that of the 44 properties sold in the 4 m storm-surge risk zone, 98 % of these transactions were within the last 10 years (>2000). Of these, 52 % and 36 % were located in the marina, and recreational open space zones, respectively, and these were located at a distance of 130–700 m from the original shoreline of Exmouth Gulf.

To determine the effect of Cyclone Vance, three separate regressions were run to test if there was any effect on property prices within 6-months, 1-year and 2-years after the event (R_K). Cumulatively, there were 52 sales transactions 6-months after cyclone Vance, 93 within a year after, and 150 sales in the 2 years following the event.

The log-log model showed the highest R-square value, where the variables used accounted for 83 % of the variation in house prices and, given that its Durbin-Watson statistic (that tests for the presence of auto-correlation) was within the required range, this was considered the most appropriate functional form. The variable indicating if a property was sold within 6 months after Vance was not significant. This showed that there was no effect on sales prices if a property was sold within this period. An HPM was also carried out to determine if there was any effect on properties sold within 1 year and 2 years after Vance. Results again indicate that there was no effect.

The location of a property within a 100-year ARI flood zone was the second risk characteristic (R_K) tested in a HPM. There were 180 property sales transactions in the 100-year ARI area during the study period. Summary statistics indicated that the log-log was the best functional form, with an R square-value of 0.84. The coefficient for 100-year ARI risk was not significant, indicating that this does not have an effect on sales price.

Finally, HPM analyses were carried out to determine the possibility of interaction of Cyclone Vance with the 100-year ARI risk. As with the other models, the structural characteristics (S_I) and neighbourhood characteristics (N_J) used were those given in the final equation specified for the HPM used in this study.

Three separate regressions were carried out to represent the three risk characteristics (R_K), which comprised of the interaction effect between location in a 100-year ARI zone and the property being sold within 6 months, 1 year or 2 years following cyclone Vance respectively. There were 4, 6 and 11 transactions that were located in the 100-year zone and sold within 6 months, 1 year and 2 years of Vance, respectively. The positive significance of the coefficient in the semi-log and log-log HPMs, imply, that even for properties located in the 100-year floodplain, the experience of cyclone Vance did not result in a discount on property prices.

Other Variables

Variables depicting properties with iron roofs or if the property was a duplex unit, villa, house, flat or single unit, had a positive coefficient at a 0.01 significance level, indicating preference for such houses. Variables denoting steel-framed and brick walls had positive coefficients, whereas a negative impact was indicated for houses with fibro walls. However, they were not significant at 0.01 and 0.05 levels, suggesting a lower priority, in this regard, among buyers. The variable for asbestos walls, however, resulted in lower property prices, and the significance of this coefficient at 0.01 level indicates that buyers took this into serious consideration.

Surprisingly, lot size and property age appeared to have no effect on price in the semi-log function, although the findings showed increase in price if there was a family room in the house. The coefficients of these in the linear and log-log forms show a decrease in price for age of property. Contrary to the findings of the bivariate model, the further away a property was from the shopping area, as represented by the variable indicating distance to a grocery store, the lower the price, which indicated that property owners have greater preference for properties with more access to urban amenities. This result was considered to be the most accurate finding, as multivariate regression models are considered to provide the better results than bivariate methods (Gujarati 2006).

Discussion

This study found that properties located in the 4 m storm-surge prone areas had higher property values than others. Further investigation indicated that this was as a result of the positive effect created by greater access to coastal amenity. In such a context, decisions to purchase property focus on positive coastal amenity, outweighing concern about cyclonic storm-surge inundation and flood risk. Other research has found similar effects. In an HPM carried out by the Sydney Coastal Councils Group (2011), lots located on the beachfront along the Collaroy-Narrabeen area in New South Wales were found to be 40 % more valuable. On average, buyers pay in excess of 1 million dollars for such properties. Properties with access to the ocean and beach view, near Noosa Heads in Queensland were also found to be more valuable (Pearson et al. 2002).

Since all of the properties at risk of a 4 m surge are located in the marina village, there are other associated benefits that could have been confounding the analysis. The houses are more modern, and the surrounding area is landscaped, to provide an aesthetically pleasing environment. Other areas in the town are comparably less attractive.

The results of the HPM for 100-year floodplain indicate that buyers do not factor the threat of cyclonic storm-surge and flooding due to rainfall into their property prices. These properties are distributed in other areas of the town, not only the marina, and therefore include those that do not have access to the same coastal and aesthetic amenities.

Other studies in the US, by Bin and Polasky (2004), and Morgan (2007), who also used HPM to investigate the major event versus 100-year floodplain interaction effect, found different results to that of Exmouth. This is attributed to the requirement, by law, for these properties to obtain insurance. Fridgen and Shultz found as much as an 81 % discount in house value resulted from the monetary signal created by higher insurance premiums. However, there is no compulsory insurance requirement for Exmouth, or other parts of northern WA. The absence of a monetary signal via insurance to create awareness of risk in Exmouth could therefore be another reason buyers failed to factor risk into the prices they paid for coastal property. It also suggests that knowledge and awareness of risk alone, may be insufficient to change economic behaviour.

A survey of local risk perceptions in the US found that the presence of shoreline armouring, seawalls, and other structural devices created a sense of safety from coastal risk, to the extent that it suppressed the need to purchase flood insurance (Kriesel and Landry 2004). This is another element which may also explain the findings of this study. At Exmouth, the flood drainage channels established in the marina village were built to direct excess flooding into the ocean. However, since their establishment, there has not been an opportunity to test the ability of these channels to mitigate flooding in the townsite. The presence of these may be instrumental in creating a perception of safety among those purchasing property in this area.

Cyclone Vance Effect

HPM analyses carried out to assess prices following cyclone Vance, and the increased awareness and concern about potential danger, revealed no effect. This finding is different to that of Hallstrom and Smith (2005), who found a decrease in property prices in Lee County, Florida, following hurricane Andrew in 1992, even in areas that were not affected. This may be because property owners in Lee County, and other parts of the US, living in risk-prone areas, are compelled to pay higher insurance premiums following such major events. This is not, however, the case in Exmouth, and possibly other similar regional towns in northern Australia. This can, on the one hand, imply lower sensitivity to potential risk among buyers at Exmouth. On the other hand, and more importantly, it indicates the importance of the absence of a monetary signal of risk, and its translation into economic behaviour.

Analyses carried out to assess the interaction effect of properties sold after cyclone Vance, located in 100-year ARI flood areas, found an increase, rather than a decrease in sales price at a 0.05 significance level. This is similar to findings by Skantz and Strickland (1987), for Houston, Texas, who found that the experience of a major event did not affect prices even for properties located in flood prone areas. They attributed this to the availability of subsidized insurance, where owners did not have to factor in the full cost of risk into their purchase decisions. This was corroborated with their finding that a sharp increase in insurance premiums a year later, however, resulted in a drop in property prices. Babcock and Mitchell (1980) also reported a similar finding for Ontario in Canada, where the experience of a major event did not affect prices, even for those located in the floodplain. A similar finding was reported by Bartosova et al. for Wisconsin in the US.

Another reason why buyers have not discounted for risk-prone property in Exmouth could be the expectation of compensation from the government, a classic example of moral hazard As was demonstrated with the relief provided through a nation-wide levy to property-owners affected by the Queensland flooding and cyclone earlier in 2011 (Australian Government, Prime Minister's Office 2011), the problem lies in an imbalance between social and private costs. While property owners enjoy the amenity gained from living so close to the coast, they do not bear the full cost of their decisions, either through insurance, or paying for damages following a major event. This creates an inequitable situation, where the rest of society, who do not partake in these benefits, is forced to fund emergency, response and long-term rehabilitation costs.

An HPM analysis of the interaction between risk of 4 m storm-surge and the Vance effect was not carried out, because at the time of cyclone, the construction of the marina had not yet commenced. With findings that buyers did not discount for property prices even directly following the event, it was not expected that there would be any effect for at-risk properties in the marina area, which came on the market 2 years after the event, by which time memories of the event had faded.

Policy Implications

The lack of the impact of disaster risk on property sales requires policy that will serve to internalise the social costs of disasters into the private calculations of property owners and developers. This can be done in various ways where, for example the local government can implement land-use control measures. Implementing and enforcing this may require various economic instruments to encourage developers and local communities to behave more in accordance with federal and state controls. Monetary obligations to incorporate the threat of flood-risk may be the only way that more judicious land-use controls are enforced.

It is 15 years since cyclone Vance (1999), and over time the memories of such catastrophic events fade, and other priorities take precedence. Several of the marina properties have come on the market in the last few years, and buyers may not be aware of the extent of risk to which they are exposed to. Marina property sales are targeted at employees of major oil and gas companies working in the area (Ray White 2011). It is possible that the nature of such industries may attract people from capital cities of Australia, who are not aware of the extent of cyclonic and other coastal risk of properties developed in towns such as Exmouth. While Exmouth is generally well-known as a cyclone-prone area, there is a case for additional information to be provided to buyers on the specific risks to which their property is exposed.

Another management pathway associated with cyclonic storm-surge inundation and flooding risk is the designation of high-risk areas through mapping, and making this information publicly available to future property buyers in the area. The basis of this management action is that if people are aware of risk, they will translate this knowledge into action, by paying lower prices for properties located in high-risk zones. The rationale is that the lower price reflects the potential costs of damage they may bear in the future. Alternatively, they may undertake risk mitigation measures to minimize potential damages.

From the perspective of property buyers, awareness of risk alone may not, however, always translate into action. For instance, an investigation into flood insurance purchase for coastal properties in the US, found that only 49 % of households maintained flood insurance, despite mandatory purchase requirements for federally backed mortgages (Kriesel and Landry 2004). This implies that regulatory measures alone are not sufficient to change economic decisions to buy risk-prone property. Given that Australia has no mandatory insurance program such as the National Flood Insurance Program in the US, this begs the question, will such legal requirements change the way the Australian property buyers perceive and make decisions regarding coastal risk? Monetary signals through higher insurance premiums or other forms of economic incentives may therefore be more effective in influencing property purchase behaviour. This was found to be the case in the US, where studies have shown that buyers who pay high insurance premiums for flood risk-prone properties pay lower prices at the time of sale (Troy and Romm 2003).

Natural disaster insurance is the most commonly explored economic strategy applied to risk. It serves to correct market failures arising from the externalities created by developers and property buyers, and failure on the part of the state and local governments. Following the 2011 Queensland floods, in eastern Australia, a national disaster insurance review recommended flood insurance for riverine events. This approach could be extended to include cyclonic storm-surge inundation, whereby a compulsory private insurance scheme, possibly subsidised in the short-term through a federal government fund, could be used for high-risk areas. Conditions could be built in the future to alleviate any perverse incentives that might arise from such a program.

Natural disaster insurance, however, cannot internalise the risks created to other parts of the community through development in high-risk areas. It also fails to internalise the social cost of disasters, created by property developers. The use of subsidised insurance in the long-term for high-risk properties could also create distortions in the market, resulting in reduced participation and under-insurance, thereby subverting the original goals.

Conclusions

The study findings indicated that property owners did not perceive coastal disaster risk as a significant factor in their decision of where to build houses. Amenity and other benefits of living close to the ocean and the expectation of government compensation in the event of a disaster could be factors affecting individual responsibility for risk mitigating actions.

To make judicious land-use decisions, state and local governments concerned with natural disaster management, must be cognizant of the long-term social cost of real-estate development decisions. Coastal development cannot be driven by economic interests alone, but must balance environmental goals and equity considerations, especially with regard to who pays for the cost of disaster risk.

While the study focused on Exmouth in the north of WA, the need to assess the economics of planning and coastal property development in disaster-prone areas is equally germane for other coastal disaster prone countries in the Asia Pacific such as the Philippines, Bangladesh, Myanmar and China, where unbridled development, growing coastal populations and injudicious land planning amplifies the predicted disaster risk due to climate change and extreme weather events.

References

Anselin L (1988) Spatial econometrics: methods and models. Kluwer Academic, Dordrecht

Australian Government, Prime Minister's Office (Hon. Julia Gillard MP) (2011) Rebuilding after the floods. Media release, Press Office, Canberra, 27 January, http://www.pm.gov.au/press-office/rebuilding-after-floods.

Babcock M, Mitchell B (1980) Impact of flood hazard on residential property values in Galt (Cambridge), Ontario. Water Resour Bull 16(3):532–537

Bin O, Polasky S (2004) Effects of flood hazards on property values: evidence before and after hurricane floyd. Land Econ 80(4):490–500

Bin O, Kruse JB, Landry CE (2008) Flood hazards, insurance rates and amenities: evidence from the coastal housing market. J Risk Insur 75(1):63–82

Bowman Bishaw Gorham (1997) Exmouth marina, resort and residential development. Public environmental review, prepared for Landcorp, Perth.

Gujarati DN (2006) Essentials of econometrics. US Military Academy, West Point

Hallstrom DG, Smith VK (2005) Market responses to hurricanes. J Environ Econ Manage 50 (3):541–561

Intergovernmental Panel on Climate Change (IPCC) (2007) Climate change 2007: the physical science basis. Contribution of working group I to the fourth assessment report of the intergovernmental panel on climate change. Cambridge University Press, Cambridge

Keskin B (2008) Hedonic analysis of price in the Istanbul housing market. Int J Strateg Prop Manage 12(2):125–138

Knutson TR (2011) Tropical cyclones and climate change. Extended abstract for GFDL/NOAA, http://yosemite.epa.gov/ee/epa/eerm.nsf/vwAN/EE-0566-103.pdf/$file/EE-0566-103.pdf

Kriesel W, Landry C (2004) Participation in the national flood insurance program: an empirical analysis for coastal properties. J Risk Insur 71(3):405–420

MacDonald DN, Murdoch JC, White HL (1987) Uncertain hazards, insurance and consumer choice: evidence from housing markets. Land Econ 63(4):361–371

MacDonald D, White H, Taube P, Huth WL (1990) Flood hazard pricing and insurance premium differentials: evidence from the housing market. J Risk Insur 57(4):654–663

Morgan A (2007) The impact of hurricane ivan on expected flood losses, perceived flood risk and property values. J Hous Res 16(1):47–60

Nakagawa M, Saito M, Yamaga H (2007) Earthquake risk and housing rents: evidence from the Tokyo metropolitan area. Reg Sci Urban Econ 37(1):87–99

Nakagawa M, Saito M, Yamaga H (2009) Earthquake risks and land prices: evidence from the Tokyo metropolitan area. Jpn Econ Rev 60(2):208–222

Pearson L, Tisdell C, Lisle A (2002) The impact of Noosa national park on surrounding property values: an application of the hedonic price method. Econ Anal Policy 32(2):155–171

Ray White Exmouth (2011) Ray White Exmouth. http://www.raywhiteexmouth.com/display/xml_listing/ressale. Accessed 22 Jan 2011

Roberts RS (2012) Economic strategies for coastal disaster risk-reduction: a case study of Exmouth, Western Australia. Unpublished PhD dissertation, Murdoch University, Perth, Western Australia

Rosen S (1974) Hedonic prices and implicit markets: product differentiation in pure competition. J Polit Econ 82(1):34–55

Skantz TR, Strickland TH (1987) House prices and a flood event: an empirical investigation of market efficiency. J Real Estate Res 2(2):75–83

Sydney Coastal Councils Group (2011) Quantifying the value of Sydney's beaches. http://www.sydneycoastalcouncils.com.au/Project/Quantifying_Value_of_Sydney_Beaches

Taylor LO (2003) The hedonic method. In: Champ PA, Boyle KJ, Brown TC (eds) A primer on nonmarket valuation. Kluwer, Dordrecht, pp 331–393

Troy A, Romm J (2003) Assessing the price effects of flood hazard disclosure under the California natural hazard disclosure law. J Environ Plan Manage 47(1):137–162

Walk Score (2013) Walk Score. http://www.walkscore.com. Accessed 15 Jan 2011

Western Australian Government, Planning Commission (2013) State coastal planning policy guidelines. Western Australian Government, Perth

Western Australian Government, Planning Commission, Shire of Exmouth (2004) Exmouth townsite structure plan. Western Australian Government, Perth

Western Australian Government, Department of Water, Sinclair Knight Mercz (SKM) (2007) Floodplain management study. Unpublished report, accessed 12 Feb 2012

Chapter 10
Understanding the Role of Trust in Network-Based Responses to Disaster Management and Climate Change Adaptation in the Asia-Pacific Region

Victoria L. Ross, Elizabeth L. Malone, and Susan Kinnear

Abstract Effective communication and strong collaboration amongst organisations are essential tools for disaster management responses and climate change adaptation. Complex situations such as disaster preparation and response require trust-based and collaborative networks. Given that mutual trust and willingness to share information are central to successful disaster management, it is crucial to increase our understanding of how to build and sustain network trust and collaboration. Based on a comprehensive literature review, including both theoretical and empirical studies, we explore the role and impact of trust on facilitating improved network co-ordination in emergency responses to climate-driven events. Social identity theory and cultural cognition theory are applied as a framework for understanding trust development, particularly in the context of the Asia-Pacific region. The paper identifies obstacles to trust development, and then outlines key methods by which trust can be established and enhanced to improve communication and network effectiveness in disaster management. These findings provide critical information for policy makers and network participants with regard to building network trust to better underpin disaster management and climate change adaptation in the Asia-Pacific region.

V.L. Ross (✉)
Smart Water Research Centre, Griffith University, Gold Coast Campus, Southport, QLD 4215, Australia
e-mail: victoria.ross@griffith.edu.au

E.L. Malone
Pacific Northwest National Laboratory, Joint Global Climate Change Institute, College Park, MD 20740, USA
e-mail: e.malone@pnnl.gov

S. Kinnear
Central Queensland University, Rockhampton, QLD 4700, Australia
e-mail: s.kinnear@cqu.edu.au

W. Leal Filho (ed.), *Climate Change in the Asia-Pacific Region*, Climate Change Management, DOI 10.1007/978-3-319-14938-7_10

Introduction

Climate change is predicted to increase the frequency, severity and intensity of natural disasters in the decades to come (Djalante 2013; WMO 2014). Globally, the increasing incidence of extreme events presents a significant threat to human society as well as the built environment (Cadag and Gaillard 2012; Hidayat and Egbu 2010; Kostoulas et al. 2008), with casualties and damage to vital infrastructure now being a major global humanitarian concern (Haque 2003). This is a particularly vexed issue in the Asia-Pacific region, given that the area currently experiences over 70 % of the world's natural disasters; and scientists have warned that the intensity and/or frequency of disasters in the region are likely to increase further (APEC 2014; Haque 2003). The recent IPCC fifth assessment report (AR5) noted that in the Asia region, extremes of precipitation will be characterised by strong variability, with both increasing and decreasing trends being observed spatially across Asia, and temporally across seasons (IPCC 2014). The IPCC also reports a high degree of confidence that 'extreme climate events will have an increasing impact on human health, security, livelihoods and poverty' (IPCC 2014, p. 3).

Given the clear link between climate change, disaster vulnerability and disaster management (Bijay et al. 2013), disaster response and mitigation are becoming increasingly important as strategies for adaptation to climate change and regional sustainability (Malone and Kinnear 2015). Better preparation for, and responses to, emergencies and disasters are a priority across the Asia-Pacific, as this helps to reduce the risk of disasters and creates the ability to build business and community resilience (APEC 2014). However, tackling this agenda is complex and multifaceted, reaching across multiple tiers of government, communities, organisations and individuals.

This paper is innovative in exploring the impact of trust on facilitating improved network co-ordination in emergency responses, particularly in the Asia Pacific region. Existing research has demonstrated the link between effective response and recovery activities and well-coordinated, inter-organisational networks and trust (Cadag and Gaillard 2012; Kapucu 2005). However, no studies to date have established how and why this occurs, and what the implications are for those organisations working in climate change adaptation. In this paper, a theoretical framework is presented to provide an understanding of trust development, obstacles to trust are identified, followed by an outline of strategies by which trust can be established and enhanced to improve communication and network effectiveness in disaster management. Finally, practical recommendations for trust building strategies are provided for disaster management network managers. In applying these, however, it must be acknowledged that this review is constrained by the available literature on social networks and trust, particularly the paucity of literature that examines these concepts using field examples from the Asia-Pacific. The understanding and measurement of network-based trust is a nascent field, and as such this paper is an exploratory, and offers a review of insights-to-date in this developing field.

Understanding Trust and Networks: An Historical Perspective

Trust and networks have been studied for decades as elements of social capital, economic transactions, and organizational performance. Jane Jacobs emphasized the important role trust plays the dense urban networks that constitute social capital (Jacobs 1961). Trust among the shopkeepers and residents of neighborhoods meant that people networks effectively addressed problems such as crime, whereas the lack of trust among groups in the Hausmann-style "City Beautiful" effectively prevented social capital of this type, instead fostering distrust among groups estranged from each other, necessitating strong police forces and other formal organizations. These themes were developed with regard to business development by ethnic groups in the 1970s (see Fukuyama 1997 for this history). Later, sociologists in the 1980s and 1990s, notably Robert Putnam (1993, 1995) emphasized trust as a necessary component of the shared values that promote or prevent economic prosperity.

Trust, of course, may facilitate action within a group but hinder actions that must be taken with outsiders; that is, trust may be based on strong, shared values that are the very core of the organization and thus include distrust of outsiders who, by definition, do not share those values (Fukuyama 1995). This is the dilemma of organizations that are joined in a network without any or much of a basis for trust among the organizations. Their "radius of trust" may not extend beyond their own organizations, so trust needs to be consciously built. Unfortunately, the theoretical and empirical work on trust and networks does not examine the case of organizations that are directed to work together, i.e., form a network to perform specific tasks or functions. Rather, research and analysis has been targeted at firms who voluntarily work together, knowledge networks in the case of research groups (e.g., Hardwig 1991), or at face-to-face networks (families, kin, ethnic groups, civil society organizations). This leaves a conspicuous gap in the literature, particularly in the context of disaster management, which typically calls for numerous organisations to co-deliver responses to acute climate events.

Understanding Trust and Collaborative Networks in Disaster Management

To achieve a rapid and effective response to unexpected events, multi-agency disaster management is often comprised of collaboration amongst geographically distributed organisations, both public and private (Janssen et al. 2010). Within this, a key challenge for disaster relief efforts is the need for these numerous organisations to collaborate effectively, despite often having had no previous interactions with one another (Kapucu and Garayev 2011). Furthermore, participating organisations often have qualitatively different ideas about how collaboration should work and what elements and outcomes of disaster relief are the most important (Malone and Kinnear 2015). Not surprisingly, research studies have thus shown that inter-organizational communication is not always effective during a crisis (Kinnear et al. 2014; Longstaff and Yang 2008; Palttala et al. 2012; Moe and Pathranarakul 2006).

Palttala et al. (2012) suggest that differing organisational cultures, structures and approaches are a hindrance to cooperation during a crisis, although, at least theoretically, these differences could also complement each other and provide a basis for more comprehensive responses. Janssen et al. (2010) stated that many disaster management systems lack the capacity to cope with complexity and uncertainty; possibly driven by a lack of human-centred approaches to improve multi-agency disaster management. In particular, trust-based relationships have been identified as critical for effective emergency responses (Janssen et al. 2010; Kapucu 2005; Kapucu and Garayev 2011; Longstaff and Yang 2008).

Why do complex situations such as disaster response and recovery require trust-based and collaborative networks? A strong body of research has demonstrated the important role of trust in network collaboration, communication and co-operation (e.g., Beratan 2007; Fisher 2013; Janssen et al. 2010; Kapucu and Garayev 2011; Longstaff and Yang 2008; Moe and Pathranarakul 2006; Vasavada 2013). In interpreting this material, it is worthwhile to note that trust is important between emergency responders; but so too it is important for those organisations to be trusted by the public they serve (Longstaff and Yang 2008). According to Beratan (2007), it is necessary to establish an interpersonal culture of trust and pragmatism in order that collaborative social networks can operate effectively. Trust is also critical for network performance and sustainability (Provan and Kenis 2008), and has been demonstrated to accelerate coordination in emergency management response operations (Kapucu 2005).

Longstaff and Yang (2008) conducted research to test the role of trust in the planning and management of communication in crisis situations; and concluded that trust is one of the most important variables in effective communication management in times of "surprise."

The authors analysed reports from a wide variety of global crisis incidents and found a direct correlation between trust and the level of an organization's preparedness and internal coordination of crisis communication. This provided strong evidence that a higher level of trust led to better internal coordination of crisis communications with the organisation's stakeholders as well as with external media. However, the authors stressed that trust must be two-directional (i.e., "I will trust local government if it trusts me enough to be prepared to tell me the whole story."). Their findings also indicated that more pre-crisis preparedness can lead to significantly less attribution of blame to other parties during and after a crisis event.

While network structures are often hierarchical (government bureaucracy, for example), the dynamic situations presented by disaster responses actually require the simultaneous operation of hierarchies as well as decentralised nodes, in order to facilitate an effective response. Here, trust-based networks are preferable to other network architectures, including those based on market mechanisms (which do not function in disaster response, except possibly in a predatory way) and those based on regulations and sanctions (which may be difficult to enforce and inadequate for disaster conditions) (Gausdal and Hildrum 2012).

Trust and Disaster Management in the Asia Pacific Region

A paucity of research exists regarding the role of trust in disaster management specifically within the Asia Pacific region. However, several key research papers relevant to the region are outlined below.

Moe and Pathranarakul (2006) conducted a detailed case study of Thailand's tsunami responses, examining the critical success factors for successful disaster management. They argued that many developing countries prone to natural disasters appear to lack proactive strategies for early warning, mitigation and preparedness. Moe and Pathranarakul reported that project success was strongly linked to communication and co-operation between stakeholders, and concluded that trust facilitated by effective communication between key stakeholders was the key factor in successful disaster management (Moe and Pathranarakul 2006).

According to Nakagawa and Shaw (2004), broader social capital (comprising trust, social norms and networks) also plays an important role in disaster recovery. These scholars examined the role of social capital in the post-earthquake rehabilitation and reconstruction programs across two case study regions: Kobe, Japan and Gujarat, India. Despite differences in socio-economic and cultural contexts, the authors demonstrated that the social capital elements of trust, social norms and networks were positively associated with the highest satisfaction rates with post-disaster planning as well as the fastest recovery time (Nakagawa and Shaw 2004).

Consistent with these findings, Vasavada (2013) asserts that high levels of trust are a crucial mechanism for effective networked responses for disaster management. Vasavada studied a disaster management network in the state of Gujarat, India. Due to its geography, Gujarat has experienced numerous natural disasters, and its underdeveloped infrastructure makes the area particularly vulnerable to the effects of these events. Using social network analysis and interviews, the research examined the structure of the disaster management network and identified the key factors for network effectiveness: these included trust; the number of participants in the network; goal consensus and the need for network-level competencies. Results showed a high density of trust among the central actors who shared an interdependence for financial resources and outreach efforts; there were also higher levels of trust among peripheral actors who had worked together prior to the disaster and thus had pre-established relationships. By contrast, peripheral actors had a low level of trust in the central actors. Vasavada stressed that it was important for network members to focus on trust and goal consensus as being necessary to underpin effective recovery efforts; as these factors are particularly helpful in managing the dynamics of the network relationships.

In Australia, Kinnear et al. (2014) conducted research on network governance and climate change adaption in response to the widespread 2010/2011 Queensland floods, which affected 70 % of the State and had large personal and social impacts (Kinnear et al. 2014). Primary data collected from organisations involved in disaster management and water management were analysed to examine collaboration and communication patterns; changes in the network structure from routine management to flood operations; similarities and differences between the geographic regions, and whether collaboration was correlated with trust. Results showed that having trust in other organisations and using personal networks to establish good working relationships were highly valued by respondents. High value was also placed on collaborators who were experienced, flexible, trustworthy, and had good knowledge and understanding of local contexts. Providing and receiving resources and information tended to be associated with trust, echoing the finding of Vasavada (2013). Some of the key implications for policy and practice included the need for stakeholders to drive adaptation to climate change through collaboration and communication; the need for stakeholders to share a common goal and language; and the need to establish collaboration outside of disaster events (Kinnear et al. 2014).

Taken together, this series of research studies from the Asia Pacific region show that trust is essential for establishing good working relationships, and in turn, effective social networks in disaster management. It is likely that trust is so important in this context because people (and hence, organisations) have an inherent desire for human-centred approaches and for inter-personal relationships when coping with complexity and uncertainty. In considering this, the theoretical framework described in the next section provides an understanding of peoples' needs for social relationships to reduce perceptions of risk and generate trust.

A Theoretical Framework for Understanding Trust

Given the critical importance of trust to effective disaster response operations (Janssen et al. 2010; Kapucu 2005; Kapucu and Garayev 2011; Longstaff and Yang 2008), it is crucial to increase the current understanding of how to build and maintain network trust and collaboration. Examination of the trust literature shows that definitions of trust vary widely (Twyman et al. 2008). Notwithstanding this, the most common elements for the existence of trust appear to be, vulnerability, risk, and positive expectations regarding the behaviour of others. A widely accepted definition of trust is 'a psychological state comprising the intention to accept vulnerability based upon positive expectations of the intentions or behaviour of another' (McEvily et al. 2003; Rousseau et al. 1998). According to researchers, personal vulnerability lies at the heart of trust—if we did not feel vulnerable, we would not feel the need to trust people (e.g., Hogg 2007; Marar 2003).

Two contrasting orientations have been advanced to explain the development of trust. The instrumental or rational model views trust as a subjective calculation of the potential costs and benefits of future interactions. However, this approach has been criticised for affording too little role to emotional and social influences on trust decisions (Kramer 1999). In contrast, the relational model emphasises social identification and social relationships in relation to trust (Tyler and Degoey 1996). The relational framing of trust suggests that trust is related to the nature of social bonds and shared identities between group members (Tyler and Degoey 1996). According to this perspective, shared social identification (a sense of "we"--ness, or belonging to a social group) enhances trust in other community members, thus increasing cooperative behaviour (Tyler and Degoey 1995). The relational model of trust provides a valuable framework for understanding why trust appears to be linked to peoples' needs for social relationships. It is particularly useful when applied to broader social network analysis, which is increasingly being used to examine relationships amongst actors in disaster management settings.

The relational framing of trust accords with social identity theory, which posits that social groups provide members with a social identity: a definition of whom one is and an evaluation of what that entails (Hogg and Abrams 1998; Tajfel and Turner 1986). Research from the social identity perspective has shown that people who belong to our groups (i.e., in-group members) are perceived in more desirable ways—such as being more trustworthy or honest—compared with outsiders; and that shared social identity reduces risk and generates in-group trust (Brewer 2007; Hogg 2007; Ross et al. 2014; Tanis and Postmes 2005). According to Hogg (2007), we trust fellow group members to do us no harm if they are "one of us" and behave in ways that protect and promote the group. However, group trust also features an inter-group dimension, where members of other groups can be viewed with suspicion and distrust (Hogg 2007).

Within the context of disaster management in the Asia-Pacific, and particularly in the case of networked governance studies, social identity theory can help us understand why distrust may exist between different organisations. This also raises the interesting phenomenon of some organisations essentially being perceived as 'out-group' although they are in fact actors *within* the social network, and may record a high density of social ties. As will be discussed in a later section, social identity theory can also be applied to develop methods to build trust between the different organisations involved in disaster management.

Another theory that recognises the influence of social and cultural factors on trust and risk perceptions is that of cultural cognition. According to this theory, when people have to make decisions in an area in which they do not have expertise, their only option is to rely on experts in whom they trust (Kahan and Braman 2006). Here, the trusted people are inevitably ones whose cultural values they share, and who are inclined to credit or dismiss scientific evidence based on its conformity to their cultural values. In other words, cultural cognition influences perceptions of credibility (Kahan et al. 2011).

According to cultural cognition theory, individuals reflexively reject information that is inconsistent with their predispositions, especially when they perceive that it is being advocated by experts whose values they reject and opposed by alternative people whose values they share. Kahan et al. therefore assert that to overcome the cultural cognition effect, communicators must attend to the cultural meaning of information, and take care when 'framing' messages for intended audiences. Here, framing needs to consider that people interpret information through their ideological filters and are strongly influenced by their own identity, worldview, and belief systems (Kahan et al. 2011).

The approaches outlined above share a commonality in suggesting that people are not always 'rational' in their perceptions and decision-making due to the influence of perceptual biases. Rather, all decisions are made through a complex process of emotion, personal and social filtering. Audiences filter information through the lenses of their value and belief systems, as well as cultural experiences and norms (Vaughan and Hogg 1995). This is an important message given that all organisations are comprised of people, each bringing their own dynamics. Hence, it is reasonable to conclude that communication amongst people and organisations will need to be complex and nuanced, if the development of trust is to be fully realised.

Obstacles to Trust Development

Establishing trust is a major challenge in extreme situations due to the involvement of diverse response teams and the frequently changing actors that emergency responders have to deal with during disasters (Kapucu and Garayev 2011; Kostoulas et al. 2008). According to Kostoulas et al. (2008), first responders can be hesitant to communicate and interact with those outside of their organization due to a lack of prior interactions. As foreshadowed earlier, this tendency to be distrustful of those outside of one's group is consistent with social identity theory and the concept of inter-group suspicion and distrust. Kostoulas et al. (2008) assert that it is the task of first responders to assess the trustworthiness of others in order to reduce uncertainty. However, there is often considerable pressure to do this quickly, which may distort the otherwise gradual sequence of trust development. For example, according to Beratan (2007), trust has to be earned over time, through repeated experience of positive interaction.

It is not yet clear whether trust-connections built at unnaturally rapid speed are more tenuous compared than those grown gradually; or if they in fact have a similar durability once formed, but that the likelihood of genuine formation is decreased under time-pressured circumstances. What is clear, however, is that organisations tend to be reluctant to commit to relationship and trust-building activities because of the time imposition, the possibility of failure, and the difficulties in recognising progress (Beratan 2007). The latter is an especially acute problem because it is difficult to measure the level of trust among participating organizations during emergencies, especially where different and frequently changing actors are involved (Kapucu and Garayev 2011). In the absence of an agreed metric, this may lead to nonchalance amongst network actors with regard to the basic value of trust, and less propensity to invest in building trust assets. For example, Beratan

(2007) claims that few organisations, particularly public sector agencies, feel that they can afford to persist with activities that do not produce visible short-term progress. However, in the Asia Pacific region, where climate events are sudden, and where there may be rapid 'churn' of management in government and/or other key disaster response organisations, it could be argued a time-lag to allow for trust development is a luxury that cannot often be afforded; thus intensifying the need to dedicate resources to trust development in the pre-preparation phase.

The second key barrier to trust-building is the tendency for organisations to have differing cultures, structures and approaches. The many different groups of responders involved in relief operations (e.g., police officers, firefighters, medical personnel, civil engineers, public and private agencies) have distinct roles, skills, knowledge and experience, as well as policies and protocols on how to respond to emergency situations (Kostoulas et al. 2008). Palttala et al. (2012) claim that communication problems occur when actors need to cooperate, but are in fact from organisations that have differing ways of handling a crisis. Given that cultures develop slowly, changing these in favour of greater integration can be a challenge (Palttala et al. 2012).

Building Trust

The adage "trust arrives on foot but leaves on horseback" by the nineteenth century Dutch politician, Thorbecke, aptly describes the concept of trust being difficult to earn and easy to lose (Nooteboom 2002). It is well established that trust building is a long-term and time consuming process (Beratan 2007; de Bruijn and ten Heuvelhof 2008; Longstaff and Yang 2008; Tummers et al. 2012; Vasavada 2013). Within the context of disaster management, Longstaff and Yang (2008) assert that trust cannot simply be established on the first day of a disaster. Trust has to be earned over time, and can occur only if there is a long-term commitment to the process. Moreover, the collaborative engagement needed for trust building requires a sincere and visible commitment of time and resources (Beratan 2007). An interesting question amongst this is the impact that repeated, and more frequent, disaster events may have on the trust dynamics of response organisations across the Asia-Pacific: one possibility is that organisations will become more comfortable and familiar with each other after having partnered on several disaster events. Equally, it may be that organisational values become more fluid, and individuals become more accepting of those from different cultural bias. On the other hand, it is also possible that trust may break down over time, because of negative experiences in working with others, particularly if those organisations are poorly prepared or ineffective in communicating.

Longstaff and Yang (2008) assert that perhaps the most important and least understood role for policy makers is insuring that emergency communications can be trusted by other emergency responders and by the public. Longstaff and Yang suggest that as a minimum, ongoing communication should be provided to help build trust; accompanied by mandated communications plans to guide organisational responses in times of surprise. Organisations that have previously been involved in collaborative efforts with dialogue between actors appear to have a greater ability to build trust (Ansell and Gash 2008; Kapucu and Garayev 2011). Gaillard (2010) states that a concerted effort is needed to open dialogue and thus build trust between scientists, non-governmental organisations, disaster reduction practitioners and local communities. Palttala et al. (2012) also found that organizations with a good reputation are considered trustworthy, and other actors are willing to share information with these. In a practical vein, technologies such as mobile phones, the internet, geographical information systems, and community radios can help to strengthen networks and promote effective disaster response by keeping those who trust one another in communication (Ospina 2011).

Trust building in disaster management requires not just professional expertise to rescue people and mitigate harmful consequences, but also openness, empathy, and explaining decisions and alternatives (Palttala et al. 2012). Palttala et al. conducted a study to identify gaps in communication in disaster management; asking experts from governmental organizations and NGOs about their experiences in disaster management and communication. The results stressed the importance of honesty, candour and openness in crisis communications. Clarity is crucial in the construction of crisis messages, as inadequate communication leads to confusion and public mistrust (Palttala et al. 2012; Seeger 2006). Palttala et al. concluded that the use of accurate, widely distributed, and timely information reduces anxiety and strengthens people's sense of self-efficacy. In this way, trust is a result of actions and communication flow throughout all the crisis phases (Longstaff and Yang 2008; Palttala et al. 2012).

Building Trust in Disaster Response Networks in the Asia-Pacific Region

Unfortunately, within the Asia-Pacific region, the distribution of accurate and timely communications is not always possible. In developing nations, science and predictive abilities may not be as well advanced, meaning accurate information may not always be available. The large and dispersed population of the Asia-Pacific, coupled with poor baseline communications infrastructure in some areas, could also be a hindrance to achieving mass messaging effectively. Lebel (2014) has reported that 'major knowledge-action gaps exist with respect to adaptation to climate change in the Asia-Pacific', due to missing, inaccessible or unused information. In this landscape, it may be very difficult to build trust through accurate and timely communications.

So, how can organisations effectively build trust when many of the factors on which trust development is influenced, are external to their control? Below we outline key recommendations for policy makers and disaster management network participants for facilitating trust building in the Asia Pacific region. These strategies should prove particularly valuable for disaster management network managers.

Recommendations

Emphasising a Common Goal and Shared Identity

A key strategy for facilitating trust building is through an emphasis on striving to achieve a common goal (i.e., effective disaster response). As discussed previously, social identity theory asserts that we trust others if they are perceived to be a member of our group, or "one of us". Consistent with both social identity theory and cognitive cognition theory, it is well-established that we also trust people whose values are compatible with our own goals or are perceived to have shared values (Earle and Cvetkovich 1995, Ross et al. 2014; Vaske et al. 2007). Therefore, in order to develop a sense of shared social identity and network trust it will be vital for disaster management communicators to emphasise that all organisations are "working together for a common goal". In addition, inter-group suspicions and distrust could be overcome if network organisations perceive themselves as belonging to a larger group of emergency responders all working to an effective disaster response.

As trust is based on the communicator and target persons sharing similar values, Siegrist et al. (2007) stresses that communicators should therefore not only be knowledgeable about the facts of the case, but should primarily be knowledgeable the concerns and values of the target audience. In the context of building network trust in the Asia Pacific region, it will be important to be cognizant of peoples' inherent desire for human-centred approaches when coping with complexity and uncertainty. In this way trust can be facilitated through developing a sense of shared values and concerns, and thus 'we-ness' among disaster response networks. This sense of shared social identification through understanding values of network partners will enhance trust and facilitate co-operative behaviour (Tyler and Degoey 1995).

Understanding Cultural Values

The research reviewed in this chapter shows that people are not necessarily rational in their perceptions and decision-making due to the influence of perceptual biases. Cultural cognition theory explains the tendency of individuals to trust those who they perceive to have shared values, and distrust or reject information from those they perceive to oppose their values (Kahan et al. 2011). In the Asia Pacific region, where there are a wide range of differing cultures, it will be vital for communicators to attend to the cultural meaning of information and messaging in order to overcome the cultural cognition effect. Understanding differences in opinions and working towards building consensus will be central to facilitating (and sustaining) network trust.

Attending to Message Framing

When framing messages it should also be considered that people interpret information through their ideological filters and are strongly influenced by their identity, worldview, and belief systems. It will also be important to take into account the cultural values and endpoint goals, as already described above. It will be critical to convey messages that do not directly challenge people's embedded values, but instead encourage consensus building among diverse groups. This approach will be particularly important when establishing network trust in the geographically and culturally diverse expanse of the Asia-Pacific.

Build Trust Early and Repeatedly

The extant literature clearly shows that trust-building requires an investment of time, yet conversely, climate impacts are typically sudden and acute, leaving little time for the establishment of trust. Hence, the importance of pre-disaster planning and scenario-based training is likely to increase, as the Asia-Pacific confronts the reality of climate change. Furthermore, Fukuyama (1995) categorically states that trust is essential to functioning networks: trust may begin with trust in professional standards, reputation, ethnicity, or some other basis, but it must be fostered for the tasks of the network. Hence, the regular flux (substitution, replacement, or entry/exit) of actors throughout disaster networks also signals the need to commit resources to ensure the sustainability of trust-based partnerships.

Explore Metrics for Trust

Future research work in this field may concentrate on identifying both qualitative and quantitative measures of trust. Based on his analysis of social capital and the role of trust, Fukuyama (1997, p. 429) proposed a "radius of trust" measure of social capital, which measures "the degree to which individual members are capable of collective action on the basis of mutual trust." However, a practical application of this theory has not been brought forward: this is a necessary development, as without it, organisations will be without a tangible appreciation of their existing levels of trust, and whether these are being maintained or degraded over time.

Conclusions

The Asia Pacific region, with a large and dispersed population and difficulties in distributing accurate and timely communications, presents unique challenges for building network trust. This work makes a significant contribution to the climate change adaptation literature by providing a insight into how network trust can be facilitated, specifically in the Asia pacific region. The relational model of trust, as well as both social identity and cultural cognition theories have provided useful theoretical frameworks for understanding peoples' needs for social relationships to reduce perceptions of risk and generate trust. In applying these models, and integrating the findings of existing case studies from the Asia-Pacific, this review identified at five key learnings to assist future climate change adaptation initiatives: emphasise a common goal and develop shared identity; understand cultural values; attend to message framing; build trust early and repeatedly, and explore the development of metrics for trust, to guide further development of new and existing social networks.

References

Ansell C, Gash A (2008) Collaborative governance in theory and practice. J Public Adm Res Theory 18(4):543–571

Asia-Pacific Economic Cooperation (APEC) (2014) http://www.apec.org/Groups/SOM-Steering-Committee-on-Economic-and-Technical-Cooperation/Working-Groups/Emergency-Preparedness.aspx. Retrieved 5 March 2014

Beratan KK (2007) A cognition-based view of decision processes in complex social–ecological systems. Ecol Soc 12(1):27

Bijay P, Leal Filho W, Schulte V (2013) Understanding the links between climate change and disaster management in Pacific Island countries. In: Leal Filho W (ed) Climate change and disaster risk management. Springer, Berlin, pp 55–69

Brewer MB (2007) The importance of being we: human nature and intergroup relations. Am Psychol 62:728–738

Cadag JRD, Gaillard J (2012) Integrating knowledge and actions in disaster risk reduction: the contribution of participatory mapping. Area 44:100–109

de Bruijn H, ten Heuvelhof E (2008) Management in networks: on multi-actor decision making. Routledge, New York

Djalante R (2013) Identifying drivers, barriers and opportunities for integrating disaster risk reduction and climate change adaptation in Indonesia: an analysis based on the earth system governance framework. In: Leal Filho W (ed) Climate change and disaster risk management. Springer, Berlin, pp 131–147

Earle TC, Cvetkovich GT (1995) Social trust: towards a cosmopolitan society. Praeger, London

Fisher R (2013) A gentleman's handshake: the role of social capital and trust in transforming information into usable knowledge. J Rural Stud 31:13–22

Fukuyama F (1995) Trust: the social virtues and the creation of prosperity. Free Press, New York

Fukuyama F (1997) Social capital: the tanner lectures on human values, pp 377–484. http://tannerlectures.utah.edu/_documents/a-to-z/f/Fukuyama98.pdf

Gaillard JC (2010) Vulnerability, capacity and resilience: perspectives for climate and development policy. J Int Dev 22:218–232

Gausdal AH, Hildrum JM (2012) Facilitating trust building in networks: a study from the water technology industry. Syst Pract Action Res 25:15–38
Haque CE (2003) Perspectives of natural disasters in East and South Asia, and the Pacific Island states: socio-economic correlates and needs assessment. Nat Hazards 29:465–483
Hardwig J (1991) The role of trust in knowledge. J Philos 88(12):693–708
Hidayat B, Egbu C (2010) A literature review of the role of project management in post-disaster reconstruction. In: Egbu C (ed) Proceedings of 26th annual ARCOM conference, Association of Researchers in Construction Management, Leeds, pp 1269–1278, 6–8 September 2010
Hogg M (2007) Social identity and the group context of trust: managing risk and building trust through belonging. In: Siegrist M, Earle TC, Gutscher H (eds) Trust in risk management. Earthscan, London
Hogg MA, Abrams D (1998) Social identification. Routledge, London
IPCC (2014) Working group II, fifth assessment report, volume II: regional aspects – Chapter 24: Asia. Intergovernmental panel on climate change. http://www.ipcc.ch/report/ar5/wg2/
Jacobs J (1961) The death and life of great American cities. Vintage Books, New York
Janssen M, Lee J, Bahrosa N, Cresswell A (2010) Advances in multi-agency disaster management: key elements in disaster research. Inf Syst Front 12:1–7
Kahan DM, Braman D (2006) Cultural cognition and public policy. Yale Law Policy Rev 24:149–172
Kahan DM, Jenkins-Smith H, Braman D (2011) Cultural cognition of scientific consensus. J Risk Res 14:147–174
Kapucu N (2005) Interorganizational coordination in dynamic context: networks in emergency response management. Connections 26:35–50
Kapucu N, Garayev V (2011) Collaborative decision-making in emergency and disaster management. Int J Public Adm 34(6):366–375
Kinnear S, Patison K, Mann J, Malone E, Ross V (2014) Network governance and climate change adaptation: collaborative responses to the Queensland floods. National Climate Change Adaptation Facility, Gold Coast
Kostoulas D, Aldunate R, Pena Mora F, Lakhera S (2008) A nature-inspired decentralized trust model to reduce information unreliability in complex disaster relief operations. Adv Eng Inform 22:45–58
Kramer DM (1999) Trust and distrust in organisations: emerging perspectives, enduring questions. Annu Rev Psychol 50:569–99
Lebel L (2014) Closing knowledge-action gaps in adaptation to climate change in the Asia-Pacific region. Int J Environ Sustain Dev 13(2):204–222
Longstaff PH, Yang S (2008) Communication management and trust: their role in building resilience to "surprises" such as natural disasters, pandemic flu, and terrorism. Ecol Soc 13(1):3
Malone E, Kinnear S (2015) How and why: complementary analyses of social network structures and cultural values: improving flood response networks in Queensland, Australia. Qual Quant, 49:203–220
Marar Z (2003) The happiness paradox. Reaktion, London
McEvily B, Perrone V, Zaheer A (2003) Trust as an organizing principle. Organ Sci 14:91–103
Moe TL, Pathranarakul P (2006) An integrated approach to natural disaster management: public project management and its critical success factors. Disaster Prev Manage 15(3):396–413
Nakagawa Y, Shaw R (2004) Social capital: a missing link to disaster recovery. Int J Mass Emerg Disasters 22(1):5–34
Nooteboom B (2002) Trust: forms, foundations, functions, failures and figures. Edward Elgar, Cheltenham
Ospina AV (2011) The role of trust in climate change adaptation and resilience: can ICTs help? Nexus for ICTs, climate change and development (NICCD). http://niccd.wordpress.com/2011/02/27/the-role-of-trust-in-climate-change-adaptation-and-resilience

Palttala P, Boano C, Lund R, Vos M (2012) Communication gaps in disaster management: perceptions by experts from governmental and non-governmental organizations. J Contingencies Crisis Manage 20:2–12
Provan KG, Kenis P (2008) Modes of network governance: structure, management, and effectiveness. J Public Adm Res Theory 18(2):229–252
Putnam RD (1993) Making democracy work: civic traditions in modern Italy. Princeton University Press, Princeton, NJ
Putnam RD (1995) Bowling alone: America's declining social capital. J Democr 6:65–78
Ross VL, Fielding KS, Louis WR (2014) Social trust, risk perceptions and public acceptance of recycled water: testing a social-psychological model. J Environ Manage 137:61–68
Rousseau DM, Sitkin SB, Burt RS, Camerer C (1998) Not so different after all: a cross discipline view of trust. Acad Manage Rev 23:393–404
Seeger MW (2006) Best practices in crisis communication: an expert panel process. J Appl Commun Res 34(3):232–244
Siegrist M, Earle TC, Gutscher H (2007) Trust in cooperative risk management. Earthscan, London
Tajfel H, Turner JC (1986) The social identity theory of inter-group behaviour. In: Worchel S, Austin LW (eds) Psychology of intergroup relations. Nelson-Hall, Chigago
Tanis M, Postmes T (2005) A social identity approach to trust: interpersonal perception, group membership and trusting behaviour. Eur J Soc Psychol 35:413–424
Tummers L, Steijn B, Bekkers V (2012) Explaining the willingness of public professionals to implement public policies: content, context and personality characteristics. Public Adm 90 (3):716–736
Twyman M, Harvey N, Harries C (2008) Trust in motives, trust in competence: separate factors determining the effectiveness of risk communication. Judgm Decis Mak 3:111–120
Tyler TR, Degoey P (1995) Collective restraint in social dilemmas: procedural justice and social identification effects on support for authorities. J Pers Soc Psychol 69:482–497
Tyler TR, Degoey P (1996) Trust in organizational authorities. In: Kramer RM, Tyler TR (eds) Trust in organizations: frontiers of theory and research. Sage, London
Vasavada T (2013) Managing disaster networks in India. Public Manage Rev 15(3):363–382
Vaske JJ, Bright AD, Absher JD (2007) Salient value similarity, social trust, and attitudes toward wildland fire management strategies. Hum Ecol Rev 14(2):223–232
Vaughan GM, Hogg MA (1995) Introduction to social psychology, 4th edn. Prentice Hall, Sydney
World Meteorological Organization (WMO) (2014) http://www.wmo.int/pages/prog/drr/documents/Climate_DRR-FactSheet_EN.pdf. Retrieved 1 July 2014

Chapter 11
Local Adaptation to Climate Change: A Case Study Among the Indigenous Palaw'ans in the Philippines

Denise Margaret Matias

Abstract Different cultural, economic, political and social forces shape adaptive capacity. In addition, spatial and social differentiations occurring at sub-national levels also result to differences in levels of vulnerability in one country. One social group often excluded in the discussion of climate change is the indigenous peoples. Traditionally subsisting and living on very minimal assets, they shape and are being shaped by the different ecosystems that they live in and depend on. A group of indigenous Palaw'ans in Palawan, Philippines exhibit social-ecological dynamics with their ancestral domain, part of which is declared a protected area under the Mount Mantalingahan Protected Landscape. Through research data from qualitative methods of key informant interviews, focus group discussions, and participant observation, this paper takes a look on how Palaw'ans perceive climate change and ascertains their adaptive capacity based on their transformability, resilience, and adaptability as well as on their local institutions as social networks. The research finds that the several local adaptations to climate change of Palaw'ans are a function of their transformability, resilience, adaptability, and, to a certain extent, to the social learning gained from their local institution.

Introduction

Traditionally, scientific and policy perspectives on climate change have focused on mitigation rather than adaptation (Füssel and Klein 2006). However, in recent years, adaptation has been receiving increased attention as a response to climate change impacts. Several developing countries with high vulnerability and limited adaptive capacity have already made climate change adaptation a priority. The Philippines, known as a climate hotspot (UNU-EHS 2011), stipulated adaptation as a priority in its National Framework Strategy for Climate Change for the years 2010–2022.

D.M. Matias (✉)
Department of Environmental Sciences and Policy, Central European University, Nádor utca 9, 1051 Budapest, Hungary, Philippines
e-mail: denise.matias@uni-bonn.de

W. Leal Filho (ed.), *Climate Change in the Asia-Pacific Region*, Climate Change Management, DOI 10.1007/978-3-319-14938-7_11

It is important to note, however, that there is no standard level of vulnerability for the different strata of governance and society in one country. Kelly and Adger (2000) point out that, within one country, enormous differences in levels of vulnerability can occur. This is related to what Saunders (1990 as quoted in Marino and Ribot 2012) says that bio-physical changes in the earth system is not only received by various geographies but also by different social strata. Thus, country-level analyses fail to capture not only spatial but also social differentiation of vulnerability occurring at sub-national levels (Adger 2006).

At the heart of social stratification discourse in climate change are those who subsist and live on very minimal assets; they are said to be most at risk due to their proximity to the threshold of disaster (Marino and Ribot 2012). One group in most societies that subsist and live on very minimal assets are indigenous peoples. They and other traditional peoples are often excluded in academic, policy, and public discussions on climate change, in spite of the impacts that they are and will further be experiencing from climatic change (Salick and Ross 2009). Studying indigenous peoples in light of climatic changes is of interest, not only because there is a need to address lack of research on vulnerable social groups and social-ecological systems (Adger 2006) but also due to the active role indigenous peoples play in shaping the very same environment that shapes their day-to-day living. Indigenous peoples are not only victims of climate change but also primary actors in climate change monitoring, adaptation, and mitigation due to their active and significant roles in different ecosystems (Salick and Ross 2009).

The complex social, ecological, and political dynamics existing in the research site make for an interesting case study for climate change adaptation and resilience of social-ecological systems (SES). Situated on a tropical archipelago, the main island of Palawan is highly dependent on ecosystem services from both terrestrial and coastal ecosystems (Tompkins and Adger 2004). Due to the island's high biodiversity, it has also been declared as a United Nations Educational, Scientific and Cultural Organization (UNESCO) Man and Biosphere Reserve in the year 1990. The high migration rate of non-indigenous peoples are affecting both the ecology and sociology of the area, resulting to increased marginalization of the indigenous peoples traditionally living on the island (Novellino 2000). Lastly, the research site is constituted by cross-scale boundaries. The approval of the ancestral domain claim of the Palaw'ans in the village (*Barangay*) Panalingaan has been given in the year 2009, and part of this land has been declared as a protected area under the Mount Mantalingahan Protected Landscape (MMPL).

This study aimed to gain insight into the vulnerability of indigenous Palaw'ans in the village of Panalingaan by looking into the interrelationships of the ecosystem services in the research site, the community's perception of environmental change and their corresponding responses, and the approval of the ancestral domain of the Palaw'ans, which happened on the same year that Mount Mantalingahan was declared a protected area. The research focused on the adaptive capacity aspect of vulnerability, while discussing exposure and sensitivity to some extent. These allowed the research to make empirical assessments of several statements on resilience and vulnerability such as the one of Adger (2006) which says that country-level analyses fail to take into consideration the spatial and social differentiation of vulnerability at sub-national levels and the local conditions that affect adaptive capacity.

The Village of Panalingaan, Rizal, Palawan

The study was conducted in the village of Panalingaan in the municipality of Rizal in Palawan (Fig. 11.1), located at the foot of Mount Mantalingahan. It is part of the ancestral domain of the Palaw'ans, which has been approved in July 2009 under Certificate of Ancestral Domain Title (CADT) number R04-RIZ-0709-129 and has been formally awarded in October 2012. The ancestral domain area covers 69,735.23 hectares and includes the villages of Panalingaan, Taburi, Latud, and portions of Canipaan and Culasian under the municipality of Rizal. The population of indigenous Palaw'ans in this ancestral domain is 7,651 as of the approval date of the ancestral domain. They are one of three main indigenous groups living on the main island of Palawan, mostly concentrated in Southern Palawan (Cayron 2011). There are several settlements along the coast, but the arrival of Muslim groups from the Sulu archipelago as early as 1700s caused the coastal Palaw'ans to move inland (Cayron 2011; Macdonald 2003). The Palaw'an community in *Barangay* Panalingaan are two distinct sub-groups of *netibo* or natives and Panimusaan, which are cross-bred Muslim Palaw'ans. Most of the Muslim Palaw'ans live below the mountain area, while the rest of the Palaw'ans live in the mountains. The two sub-groups interact, albeit the *netibo* seem to be more reserved than the Panimusaan. In this research study, the *netibo* and the Panimusaan will be collectively called Palaw'ans.

The Palaw'ans mainly subsist on upland agriculture, where rice is grown along with other crops such as *kamoteng kahoy* or cassava (*Manihot esculenta*) (Cayron 2011; Macdonald 2003). Macdonald (2003) points out that rice is the most valued type of food by the Palaw'ans but they also subsist on other cultigens and root crops such as cassava. Cayron (2011), likewise, states that rice is the main crop, with cassava, corn, and vegetables as supplementary crops. From interactions with the Palaw'an community, they mention that when there is no rice, they eat cassava.

Methodology

The research made use of a qualitative approach, which employed key informant interviews, focus group discussions, and participant observation. These enabled the identification of processes and captured interdependencies of the different units of analysis, namely, multi-level governance, social-ecological dynamics, and local perceptions of environmental change.

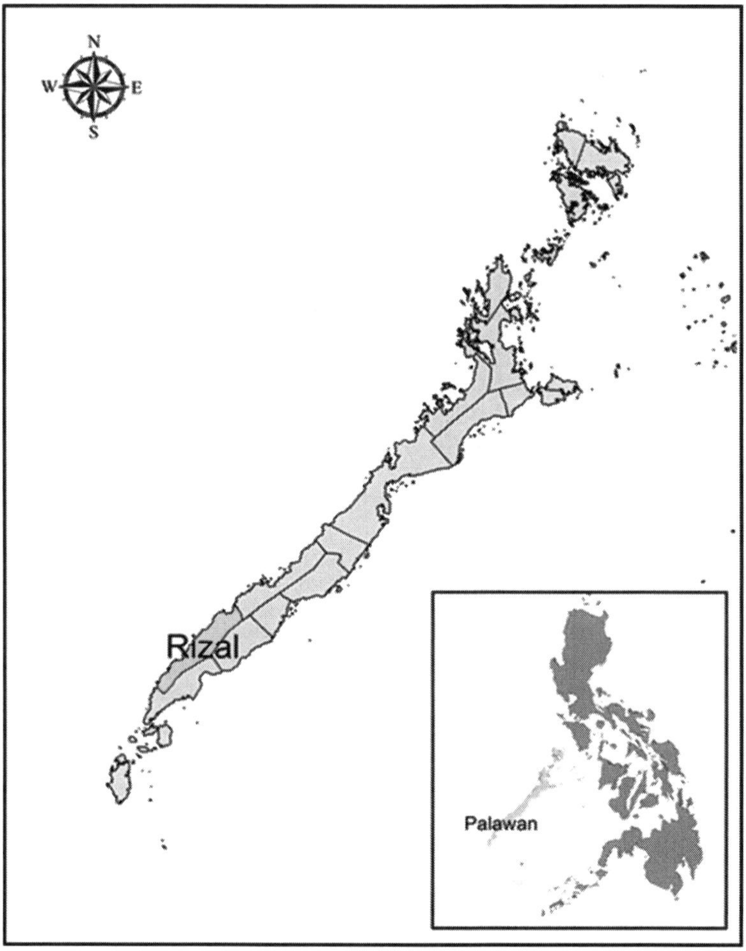

Fig. 11.1 Map of Rizal, Palawan, Philippines (Data source: GADM 2012)

The study employed both primary and secondary data collection. Primary data collection entailed field visits in the year 2012 that used the methods of key informant interviews, focus group discussions (FGD), and participant observation. Field visits were done in three locations: Manila (the capital of the Philippines), Puerto Princesa (the capital of the island of Palawan), and the research site (*Barangay* Panalingaan, province of Rizal, southern Palawan).

Key informant interviews were conducted with relevant institutions from the national, regional, and local levels. Upon commencement of key informant interviews, referral (or chain) sampling was employed, where interviewees were also consulted in identifying other relevant interviewees. Interviews were held with relevant officials and personnel of the national level Department of Environment and Natural Resources (DENR) Protected Areas and Wildlife Bureau (PAWB),

Philippine Atmospheric, Geophysical and Astronomical Services Administration (PAGASA), National Commission on Indigenous Peoples (NCIP), regional level Palawan Council for Sustainable Development (PCSD), DENR—Provincial Environment and Natural Resources Office (PENRO), DENR—Protected Areas Superintendent (PASu), Conservation International—Philippines (CIP), and local level town (*barangay*) council, indigenous people's organization (PO) Bangsa Palaw'an Philippines, Inc. (BPPI), and field level CIP.

Three FGDs were requested from the community. One FGD was done per gender group of men and women and one FGD of mixed genders was done at the end of the researcher's stay in the community. Separate FGDs for men and women were done to allow for in-depth discussion and avoid male dominance in discussion as the Palaw'an community is culturally patriarchal. The separate FGDs also allowed a more comprehensive inquiry into the socio-ecological dynamics of the community, as each gender had different roles in the management of their natural resources. The final FGD was a validation discussion, where initial results of the study were presented to the president and board member of the BPPI, field staff of CIP, and to some Palaw'an Panimusaan community members.

Secondary data collection was done through desk-based archival and documentary research and meetings and correspondences with relevant organizations. Relevant laws, published studies and grey literature were accessed through desk-based research and correspondences with relevant organizations and persons. The international non-government organization (NGO) International Union for Conservation of Nature (IUCN) Commission on Ecosystem Management (CEM) assisted with identifying literature on ecosystem-based adaptation and indigenous peoples. Important biodiversity and climatology documents for use in the research study were identified through several visits to the government offices PAWB and PAGASA. Internal documents related to the research site were accessed through visits to CIP national and local offices.

The study only included analysis of present-day adaptive capacity (as opposed to speculating on future adaptive capacity) of the Palaw'an community and was limited to examination of exposure and sensitivity, which are other components of vulnerability. The study was not meant to predict future biophysical or social changes in the community and their area. Individual household surveys and interviews were not carried out due to the short research timeframe. As a consequence, the research results contain limited demographic data on the Palaw'ans in the research site. The research site was only limited to Sitio Cadulan, which is only a part of the ancestral domain and of the Mount Mantalingahan Protected Landscape. The study was conducted in the national language of the Philippines, Filipino (Tagalog). Translations from the indigenous language, Pinalawan, were facilitated by other community members when a respondent does not speak Filipino.

Results

Local Perceptions of Climatic Change

The key informant interviews and the gender-specific FGDs (historical timeline and seasonal calendar) asked members of the local community if they have noticed any changes in the season or their environment throughout these past years. Caution was exercised not to use the term "climate change" during the inquiry. However, some of the community members, especially those who are part of the BPPI BOT, already heard of the term and articulated the term as a response to the inquiry, "Did you notice any changes in the season during these past years?" This may be attributed to their attendance in a climate-related seminar conducted by CIP in late 2011. Nonetheless, two common answers from the community point to a changing climate: there is not good enough heat anymore for preparing their swidden plots due to erratic weather patterns and, incongruously, there is increased intensity of heat and rain.

Increased Intensity of Heat and Rain

During interviews with the PASu and the CIP field officer, they already referred to a changing climate by citing extreme and out-of-season weather events. The PASu mentioned that the changing climate is certainly felt in the area and that the weather pattern has changed. She mentioned that the seasons are no longer being followed, saying that when it is supposed to be rainy season, it will be dry. Similarly, during the dry season, there will be sudden episodes of rain.

The community articulated the same observations with regards the changing weather patterns and extreme weather. They mentioned that the heat of the sun has become unbearable, with one woman (in her late 40s) saying, "In the years before, I can work in the fields until noon. Now, the heat has become unbearable that I can only work until 10:00 in the morning. The sun is already very painful on the skin at around 09:00 in the morning." She added, as an example, that she planted *kamoteng kahoy* or cassava (*Manihot esculenta*) working under the sun for three days and, as a consequence, became sick for one week. She concluded her narrative by saying that, "Maybe the earth is just really old." Another member of the community (male, in his 40s) said that, "Now, when it rains, it really pours very hard. And when the sun is shining, it is very hot and stinging your skin." Other members of the community say the same, recalling that the sun has not been that harsh on the skin in the earlier days.

The women also shared that every summer, the deep water wells dry up. However, during the summer seasons of 2011 and 2012, the deep water wells no longer dried up. This may also be related to the next section on abandonment of swidden practices as local adaptation to climate change. As one woman (in her 40s) in the FGD shared, she abandoned her swidden plot in the year 2011.

Decreasing Forest Products Productivity

The PASu and the PCSD chief officer of the ECAN-PRD mentioned that the changing climate has decreased productivity of forest products such as wild fruits, honey, and agricultural products. The PCSD chief officer even went on further to recount their hypothesis that climate change is affecting flowering plants, which the honey bees visit. The concern of the PASu and the PCSD ECAN-PRD chief officer on decreased forest products productivity included the loss of commercial opportunities of these forest products. The PCSD chief officer mentioned that there is an international demand for the honey and their office is interested in finding the chemical composition of honey in order to satisfy the international organic market. The CIP field officer, on the other hand, mentioned that the decrease of honey production depends on the area. The Palaw'an community, however, only mentioned the decreasing production of honey when they were asked about it during the final FGD. Most of the time, they would mention the decreasing productivity of swidden crops and associated this with increased pest infestation.

Local Adaptation to Climate Change

Key informant interviews with CIP staff led to discussions on the swidden practices of the Palaw'ans. One CIP staff mentioned the swidden cultivation is prohibited especially within the limits of the protected area. It was, however, stressed that indigenous peoples are still allowed to practice swidden cultivation in old plots, but opening up of new plots is prohibited. A discussion with another CIP staff, on the other hand, revolved around the productivity of the swidden cultivation of the Palaw'ans. It was mentioned that in recent times, there has been a lot of pests attacking the cultivated crops, specifically rice, of the indigenous peoples. It was narrated that, prior to pest infestation, the IPs did not use insecticides.

Separate informal narratives from the community confirmed that swidden cultivation is no longer productive and is no longer worth the effort of long preparation of the soil for planting. Most of their reasons stem from the lack of proper heat needed to burn the plants that were cleared or "slashed." They related this to unpredictable weather patterns, with one of the community members (male, in his 60s) saying that, "The season does not seem to have any direction anymore." This may be related to what another community member (male, in his 30s) says that more weeds seem to grow now in their swidden plots, thus their amount of work in cultivating their plots is more than what is needed before.

The chieftain (male, in his 50s) and the head of the host family (male, in his 60s) explained the seasonal calendar of their swidden practice. They narrated that their swidden cultivation starts in January. They spend 2 weeks in cutting the understory and then another 2 weeks in cutting the bigger plants. Two weeks in the month of February would be spent in burning the plants that were cut down. Reminiscent of bioenergetic agriculture, the chieftain mentioned that they await the appearance of seven stars in the sky (referred to as *marupuro*) prior to the commencement of burning. The months of March and April are spent planting (*panggas*), where the man punches a hole on the soil with a stick and the woman follows with planting the seeds. The seeds are placed in a small basket called *baka-baka*, which they explained as the appropriate size for bringing seeds since it is not too heavy. By the month of May, planting is finished and weeding or "grass" removal is the next occupation in the swidden plots. The chieftain said that if by the month of May, they have not commenced with planting (e.g. due to wet conditions), it would no longer be possible to have a successful harvest of good-tasting rice. They allow 4 months of cultivation before they harvest in August. The chieftain added that after harvesting, they first offer the harvested products to the gods through a ceremony before consuming the produce.

The women also described the challenges associated with practicing swidden cultivation in recent times. One of the women present during the FGD (aged 48) mentioned that in the year 1975, there were no weeds if they plant in swidden plots or no water lilies if they plant in paddies. Now, she added, apart from having more weeds, there are also pests. The women mentioned that in order to increase productivity of their cultivated plants, they need to apply both fertilizers and insecticides. They narrated that a sack of chemical fertilizers that they call "1620-triple-40-complete" and "urea" cost around PHP 1,500 up to 2,500 (EUR 30–50) per sack. They mentioned that among all crops, only *kamoteng kahoy* or cassava (*Manihot esculenta*) and *kamoteng baging* or sweet potato (*Ipomoea batatas*) are the ones that do not need any fertilizers. However, they mentioned that these are also in need of insecticides, which are applied to the above-ground portion (leaves) of the root crops.

The head of the researcher's host family mentioned that the arrival of the non-indigenous peoples and the commencement of mining operations brought different changes to their practice of swidden cultivation and to their harvests. He mentioned that the indigenous peoples have their own way of doing swidden cultivation, which the non-indigenous peoples have modified through damaging practices such as overturning the soil (tillage). He said that other indigenous peoples copied this practice until they lost the knowledge of the traditional way of doing swidden. The women, on the other hand, mentioned that abandonment of swidden cultivation led to paddy farming or tillage, which also has its costs. While paddy farming is rain-fed, tillage entails rentals of tractor and carabao, which costs PHP 1,500 (EUR 30) per hectare of tillage and PHP 200 (EUR 4), respectively. Usually, they borrow money in order to pay these costs. They pointed out that should there be proceeds from the harvest, these are just enough to pay the debt they incurred in obtaining inputs and services.

The commencement of mining operations at the other side of the mountain is associated with the arrival of pests that are eating the crops in their swidden plots. The community as well as the former and current presidents of BPPI believe that the black bug (*Scotinophara coarctata*), one of the first pests feeding on rice crops, was brought by one of the ships of the mining company. They believe that the black bug was attracted to the light of the ship and went all the way with the ship to the Philippines. Someone in the community (male, late 60s) identified the ship as coming from Japan. Since then, the community has had its share of different pests that were infesting crops and food sources. They have identified several insect pests (specific species unknown), mammalian pests such as rats, amphibious pests such as the bullfrog (scientific name unknown), and invertebrate pests, such as the golden *kuhol* or the golden apple snail (*Pomacea canaliculata*) and earthworms (specific species unknown). One member of the community (male, in his 30s) lamented that all their crops have their respective pests, even the root crops *kamoteng kahoy* or cassava (*Manihot esculenta*) and *kamoteng baging* or sweet potato (*Ipomoea batatas*) which did not have any pests before.

The spread of pests or invasive species was consulted with the PENRO, who mentioned that there have yet to be scientific studies on the invasive species in the area and whether these are alien species. He suggested that land use change may be driving the proliferation and distribution of the invasive species, citing that the current increase in palm oil plantations in the province may be encouraging these pests. The former president of the BPPI surmised that rats live within the palm oil plantations. Whether it is a changing climate or land use change driving the proliferation of invasive species in *Barangay* Panalingaan is a research question that urgently needs to be addressed. While this research need arises from the current research, it is, however, beyond the scope of this research.

The chieftain and other members of the community recalled how life before was easy compared to their lives now, which one member of the community (male, in his early 30s) attributed to "constant hunger."One of the women (in her late 30s) shared that now they go deeper into the forest in search for food. She stated that she would go and look for root crops like *gabi* or taro (*Colocasia esculenta*) despite the itchiness that the plant brings. She further added if she is not able to find food, she would just stay at home hungry instead of stealing from other people's food supply.

The chieftain also mentioned that life was "abundant" up until the birth of his third child (between 20 and 25 years old). He also narrated that during the time when cutting trees or *rattan* palm (*Calamus* spp.) were still not widespread, produce from his swidden plots could fill his *kamalig* or storage area with a height of an arm span and a half. Likewise, the women recalled that the amount of rice they harvested from their swidden plots in 1975 could surpass the height of an average man. Now that swidden is no longer productive, they have turned to other means (see Fig. 11.2). One is rain-fed paddy-rice farming (rented paddies) or tilled swidden, which some women say as better because the soil is softer and weeding is easier. Yet another option is to subsist on root crops or to buy food, if there are any financial resources at hand. In order to have income, some of them are engaged in copra-making or selling native agricultural products on weekly market days called *tabuan*.

From the narratives of the community, it was during the 1970s when their life completely changed for worse. Prior to 1975, they narrated that they were still engaged in a barter system, where the terrestrial and coastal members of the group would exchange commodities without any monetary valuation. One member of the

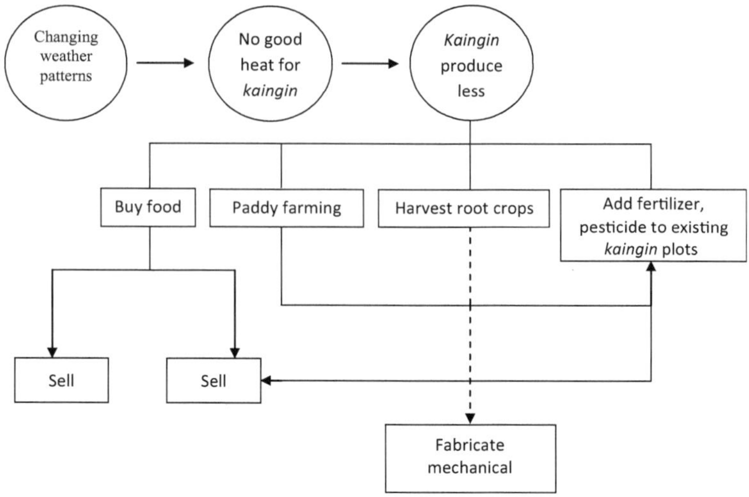

Fig. 11.2 Impact of changing climate on subsistence and local responses

community recalled that it was in 1975 that they became aware of what money is. According to the community, this is also the year that the mining company at the other side of the mountain started their operations. The community recalled that exploration began in the year 1967.

Along with their integration to the monetary system is their exposure to unnatural products, such as chemically-processed foods like canned goods or soy sauce. The chieftain recalled that one of their elders in the earlier times cannot eat meat seasoned with salt or soy sauce. Likewise, the BOT member of the BPPI noticed that their life expectancy has lowered compared to the average age of 100–120 in the earlier days. He attributed this to their consumption of unnatural food, which they still continue to do so despite this awareness.

Discussion

Transformability, Resilience, and Adaptability

Ecological, economic, and socio-political factors have affected the living conditions of the Palaw'ans. Ecologically, a changing climate coupled with the invasion of pests has made their indigenous practice of swidden agriculture impractical. Economically, their former system of trade or barter is no longer honoured especially with non-indigenous peoples. The Palaw'ans have then shifted to a current undesirable state of constant hunger. While they have been in this state or basin of attraction (Walker et al. 2004) for a number of decades now, the Palaw'ans have shown they have the capacity to find new ways of making a living. An example would be their plan of fabricating a mechanical cassava grinder as narrated in the results section. As the Palaw'ans have long lived on subsistence and have only in recent times explored copra-making or paddy rice farming as a means of livelihood, selling cassava flour is a completely new way of living. When ecological, economic, or socio-political conditions of a system make it unsustainable, a fundamentally new system may be created if there is the capacity to do so (Walker et al. 2004). This capacity is called transformability, which introduces new mechanisms and ways of making a living (Walker et al. 2004). The Palaw'ans have demonstrated their transformability through their plan for the cassava grinder.

Despite this, the resilience of the SES of the Palaw'ans in *Sitio* Cadulan remains to be seen. The study's focus on adaptability or capacity of the Palaw'ans to influence the resilience of their ancestral domain can only give a limited view of the overall resilience of the area (Walker et al. 2004). In the same vein, some Palaw'ans seem to have influenced the resilience of their SES by using agricultural inputs such as chemical pesticides and fertilizers on their swidden plots. This coping strategy may sustain their swidden cultivation for the short-term, but a longer-term study is needed to ascertain whether this will succeed in bringing back the productivity of their subsistence agriculture. This is in line with what Lal (2008) states that soil degradation, water pollution, and air contamination result from the indiscriminate use of chemicals and excessive tillage, both of which are being practiced by the Palaw'ans albeit in a non-indiscriminate and non-excessive way. Whether these chemical pesticides and fertilizers are suitable adaptation strategies in the long-term or not, the social network of the Palaw'ans provides an atmosphere conducive for social learning. The social network plays a role in the transformability of the Palaw'ans. In addition, the social network also shapes the local perceptions of change through collective thinking processes.

Local Institutions as Social Networks

Currently, Palaw'ans in the three *barangays* that the CADT covers are organized by the PO BPPI. This brings to mind what the current president of the BPPI said, "If we were not organized and did not have any system of leadership, we would not have this CADT today" [Japson RL (2010) President, Bangsa Palaw'an Philippines, Inc. (BPPI). Formal interview. Barangay Panalingaan, 28 May 2012, Personal communication]. He stressed the importance of a leader in organizing and unifying the community. He referred to other IP groups in Palawan, saying that they are in a more difficult situation because they do not have any system of leadership. BPPI typifies an informal network that developed in a bottom-up process, with no formal linkage to formal governance and management regimes (Pahl-Wostl 2009). BPPI also demonstrates the limitation of an informal shadow network, which is weak influence on policy and real implementation despite having an autonomy that increases their ability to self-organize, innovate and think creatively (Pahl-Wostl 2009).

The community constantly mentioned the lack of assistance from governmental bodies despite requests from the community. The BOT member repeatedly shared that it was his retirement pay a couple of years back that paid for seedlings of fruit-bearing trees such as coconuts that have been distributed to his fellow Palaw'ans for planting. They are now benefiting from these trees both for subsistence and livelihood. This BOT member serves as both "maven" and "connector" in the community, where his altruistic nature and influence on fellow Palaw'ans go beyond information-sharing (Gladwell 2000 as quoted in Folke et al. 2005).

The community members themselves support each other for job generation, an example of which is copra-making. The working relationship is not strictly on employment terms but that of giving assistance to both sides. While the Palaw'ans do not have a strong network of engagement with governmental bodies, they seem to have a significant capacity for collective action. A more recent demonstration of collective action is their maintenance of the water system from a sub-grant from the CIP. The Palaw'an community in *Sitio* Cadulan have formed themselves into a local water users' association called *Danum na Buwal et Mundugen*, which agreed on terms and actions on maintenance of the water system (CIP 2012). They have a monthly tree planting activity that aims to plant fruit-bearing trees as markers along the water pipes. They also specified a cleaning schedule for the water tanks.

The community's social memories on water availability and water quality helped in their self-organization as well as demonstrating framed creativity in the creation of maintenance mechanisms (Folke et al. 2005). On water availability, the women shared that, during the summer season prior to 2011 and 2012, the water wells would always dry up and it was always a race among families to the water well in the morning. Conversely, the water-related deaths in April 2011 of fellow Palaw'ans, while not believed to be caused by faecal contamination, still served as a reminder to the community about the importance of water quality.

Tompkins and Adger (2004) suggest that collective action and its preconditions may increase the resilience of a community to changes. The case of the Palaw'an community in *Sitio* Cadulan demonstrates the three principles for collective action. These principles state that there is a greater chance of success in smaller groups than in larger groups, an equitable distribution of entitlements lead to greater success and alternative institutional designs overcome failures of collective action (Tompkins and Adger 2004). In addition, the Palaw'an community demonstrate social learning in terms of turning subsistence sources (cassava) into potential livelihood sources. The Palaw'an community, as well as the BPPI, have exhibited sustained development of attitudinal and behavioural change since their direct application for CADT. The success of the cassava flour grinder is yet to be seen, as Adger (2001) points out that important subsistence farming systems do not have "developed hedonic markets." There is a possibility that the income the Palaw'ans would generate may potentially be less than what the product is really worth. One of CIP's field staff mentioned that during market days, the agricultural products of the IPs are usually bought at a price lower than the market price. This is a potential hurdle for the Palaw'ans, but this may also be a point for social learning of the community. Knowledge gained from social learning may be able to guide future decisions of the community. Having a knowledge bank or a developed social memory reinforces the adaptive capacity of the community. Apart from the knowledge gained, the process of social learning itself is important in the community's ability to respond to any changes.

Conclusion

Social networks, bridging organizations, and the demonstration of transformability by the Palaw'ans reinforce their adaptive capacity to some extent.

Transformability

Despite being in an undesirable state of poverty and hunger, the Palaw'ans have shown that they have the capacity to find new ways of making a living. The plan of the Palaw'ans to fabricate a mechanical cassava grinder reflects the transformability of the community, which seems to be able to introduce new mechanisms and ways of living. Transformability increases the adaptive capacity of the Palaw'ans. The indigenous PO (BPPI) serves is a local institution that serves as a social network among the Palaw'ans. Through social learning, knowledge that may be able to guide future decisions of the community is gained and having a knowledge bank or a developed social memory reinforces the adaptive capacity of the community. Apart from the knowledge gained, the process of social learning itself is important in the community's ability to respond to any changes.

As part of their adaptation strategy, the Palaw'ans have started using chemical pesticides and fertilizers as agricultural inputs to their farming practices. This may have aninfluence on the resilience of their ancestral domain's SES. It may be able to sustain the productivity of their swidden or rice paddy plots, but potential positive feedbacks may arise that will keep them in a cycle of pests and chemical agricultural inputs. The increasing number of pests, along with the loss of biodiversity in the area, has the potential to bring a regime shift that will alter the landscape and the social-ecological dynamics in the research site. If the present undesirable state of hunger and poverty revert to a more undesirable state, it will be difficult for the Palaw'an community to persist. Nonetheless, their demonstration of transformability is a positive reinforcement of their adaptive capacity.

Social Networks and Bridging Organizations

The BPPI, as a local institution of the Palaw'ans of the ancestral domain covering three *barangays*, serves as a social network. Despite having a weak influence on policy and real implementation, being a social network enabled the BPPI to have an autonomy, which increases their ability to self-organize,

(continued)

innovate, and think creatively. This seems to have increased the adaptive capacity of the community.

Social connectivity has been observed in the relationship of the Palaw'ans with CIP, which created room for communication, coordination and common agreement that may facilitate understanding of laws or prohibited acts. In addition, CIP can be considered a bridging organization for their work on establishing the MMPL and their current livelihood projects with communities living within the MMPL. The trustworthy relationship of CIP with the local institution of the Palaw'ans such as BPPI serves to increase the adaptive capacity of the Palaw'ans.

Future Prospects

Community transformability, social networks and bridging organizations have contributed to the strengthening of the adaptive capacity of Palaw'ans. However, these have been identified based on a scenario in a given space and time. Transformability, social networks and bridging organizations are also subject to the dynamism of times and how the indigenous Palaw'ans cope with changes in the status quo should also be looked into. The effects of climate change are varied and interventions should take differences in adaptive capacity into consideration. Indigenous peoples such as the Palaw'ans would benefit from further strengthening of their ability to adapt to climate change and its impacts.

Acknowledgements The author would like to acknowledge the support of Ms. Jeanne Tabangay, Dr. Rowena Boquiren and Conservation International Philippines, Bangsa Palawan Philippines, Inc., Mount Mantalingahan Protected Area Management Board and Superintendent, Central European University (CEU) Foundation, the CEU Department of Environmental Sciences and Policy and Dr. László Pintér.

References

Adger N (2001) Scales of governance and environmental justice for adaptation and mitigation of climate change. J Int Develop 13:921–931

Adger N (2006) Vulnerability. Glob Environ Chang 16:268–281

Cayron JG (2011) Archaeology and exchange in Palawan Island, Philippines. Doctoral thesis, Department of Southeast Asian Studies, National University of Singapore

Conservation International—Philippines (CIP) (2012) Updates First Quarter of 2012. Internal report

Folke C, Hahn T, Olsson P, Norberg J (2005) Adaptive governance of social-ecological systems. Annu Rev Environ Resour 30:441–473

Füssel HM, Klein RJT (2006) Climate change vulnerability assessments: an evolution of conceptual thinking. Clim Change 75:301–329

GADM (Global Administrative Areas). (2012). Archived from the original on 2012-04-16 http://www.gadm.org/
Gladwell M (2000) The tipping point—how little things can make a big difference. Little Brown, Boston, USA
Kelly PM, Adger WN (2000) Theory and practice in assessing vulnerability to climate change and facilitating adaptation. Clim Change 47:325–372
Lal R (2008) Soils and sustainable agriculture: a review. Agron Sustain Dev 28:57–64
Macdonald CH (2003) An anthropological investigation on suicide in Palawan, Philippines. Southeast Asian Stud 40(4):419–443
Marino E, Ribot J (2012) Special issue introduction: adding insult to injury: climate change and the inequities of climate intervention. Glob Environ Chang 22:323–328
Novellino D (2000) Recognition of ancestral domain claims on Palawan island, the Philippines: is there a future? Land Reform 1:56–73
Pahl-Wostl C (2009) A conceptual framework for analyzing adaptive capacity and multi-level learning processes in resource governance regimes. Glob Environ Chang 19:354–365
Salick J, Ross N (2009) Introduction: traditional peoples and climate change. Glob Environ Chang 19:137–139
Saunders P (1990) Social class and stratification. Routledge, London, UK
Tompkins EL, Adger WN (2004) Does adaptive management of natural resources enhance resilience to climate change? Ecol Soc 9(2):10
UNU-EHS (United Nations University—Institute for Environment and Human Security) (2011) "WorldRiskReport 2011." Bündnis Entwicklung Hilft, Germany. Archived from the original on 2012-07-19 http://www.ehs.unu.edu/file/get/9018
Walker B, Holling CS, Carpenter SR, Kinzig A (2004) Resilience, adaptability and transformability in social-ecological systems. Ecol Soc 9(2):5

Part II
Policy and Technology Responses

Chapter 12
Managing "Climate Migration" in Mongolia: The Importance of Development Policies

Benoît Mayer

Abstract Each winter, tens of thousands of destitute Mongolian herders move to the insalubrious suburbs that surround Ulaanbaatar ("*ger* districts"). This migration can partly be attributed to climate change, as a rapid warming and a slight change in the precipitation patterns (decrease in summer precipitations) reduce the yield of the grassland. On the other hand, the resilience of nomadic animal husbandry declined markedly since the collapse of the communist regime in 1990: the "Age of the Market" and the imposition of a radical neoliberal ideology led to the interruption of the services indispensable to the traditional Mongolian way of life (e.g. boarding schools, mobile health brigades, but also veterinary services and a centralized system of fodder production and distribution that mutualizes environmental risks). Thus, this chapter shows that, in the context of Mongolia's internal migration, climate change adaptation is inseparable from domestic development policies that, it is argued, need urgently to be rectified.

Introduction

The debate on climate change-induced migration has often focused on exotic "sinking" islands such as the Maldives and Tuvalu, and on costal least-developed countries such as Bangladesh. The government of the Maldives has been particularly prominent in setting the issue on the international agenda, for instance through organizing a largely mediatised under-water cabinet to call for the world's attention (BBC 2009). Yet, the effects of climate change extend well beyond those few countries and they are likely to have an impact on human migration in many other contexts. Echoing the communication strategy of the Maldives, Mongolian ministers met in the Gobi desert, at about 15 h drive from the capital city Ulaanbaatar, in order to "draw ... the attention of the world community to the fact that Mongolia's traditional nomadic civilization based on pastoral animal husbandry is likely to be

B. Mayer (✉)
Faculty of Law, National University of Singapore, Eu Tong Sen Building, 469G Bukit Timah Road, 259776 Bukit Timah, Singapore

5 impasse Sorano, 31100 Toulouse, France
e-mail: bmayer@nus.edu.sg; http://www.benoitmayer.com

W. Leal Filho (ed.), *Climate Change in the Asia-Pacific Region*, Climate Change Management, DOI 10.1007/978-3-319-14938-7_12

at risk by mid of twenty-first Century" as a consequence of climate change (Mongolia 2010b).

Nomadic livestock husbandry, which has long been Mongolia's main economic activity, continues to occupy a third of the national work force despite the rapid development of the extractive industry. Mongolian herders adapted to arid and cold climatic conditions by frequently moving their folk—which allowed the pasture to regenerate. Today, however, nomadic livestock husbandry is in crisis, and the massive migration of herders to Ulaanbaatar is a symptom of this crisis. From 2002 to 2012, Ulaanbaatar has registered about 350,000 new arrivals from the rest of the country, reaching a population of roughly 1, 3 millions, in a country that counts no more than 3 millions inhabitants. These migrants are former herders who lost their livestock; they are destitute, settle in peripheral "*ger* districts"[1] where they often lack access to basic services (health, education, sanitation and running water, etc.); most of them have been unable to find their place in the city centre's thriving economy.

This migration (the settlement of former herders in the peripheral districts of Ulaanbaatar) is related to climate change, but climate change is only a part of a complex story that also features economic (dis)incentives, developmental policies, and ideological shifts. In other words, if the impacts of climate change "cause" migration, this is only in conjunction with certain political, economic, social and cultural factors. By exploring the causes of migration in Mongolia, this chapter examines the various policy levers that could be used to address this phenomenon, in particular climate change adaptation and domestic development policies. The following discussions also illustrate the problematic causal links between climate change and its social impacts by highlighting the existence of multiple proxy factors such as political responses or economic capabilities.

This chapter is based on a series of semi-directive interviews of different stakeholders conducted in March and April 2013, completed by a documentary research. The author did not conduct any quantitative study and had to rely on a paucity of available statistical tools.

Dzud and Migration

Migrants often tell a similar story. They were herders and lost their flock. Some of them sought a job in small urban centres but were unable to make a living there; others moved directly to Ulaanbaatar in search for new economic opportunities. Attention has often been focused on the causes of the loss of their flock: the *dzud*, an untranslatable Mongolian term for a specific type of natural disaster that results from the conjunction of a dry summer with harsh winter conditions. The dry

[1] "*Ger*" (yurt) is the traditional tent in which the Mongolian nomads live. Many destitute internal migrants settle around Ulaanbaatar in their "*ger*," hence the name "ger districts."

summer decreases the yield of the pasture and the resilience of the livestock. The following harsh winter conditions may consist alternatively in extremely cold temperatures, powerful winds, heavy snowfall, freezing rain, the late arrival of the spring, or a combination of these factors (Field et al. 2012: p. 500; Batima 2006: p. 57). Whereas it is normal for Mongolian herders to lose up to 3 % of their folk during a winter, a *dzud* leads to significantly heavier losses.

Dzud is not a new phenomenon. Plenty of historical sources describe the phenomenon at different times of Mongolia's history (e.g. Lansdell 1885: p. 318; Khazanov 1978: p. 121). The Book of the Later Han (後漢書) recounts that, in 45 CE, the Xiongnu (a nomadic people living on the territory of today's Mongolia) suffered from repeated droughts, as a consequence of which "[t]wo thirds of its people and domestic animals died of hunger and illness" (at 2942–2944, cited in Fang and Liu 1992: p. 151). The 1945 *dzud* remains the worst in recent history, which led to the death of one third of the national livestock (Field et al. 2012: p. 500; Batjargal 2001: p. 41).

Yet, there is a widespread understanding that *dzud* has become more frequent and more severe over the last decades. In 2000, 2001, 2002, and again in 2010, all or most of Mongolia was hit by severe *dzud* (Field et al. 2012: p. 501; UNDP and Mongolia 2011: p. 1). At the scale of the country, 26 % of the livestock perished in 2010 (see Table 12.1). In addition to these large *dzud*, local *dzud* affect some part of the country almost every year.

These disasters have naturally had tremendous social consequences (Janzen 2005: p. 80). At least 75,000 herders families lost more than half of their livestock in 2010 (Sternberg 2010). Many more encountered a situation of great economic distress: "[t]he most critical consequences of *dzud* are increased poverty and mass migration from rural to urban and from remote to central regions" (Field et al. 2012: p. 502). Historical studies show that *dzud* affects the migratory behaviour of the Mongols. In 45 CE, the Xiongnu reportedly migrated Southward and resettled as far as Yunyang (now in Chongqing province, central China) (Fang and Liu 1992: p. 151). On a long perspective, Southward migration flows toward China were correlated to drought and extreme winter conditions (Fang and Liu 1992: p. 166; Zhang et al. 2007: p. 405). Some studies even suggest that a slight climatic change

Table 12.1 Loss of livestock in years of large *dzud* (more than 3 % of loss), 1972–2012. *Source*: data provided by the Mongolian National Statistical Office

Year	Loss of livestock
1976	3 %
1977	3 %
1980	2 %
1983	5 %
1984	4 %
1991	1 %
1993	2 %
2000	10 %
2001	14 %
2002	8 %
2010	26 %

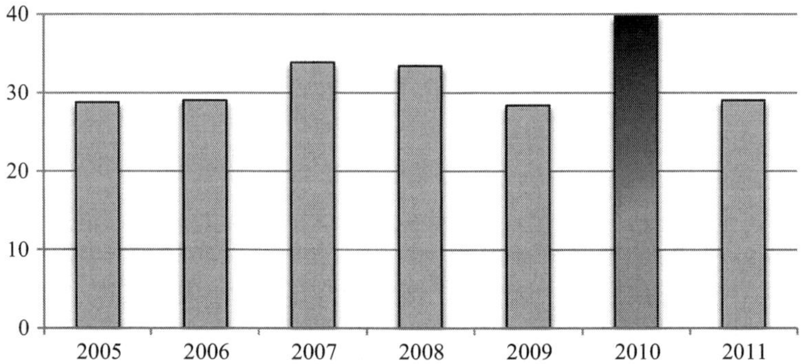

Fig. 12.1 Registration of individual newcomers from the countryside in Ulaanbaatar (in thousands), 2005–2011. *Source*: data provided by Ulaanbaatar registration office (3 April 2013)

in the early thirteenth Century may have precipitated Genghis Khan's conquest of the world, forming an empire that extended, at its peak, all the way to Hungary (Jenkins 1974; Hvistendahl 2012; May 2012). Although recent *dzud* do not seem to impact international out-migration from Mongolia, it has a discernible impact on internal migration. Following the 2010 *dzud*, the registration office of Ulaanbaatar noted an increase of the annual inflow of migrants from the country to Ulaanbaatar by 40 %, representing 10,000 additional migrants (see Fig. 12.1). Many other migrants may only move to Ulaanbaatar the following years, or may be displaced by more local *dzud*.

The Causes of Migration

The concept of *dzud* is a complex one, as it relates environmental phenomena (summer drought and harsh winter conditions) as well as to their social impact (loss of livestock). The increasing frequency and severity of *dzud* may be explained in two ways. Firstly, the physical phenomena are amplified by a change in the prevailing climatic conditions. Secondly, the social impact of these phenomena is exacerbated by the change of the political conditions since the early 1990s.

Climate Change

The Mongols are highly aware of climate change (Marin 2010; Sternberg and Chatty 2008), perhaps due to their conscience that Mongolia's "fragile ecosystems, pastoral animal husbandry and rainfed agriculture are extremely sensitive to climate change" (Mongolia 2010a: p. 7). Yet, the causal relation between climate change and migration is indirect and complex. Warming may have a direct positive impact on Mongolia's environment by prolonging the growing season, but, in most places, the ensuing lack of water will overwhelmingly dissipate this positive impact. Overall, warming and a change in the precipitation pattern that can be attributed to climate change may increase the frequency and severity of *dzud*, hence indirectly "cause" migration.

Warming is occurring in Mongolia three times faster than the global average: temperatures increased by 2.1 °C between 1940 and 2010, compared with 0.7 °C globally (Dagvadorj 2010: p. 98). The country is characterized by an arid climate situated at an ecotone between forests and steppes, which is highly sensitive to virtually "any external disturbance of the environment, natural or human" (Saizen 2013: p. 215). The country has little surface or ground water storage, and evapotranspiration accounts for 82–97 % of the precipitations. The increase in temperature has already led to an estimated 7–12 % increase of the potential evapotranspiration (Tsogtbaatar 2013: p. 90), thus significantly increasing the lack of water and causing numerous rivers and lakes to dry up, desterfication and land degradation to progress, and dust and sand storms to become more frequent (Mongolia 2010a: p. 61). Only a few places in the East of the country benefit from a temporary availability of water from the melting of glaciers (Lkhagvadorj et al. 2013: p. 88).

A change in the precipitation pattern has also had a significant impact on the viability of nomadic livestock husbandry. While the annual level of precipitations has not changed significantly, summer precipitations have decreased while winter precipitations were increasing (Batima et al. 2005: p. 20; Dagvadorj 2010: p. 99). The persistence of Mongolia's grassland depends on the concentration of the precipitations in June and July, where higher temperatures allow the vegetation to grow. The livestock also needs access to water. Herders identify water shortage as the main environmental problem with which they are faced (Marin 2010: p. 166; Sternberg 2008). On the other hand, snowfall is harmful to nomadic livestock husbandry, as the layer of snow may prevent the livestock from eating, or even from moving; snowstorms may even be fatal to herders.

Thus, by increasing the frequency of drought and snowfall, climate change affects the viability of nomadic livestock husbandry, increasing in particular the likelihood of *dzud* causing heavy loss of livestock. As a consequence, many destitute herders migrate to seek an alternative livelihood in Ulaanbaatar.

This, however, is only one part of the story. A natural disaster is never totally "natural." In the case of Mongolia, it is particularly evident that political factors have largely increased the vulnerability of the herders to *dzud*, as this will now be shown.

Regime Change

From 1924 to 1990, Mongolia was governed by a communist regime supported by the USSR, following which it entered into what the Mongols call the "Age of the Market." The brutal "shock therapy" of the early 1990s led to a profound economic upheaval (see Fig. 12.2) as the aid from the USSR, on which Mongolia largely depended, was suddenly interrupted. Within a few months, half of the state's employees were laid off; production collapsed; and unemployment, prices and corruption skyrocketed. For the Mongols, this resulted in a transformation "from a middle-income to a poor country, as if the process of development had been put on reverse" (Sneath 2006: p. 196). The economy recovered in the second half of the 1990s, and, during the first decade of the 2000S, the country witnessed a rapid economic growth on the extraction of mineral resources (except for the brief impact of the 2009 economic crisis). Yet, this economic growth has not resulted in a genuine development benefited the Mongols: half of the population was left behind. The UN Special Rapporteur on Extreme Poverty and Human Rights recently noted "the significant income inequalities affecting communities living in poverty," adding that "the gap widens, poverty is becoming entrenched, not only in rural areas, but also in urban centres" (Sepulveda 2013: para. 9). The economic gap went along an increased cultural isolation leading to the "continued denigration of rural life" by Mongolian elites. These new elites, living in the centre of Ulaanbaatar and drawing all the benefits from the mining boom, consider the herders as laid back, lazy, and responsible for their failure to adapt to a modern economy (Levin 2012; diplomatic source 2013).

This regime change contributes to explain the migration occurring in Mongolia. This is because the new regime has shown unable to maintain nomadic animal husbandry as a viable economic activity for many herders, to provide economic alternatives in the countryside, or to integrate the migrants in Ulaanbaatar.

Firstly, many herders strive to make a living, and climate change is only a part of the story. In 1990, the specific services responding to the needs of herder families were discontinued: mobile heath brigades (Medvedeva 1996: p. 182), boarding schools (Sneath 2006: p. 155), veterinary, and the system of centralized fodder production and distribution that helped herders affected by a local *dzud* (see Fig. 12.3). As the transportation costs increased, many herders concentrated close to urban centres where they could access to public services: an author estimates that one third of Mongolian herders stopped moving completely while another third reduced the frequency or distance of their displacement (Lkhagvadorj et al. 2013: p. 85). Yet, mobility is essential for nomadic livestock husbandry to sustain a fragile environment (Konagaya and Maekawa 2013: p. 11): a lesser mobility led to overgrazing—the use of the pasture beyond its capacity to regenerate. Moreover, for lack of regulation, individual profit-seeking strategies led to a dramatic increase of the number of livestock, in particular of goats (see Fig. 12.4)—certainly far beyond the carrying capacity of Mongolia's grassland. Overgrazing further decreased the resilience of Mongolian herders to *dzud*.

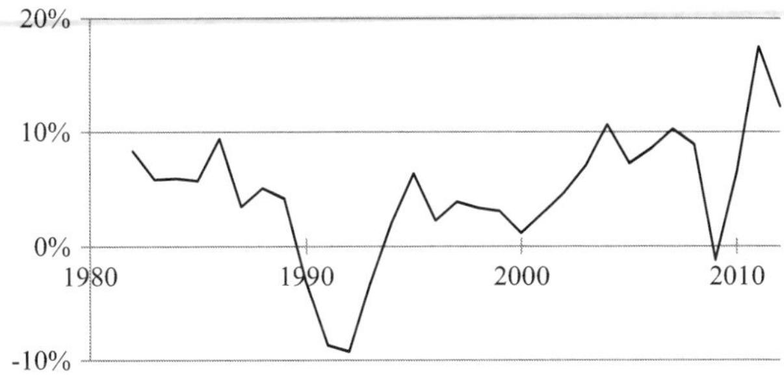

Fig. 12.2 Mongolia's annual GDP growth, 1982–2012. *Source*: data provided by the World Bank

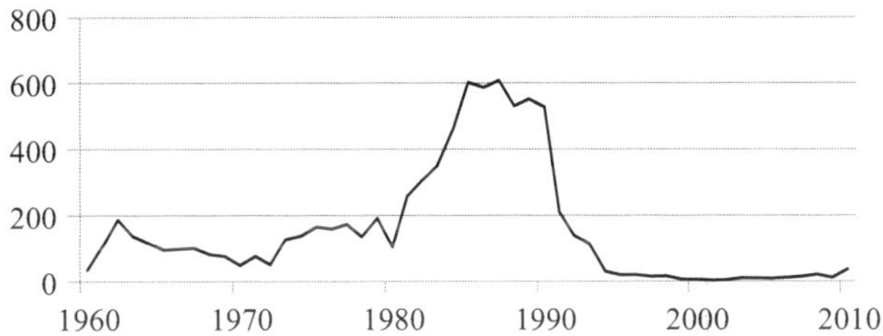

Fig. 12.3 Evolution of annual fodder crop production, 1962–2012 (thousands tonnes)

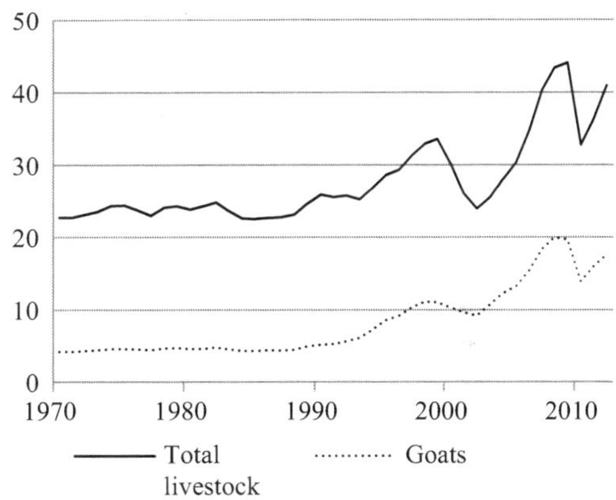

Fig. 12.4 Evolution of the Mongolian livestock, 1970–2012 (millions of animals). *Source*: data provided by the Mongolian National Statistical Office

Secondly, small urban centres are no more able to offer an alternative livelihood to herders who have lost their livestock. Herders have always interacted with small sedentary populations, which, at some time, were mostly centred around Buddhist monasteries (Fernandez-Gimenez 1999). There, destitute herders could settle for a few years and, through diverse temporary works, reconstitute a flock. Throughout the 1960s and 1970s, the communist regime organized a collective system of nomadic animal husbandry that also guaranteed that everyone would have a role to play. In particular, light industry was developed, through heavy subsidies, in small urban centres. However, the 1992 Constitution allowed freedom of movement at a time when many state employees were being laid off and where the heavy subsidies in support to light industry were discontinued. The result is that, during

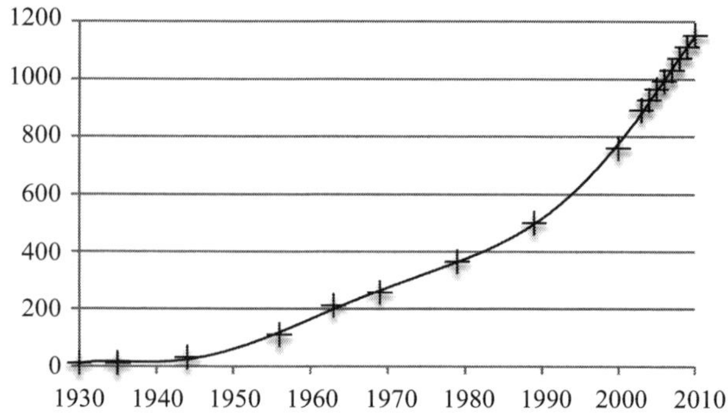

Fig. 12.5 Population of Ulaanbaatar, 1930–2010 (in thousands). *Source*: data provided by the National Statistical Office (2013)

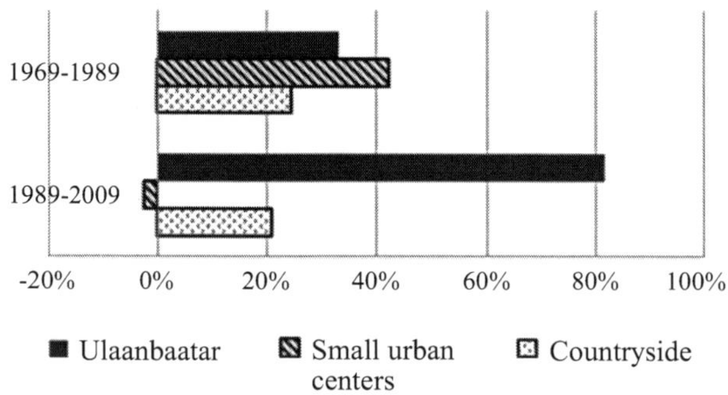

Fig. 12.6 Repartition of the population growth, during and after the democratic transition. *Source*: data provided by the Mongolian National Statistical Office

the last two decades, destitute herders have been unable to find any economic opportunity in small urban centres, and have found no alternative other than moving to Ulaanbaatar. The growth of Ulaanbaatar increased significantly in the 1990s (see Fig. 12.5). Figure 12.6 suggests that the rapid growth of Ulaanbaatar since 1990 can largely be attributed to the sharp decrease of attraction of small urban centres rather than to the slightly decreased attraction of the countryside. The population of Ulaanbaatar grew by 3,700 per year between 1930 and 1956, 12,000 between 1956 and 1989, but 24,000 between 1989 and 2000, and 39,000 between 2000 and 2010 (National Statistical Office 2013), and most of this growth is due to incoming internal migration.

Thirdly and overall, the current economic regime has been unable to integrate migrants arriving in Ulaanbaatar. Rather than necessary an issue, migration is the normal state of any society; it has its social functions, including to foster development by promoting the circulation of ideas and the construction of a social cement, and it may favour social adaptation to environmental changes. However, migration becomes a social issue when the rights of migrants are not adequately protected. In Mongolia, the geographical and economic exclusion of the migrants in Ulaanbaatar transforms rural-to-urban migration in a social issue. Current migrants are not able to find a place in Ulaanbaatar's economy as their predecessors did in previous decades. As an observer notes, "[i]t is hard to write about Ulaanbaatar's development without a sense that you are chronicling a colossal failure of city planning (or lack thereof)" (White 2012).

Policy Levers to Address Migration

Attributing "causes" to migration matters because each cause might also be a policy lever. Considering Mongolian migrants as "climate migrants" suggests that the policy responses should be included in the climate regime; attributing this migration to the policies adopted by Mongolia's government suggests different responses.

The climate regime, however, does not offer any evident policy response to Mongolian migration. Climate change mitigation should of course be an element to consider, but no amount of effort on climate change mitigation will address the needs of existing migrants or even significantly decrease future migration flows. Climate migration may be an *argument for* mitigation (whereby the migrants become, so to speak, the human face of climate change), but mitigation is at best a very partial response to the ongoing migration. Therefore, climate change adaptation may be more relevant. The *Cancun Agreements* included "measures to enhance understanding, coordination and cooperation with regard to climate change induced displacement, migration and planned relocation, where appropriate, at the national, regional and international levels" within a framework on "enhanced action on adaptation" (UNFCCC 2010: para. 14(f)). Two years later, the climate conference in Doha recognized that patterns of migration affected by climate change may be addressed as "loss and damage associated with climate change impacts in developing countries that are particularly vulnerable to the adverse effects of climate change" (UNFCCC 2012: para. 7). By including migration within adaptation (or, if it is conceived beyond the scope of adaptation, as loss and damage), these documents converge to suggest that states should address climate migration by cooperating within the climate regime, that is, "on the basis of equity and in accordance with [states'] common but differentiated responsibilities and respective capabilities" (UNFCCC 1992: art. 3). Beyond this principal approach of burden sharing, however, the climate regime does not favour any specific responses. It says little as to, for instance, whether the government of Mongolia should avoid migration by providing alternative economic opportunities in small urban centres, or whether it should rather facilitate migration by supporting social and economic insertion in Ulaanbaatar.

There is no specific legal regime applying to climate migrants, who are not recognized as refugees (McAdam 2012). In fact, it would be both impractical and arbitrary to treat migrants in a similar situation differently depending on the cause of their migration (i.e. isolating "climate migrants" among the flow of Mongolian migrants): ethical and human rights arguments suggest that all migrants in a similar situation of vulnerability should be treated in the same way (Betts 2013; Mayer 2013). Managing the partly-climate-induced migration occurring in Mongolia supposes policies that extend beyond the climate regime. The climate regime should arguably organize a financial support to domestic policies taken to address such phenomenon in application of the principle of common but differentiated responsibilities, but it should not go beyond such financial support.

International law defines some constraints as to the substances of policy responses to migration, but it does not define a unique option that the government of Mongolia or its partners must follow. The constraints include, in particular, the obligation of states to "take steps . . . with a view to achieving progressively the full realization" of economic and social rights such as the "enjoyment of just and favourable conditions of work," to social security, to "an adequate standard of living . . . including adequate food, clothing and housing," to "the enjoyment of the highest attainable standard of physical and mental health," and the right to education, among others (ICESCR 1966: arts. 2, 7, 9, 11–13). In the protection of these rights, any overt or de facto discrimination is prohibited; rather, states are "under an obligation to adopt special measures to attenuate or suppress conditions that perpetuate discriminations":

> The exercise of Covenant rights should not be conditional on, or determined by, a person's current or former place of residence; e.g., whether an individual lives or is registered in an urban or a rural area, in a formal or an informal settlement, is internally displaced or leads a nomadic lifestyle. Disparities between localities and regions should be eliminated in practice by ensuring, for example, that there is even distribution in the availability and quality of primary, secondary and palliative health care facilities (CESCR 2009: paras. 9, 34).

The government of Mongolia is arguably in breach of its obligations with regard to the non-discriminatory protection of the economic and social rights of herders and migrants. Many in Mongolia consider that their economic and social entitlements were replaced in 1990 by a protection of their civil and political freedoms. Following the rapid economic growth of the last 15 years, Mongolia has now the capacity to guarantee a reasonable level of economic and social entitlements to everyone. Yet, the constant prioritization of its development policy on the sole extractive industry creates few jobs and benefits little to unskilled herders; the benefits are reserved to the country's elite. The Mongolian government does not do enough to assist the large population of migrants living around Ulaanbaatar. It largely relies to foreign actors (international organizations and non-governmental organizations) to care for the country's poor, despite its responsibilities under international human rights law.

There are different ways for the Mongolian government to comply with its international legal obligations. One option would be to address the causes of migration. This could be done in particular through resuming the provision of basic and public services in the countryside, that were discontinued in the early 1990s; through limiting the use of the pasture to its regeneration capacities through caping the number of livestock and coordinating the geographical distribution of the herders; through developing financial mechanisms (such as insurance) and job opportunities to support the destitute herders; and through limiting the environmental impact of the many mines, in particular their use of water and their production of dust including through transportation, and preventing the development of mines in the most sensitive environmental areas. There is no reason to exclude the possibility of a "smart nomadism" that would be compatible with a modern economy, improving the living conditions of herders while preserving their mode of life, if this is the choice favoured by the Mongols through a democratic deliberation (Campi 2006: p. 50).

Another option would consist in a drastic economic transition toward a "modern" economy, through the intensive raising of livestock indoors supported by the production of fodder. This option would possibly increase the productivity of Mongolia's agriculture and, while it would also mean the end of a traditional way of life, it *may* help improve the conditions of life of the (former) herders. It is however of paramount importance, if this is the decision adopted by the Mongols, that herders be provided for with alternative livelihood, for intensive raising of livestock indoors is likely to be significantly less labour intensive. Such political orientation must therefore come along with substantial investment in education and re-training and plans for an extension of Ulaanbaatar or for the development of new urban centres in conditions that would ensure the provision of basic services to everyone with an emphasis on the needy, among others. If human rights are to be protected for everyone at all time, such transition can only be brought about progresively.

The government of Mongolia has ostensibly turned toward the latter option of modernizing Mongolia's economy. While massively desinvesting from the countryside, it has opted for a "resolute urban prioritization" (Sneath 2006: p. 162). International investments and international development assistance over the last two decades have constantly focused on the extractive industry and on urbanization to the exclusive benefit of Ulaanbaatar. Such development policy is inadequate because it does not benefit to the poorer half of the population: herders and migrants. It was seemingly justified by the widespread misconception of herders as self-sufficient, able to count on their folk to survive.

It is likely to be true that migrants are generally better off than herders. In this sense, the UN Special Rapporteur on Extreme Poverty and Human Rights asserted that, "poverty is more prevalent in rural and remote areas of Mongolia," yet immediately adding that "inequality in living standards is more pronounced in urban areas" (Sepulveda 2013: para. 83). From this observation, many concluded that Mongolian herders had to first settle if they wanted to benefit from development. This, however, is not necessary: basic services could also be effectively provided to nomadic herders, and a growing economy could support the additional expenses that this would induce. Yet, the cultural changes that accompanied the Age of the Market led the younger generation to aspire for an urban life, and the older generations to stop fighting against such aspirations (Sneath 2006: p. 177; Lkhagvadorj et al. 2013; Marin 2010). New development policies need to be conceived that would respond to these aspirations while realizing the economic and social rights of everyone.

Conclusion
There are plenty of ways to manage the impacts of climate change, but they do not necessarily fall squarely within a distinct category of measures on "climate adaptation." Partly-climate-induced migration can most adequately be addressed through development policies. In Mongolia, it seems, the impact of climate change exacerbates the effects of *dzud*; in the current circumstances, this accelerates the migration of destitute migrants to Ulaanbaatar. The causal link between climate change and migration pleads for a financial support by other states, in application of the principle of a common but differentiated responsibilities for climate change. Yet, such migration may also be seen as the symptom of a growing tension between the aspirations of herders and the economic opportunities offered to them.

The concept of climate adaptation says little about what could be done to address migration. By contrast, in the Mongolian context at least, approaching partly-climate-induced migration as the symptom of development issues suggests a set of effective measures that the government of Mongolia should take in order to promote an equitable development. In other words, it seems more politically productive to consider Mongolia's internal migrants as "development migrants" rather than as "climate migrants."On the basis of the case studied, it seems that thinking of "climate migrants" as a distinct category of migrants may be misleading and counterproductive when other, more efficient policy levers are available. While the "climate migration" rhetoric identifies migration as an issue, other perspectives may reveal that it is a symptom, and, perhaps, a solution. As a symptom, migration should be the opportunity for a discussion on the social gap between Mongolia's new rich and those left behind. As a possible solution, migration should generate public debate in Mongolia as to the future of nomadic livestock husbandry in a modern economy.

Acknowledgments This chapter is the product of a visit of Mongolia in March–April 2013. This research was permitted by the financial support provided by the McGill Hans & Tamar Oppenheimer Chair in Public International Law, by the logistical support offered by the local office of the International Organization for Migration in Ulaanbaatar and by the library of the American Center for Mongolian Studies, and by the availability and kindness of about 50 interviewees.

References

Batima P (2006) Climate change vulnerability and adaptation in the livestock sector of Mongolia. Final report of the AIACC (Assessments of impacts and adaptations to climate change) project

Batima P, Natsagdorj L, Gombluudev P, Erdenetsetseg B (2005) Observed climate change in Mongolia. AIACC (Assessments of impacts and adaptations to climate change) working paper No. 15

Batjargal Z (2001) Lessons learnt from the Dzud 1999–2000. UNDP and Government of Mongolia, Ulaanbaatar

BBC (2009) Maldives cabinet makes a splash. BBC News (17 October 2009)

Betts A (2013) Survival migration: failed governance and the crisis of displacement. Cornell University Press, Ithaca, NY

Campi A (2006) The rise of cities in Nomadic Mongolia. In: Bruun O, Narangoa L (eds) Mongols from Country to City: floating boundaries, Pastoralism and city life in the Mongol Lands. Nordic Institute of Asian Studies, Copenhagen, p 21

CESCR (2009) UN Committee on Economic, Social and Cultural Rights. Non-discrimination in economic, social and cultural rights (art. 2, para. 2). General Comment 20. U.N. Doc. E/C.12/GC/20

Dagvadorj D (2010) Mongolia Second National Communication under the United Nations Framework Convention on climate change. Ministry of Nature, Environment and Tourism, UNEP, Ulaanbaatar

Fang LQ, Liu G (1992) Relationship between climatic change and the nomadic southward migrations in Eastern Asia during historical times. Clim Change 22:151

Fernandez-Gimenez ME (1999) Reconsidering the role of Absentee Herd Owners: a view from Mongolia. Hum Ecol 27:1

Field C, Barros V, Stocker T, Qin D (2012) Managing the risks of extreme events and disasters to advance climate change adaptation: a special report of Working Groups I and II of the Intergovernmental Panel on Climate Change. Cambridge University Press, Cambridge

Hvistendahl M (2012) Roots of empire. Science 337:1596

ICESCR (1966) International covenant on economic, social and cultural rights. General Assembly resolution 2200A (XXI). U.N. Doc. A/6316 (1966)

Janzen J (2005) Mobile livestock-keeping in Mongolia: present problems, spatial organization, interactions between mobile and sedentary population groups and perspectives for pastoral development. In: Ikeya K, Fratkin E (eds) Pastoralists and their neighbors in Asia and Africa. National Museum of Ethnology, Osaka, p 69

Jenkins G (1974) A note on climatic cycles and the rise of Chinggis Khan. Central Asiatic J XVIII:217

Khazanov AM (1978) Nomads and the Outside World, 2nd edn. Cambridge University Press, Cambridge, trans. J. Crookenden

Konagaya Y, Maekawa A (2013) Characteristics and transformation of the Pastoral system in Mongolia. In: Yamamura N, Fujita N, Maekawa A (eds) The Mongolian ecosystem network: environmental issues under climate and social changes. Springer, Utrecht, p 9

Lansdell H (1885) Russian Central Asia, including Kuldja, Bokhara, Khiva and Merv. Low, Marston, Searle and Rivington, London

Levin D (2012) Wealth rises in Mongolia, as does worry. New York Times (15 July 2012)

Lkhagvadorj D, Hauck M, Dulamsuren C, Tsogtbaatar J (2013) Pastoral Nomadism in the forest-steppe of the Mongolian Altai under a changing economy and a warming climate. J Arid Environ 88:82

Marin A (2010) Riders under storms: contributions of Nomadic Herders' observations to analysing climate change in Mongolia. Glob Environ Chang 20:162

May T (2012) The Mongol conquests in World History. Reaktion Books, London

Mayer B (2013) Constructing 'Climate Migration' as a Global Governance Issue: essential flaws in the contemporary literature. McGill J Int Sustain Dev Law Pol 9:87

McAdam J (2012) Climate change, forced migration, and international law. Oxford University Press, Oxford

Medvedeva T (1996) Medical services and health issues in rural areas of inner Asia. In: Humphrey C, Sneath D (eds) Culture and environment in inner Asia: the pastoral economy and the environment. White Horse Press, Cambridge, p 176

Mongolia (2010a) Assessment report on climate change 2009. Ministry of Environment, Nature and Tourism, Ulaanbaatar

Mongolia (2010b) Message from the Gobi Desert by the Government of Mongolia on climate change. 27 August. Gobi Desert

Saizen I (2013) Change in livestock species and their spatial distribution. In: Yamamura N, Fujita N, Maekawa A (eds) The Mongolian Ecosystem Network: environmental issues under climate and social changes. Springer, Utrecht, p 215

Sepulveda M (2013) Report of the special Rapporteur on extreme poverty and human rights, Magdalena Sepulveda Carmona, on her mission to Mongolia (3–7 December 2012). U.N. Doc. A/HRC/23/36/Add.2

Sneath D (2006) The rural and the urban in Pastoral Mongolia. In: Bruun O, Narangoa L (eds) Mongols from Country to City: floating boundaries, Pastoralism and city life in the Mongol Lands. NIAS, Copenhagen, p 156

Sternberg T (2008) Environmental challenges in Mongolia's Dryland Pastoral Landscape. J Arid Environ 72:1294

Sternberg T (2010) Unravelling Mongolia's extreme winter disaster of 2010. Nomadic Peoples 14:72

Sternberg T, Chatty D (2008) Mobile indigenous peoples. Forced Migration 31:25

Tsogtbaatar J (2013) Deforestation and reforestation of degraded forestland in Mongolia. In: Yamamura N, Fujita N, Maekawa A (eds) The Mongolian ecosystem network: environmental issues under climate and social changes. Springer, Utrecht, p 83

UNDP and Mongolia (2011) Human Development Report 2011: from vulnerability to sustainability, environment and sustainable development. Ulaanbaatar

UNFCCC (1992) United nations framework convention on climate change. 1771 UNTS 107

UNFCCC (2010) Cancun agreements: outcome of the work of the ad hoc working group on long-term cooperative action under the convention. Decision 1/CP.16

UNFCCC (2012) Approaches to address loss and damage associated with climate change impacts in developing countries that are particularly vulnerable to the adverse effects of climate change to enhance adaptive capacity. Decision 3/CP.18

White B (2012) Mapping Ulaanbaatar's development. The Mongolist: politics, business, and society. 16 October. http://www.themongolist.com/blog/society/23-mapping-ulaanbaatar-s-development.html

Zhang D, Zhang J, Lee HF, He Y (2007) Climate change and war frequency in Eastern China over the last millennium. Hum Ecol 35:403

Chapter 13
Climate Change and Adaptation Challenges in the Pacific

Cyn-Young Park, David A. Raitzer, Jindra Nuella G. Samson, and Paulo Rodelio M. Halili

Abstract Pacific island countries are highly vulnerable to the effects of climate change, given their unique geography and natural-resource dependent economies. This paper reviews the potential impacts of climate change over the coming decades in the Pacific island-economies to identify measures for minimizing or mitigating these impacts. Drawing on the downscaling results of Global Climate Models (GCMs), the paper summarizes climate change effects such as changes in temperature, rainfall patterns, and sea levels. Assessment of GCMs suggests that sea level increases are likely to be substantially higher than previously accepted estimates, posing risks of inundation and salinization of freshwater in the Pacific island countries. Crop growth models suggest the potential for substantial losses of agricultural yields, while fisheries models predict changes to catches, and other modeling efforts identify substantial consequences for health and tourism. Integrated Assessment Models indicate the substantial total economic losses, if the business-as-usual growth strategy is followed. Appropriate adaptation strategies must be implemented, in order to ensure that poverty eradication and sustainable development are not impeded.

Introduction[1]

The Pacific[2] is one of the world's most vulnerable regions to climate change, yet few studies have been conducted to quantify expected climatic impacts on the economies of the region. In conceptual terms, the region's vulnerability is clear.

[1] This policy paper draws on the 2013 ADB report titled "The Economics of Climate Change in the Pacific" (ADB 2013b). The views expressed in this report are the views of the authors and do not necessarily reflect the views or policies of the Asian Development Bank (ADB), or its Board of Governors, or the governments they represent.

[2] The Pacific is defined here to include Cook Island, Republic of Fiji, Kiribati, Marshall Islands (RMI), Fed. States of Micronesia (FSM), Nauru, Palau, Papua New Guinea (PNG), Samoa, Solomon Islands, Timor-leste, Tonga, Tuvalu and Vanuatu.

C.-Y. Park (✉) • D.A. Raitzer • J.N.G. Samson • P.R.M. Halili
Economics and Research Department, Asian Development Bank, 6 ADB Avenue, 1550 Mandaluyon, Metro Manila, Philippines
e-mail: cypark@adb.org; draitzer@adb.org; jsamson@adb.org; phalili@adb.org

W. Leal Filho (ed.), *Climate Change in the Asia-Pacific Region*, Climate Change Management, DOI 10.1007/978-3-319-14938-7_13

The unique geography and environment of the region, its economic fragility, its distinctive demographics and the interactions between these different factors render it susceptible to the impacts of climate change. Following the Intergovernmental Panel on Climate Change's (IPCC 2001) framework, vulnerability can be considered as the product of three elements: i) exposure, ii) sensitivity and iii) adaptive capacity, each of which reinforces its vulnerable status.

The region is exposed to climate change, as it is subjected to volatile and unpredictable weather events, and many of its island states are geographically isolated. With the exception of Fiji, the Solomon Islands, and PNG, most of the island-countries in the Pacific have limited surface water and are highly dependent on rain and groundwater resources for their water requirement. Atoll countries and limestone islands have no surface water or streams and are fully dependent on rain and groundwater. More than half of the region's population lives within 1.5 km of the ocean, which may be at risk to sea level rise (IOC/UNESCO, IMO, FAO and UNDP 2011). Cyclone risk is high, and economic damages are extensive for many countries in the region.

The Pacific is sensitive to climate change, as in a majority of countries in the region more than a quarter of Gross Domestic Product (GDP) is from natural resources based sectors (Table 13.1). These natural resource sectors are ultimately dependent on a favorable and predictable climate, and are already at risk from over-exploitation in the context of rapidly rising population.

The region has limited resources for adaptive capacity to climate change effects, as poverty remains persistent with 32 % and 54 % of the region's population surviving under the poverty line of $1.25 and 2.00 in purchasing power parity (World Bank 2013). Moreover, most economies in the region are narrowly based with heavy dependence on tourism and other natural resource based industries under increasing pressure, limiting adaptive options.

Given the vulnerability that these elements suggest, it is pertinent to better understand how climate change will impact the economies of the region and what can be done to mitigate those impacts. This paper provides new evidence on the climatic changes expected in the region. The effects of those changes for particular sectors are modeled through an array of biophysical and socio-economic approaches to quantify impacts on principal sectors for the region, such as agriculture, marine environments, tourism and health. Finally, overall economy wide damages are appraised through the use of integrated assessment models.

While specific climatic forecasts (Australian Bureau of Meteorology and CSIRO 2011) and assessments of individual sectors have been previously performed (e.g. Bell et al. 2011), no comprehensive analysis has been conducted to quantify impacts in the Pacific across the main economic sectors and at the economy-wide level. Thus, the new evidence offered by this paper should be highly relevant to adaptation planning in an area of the world where appropriate measures to reduce vulnerability are most needed. This paper provides new evidence on the overall magnitude of expected impacts, which is useful for informing the allocation of resources for adaptation to the Pacific, as compared with other regions, and it

Table 13.1 Share of Pacific Economies Dependent on Natural Resources (GDP Share by Sector, %)

Economic Indicators	Cook Islands	Fiji Islands	Kiribati	RMI	FSM	Palau	PNG	Samoa	Timor-Leste	Tonga	Vanuatu
Agriculture	4.6	12.1	26.3	15.0	27.8	5.5	29.1	9.8	3.3	18.8	23.9
Industry	9.0	22.0	8.2	13.1	9.1	8.4	44.2	27.9	85.6	21.1	10.1
Services of which international tourism receipt	86.4 44.4	65.9 23.4	65.5 2.9	72.0 2.0	63.2 8.4	86.1 56.0	26.7 0.03	62.3 20.2	11.1 2.6	60.1 5.8	66.0 34.1
Tourism plus agriculture	49.0	35.5	29.2	17.0	36.2	61.5	29.1	30.0	5.9	24.6	58.0
Employment in agriculture	4.3	1.3	2.8	12.0	52.2	7.3	72.3	35.4	50.8	27.9	60.5

Note: GDP data is for 2011 except for Palau, PNG, Samoa, Tonga, and Tuvalu, which is 2012 data. Employment data is based on most recent year available. Tourism data is for the year 2010 except for Kiribati and Tonga (2005). Nauru and Tuvalu are not included in the table owing to the absence of tourism data. Solomon Islands lacks GDP shares data. Sources: ADB (2013a) and ESCAP (2012) for tourism data provides novel insights on which sectors within the Pacific are most in need of adaptation efforts.

Scope and Methodology

The study is based on a three component approach to assess climate impacts and their resulting economic costs for countries of the Pacific region. These include: (i) climate modeling and downscaling to understand future climatic contexts, (ii) compilation of model based evidence of expected sectoral impacts to understand specific damage and (iii) assessment of economy wide effects, drawing on understanding of sectoral damages. For climate modelling, a regional climate model (RegCM3, Pal et al. 2007) was set up for dynamic downscaling, using a 20×20 km horizontal grid resolution, driven by the fifth generation atmospheric general circulation model (ECHAM-5, Roeckner et al. 2003) developed by the Max Planck Institute for Meteorology, in combination with arrays of other GCMs for assessing El Niño– Southern Oscillation changes and sea-level rise. For the sector impact assessment, original modeling results for agriculture, coral reefs, tourism and human health are combined with other estimates for agriculture and fisheries in the existing literature. Agricultural modeling used the Decision Support System for Agro-technology Transfer (DSSAT, Jones et al. 2003) model, while tourism and health impacts are estimated using the Hamburg Tourism Model (HTM) and Version 3.6 of the Climate Framework for Uncertainty, Negotiation and Distribution (FUND3.6, Anthoff and Tol 2010), respectively. To estimate broader economy wide effects, two Integrated Assessment Models, namely, the Policy Analysis of Greenhouse Effect (PAGE09, Hope 2011) and FUND 3.6 are also applied.

National and sector specific analyses are focused on Fiji Islands, Papua New Guinea (PNG), Samoa, Solomon Islands, Timor-Leste, and Vanuatu, based on data availability. All impact modeling (climate, sector, economic) was conducted under different future emissions scenarios consistent with the Intergovernmental Panel on Climate Change (IPCC) Special Report on Emissions Scenarios (SRES) as well as policy scenarios based on negotiations under the United Nations Framework Convention for Climate Change (UNFCCC) process. IPCC's SRES scenarios are classified into four scenario families namely A1, A2, B1, and B2.[3] Climate

[3] The A1 scenario family describes a future world of very rapid economic growth, but declining global population after the mid-century, with varying assumptions of major technological progress: fossil intensive (A1FI), non-fossil energy sources (A1T), or a balance across all sources (A1B). The A2 scenario family describes a very heterogeneous world with continuously increasing global population and regionally oriented economic development. The B1 scenario family describes a convergent world with declining global population after the mid-century, rapid transition toward a service and information economy, reductions in material intensity, and the introduction of clean and resource-efficient technologies. The B2 scenario family describes a world where local solutions are emphasized for economic, social, and environmental sustainability (IPCC 2007).

downscaling was undertaken for the six selected countries and the results were presented for the periods of 2041–2060 (i.e., 2050 mean) and 2061–2080 (i.e., 2070 mean) as differences vis-à-vis the 1990 level using the A1B (a medium emissions) scenario as the reference case. Overall economic impacts have been analyzed under various emissions scenarios, with detailed sector effects presented for the six selected countries under the A2 (a high emissions) scenario.

Climate Modelling: Climate Change Extremes and Scenarios in the Pacific

The climate downscaling[4] finds that annual mean temperatures will increase for all six selected Pacific countries under the medium emissions scenario. Fiji Islands and Samoa will experience a 2.0 °C rise by 2070 on average from the 1990 level, while PNG, Solomon Islands, Timor-Leste, and Vanuatu experience a rise of more than 2.5 °C on average by 2070, with some areas in these countries experiencing an increase of nearly 3.0 °C (Fig. 13.1).

The climate modeling finds that the frequency of El Niño and La Niña could increase in the future. Several GCMs results point to an increase of over 40 % in the Pacific under A1B and A2 scenarios. ENSO-rainfall linkages may also be affected, resulting in unpredictable heavy rains and dry episodes. The effects of ENSO on rainfall could be considerable, causing either intense or limited rain, depending on the areas and the seasons. Extreme temperatures (the 99th percentile of the distribution of daily maximum data) are expected to increase in all six modeled countries. Bobonaro (Timor-Leste) would be the warmest, with the temperature reaching 44 °C by 2070 under the A1B scenario. Extreme wind and rainfall are also expected to increase. In Fiji, for example, maximum rainfall will increase from 160 to 200 millimeters (mm) per day by 2070.

Sea-level rise puts large coastal areas at risk of inundation. Under the A1B scenario, high-range estimates suggest that by 2100, all Pacific island-countries but Kiribati could face a sea-level rise, exceeding 1.0 m, from 1.2 m in Cook Islands to 1.7 m in Solomon Islands; low-range estimates suggest that sea-level rise to range from 0.5 m to 1.1 m for those two islands, respectively. Airports and seaports, road infrastructure, and local communities, all of which are highly concentrated on coastal areas, could sustain significant damage from the expected sea-level rise. Perhaps more importantly, there is risk that the already limited freshwater resources in the region may be severely impacted by increasing salinization.

[4] This analysis was performed for the periods of 2050 (mean value of 2041–2060 periods) and 2070 (mean value of 2061–2080 periods) based on the projection data from the output of RegCM3 model under SRES A1B.

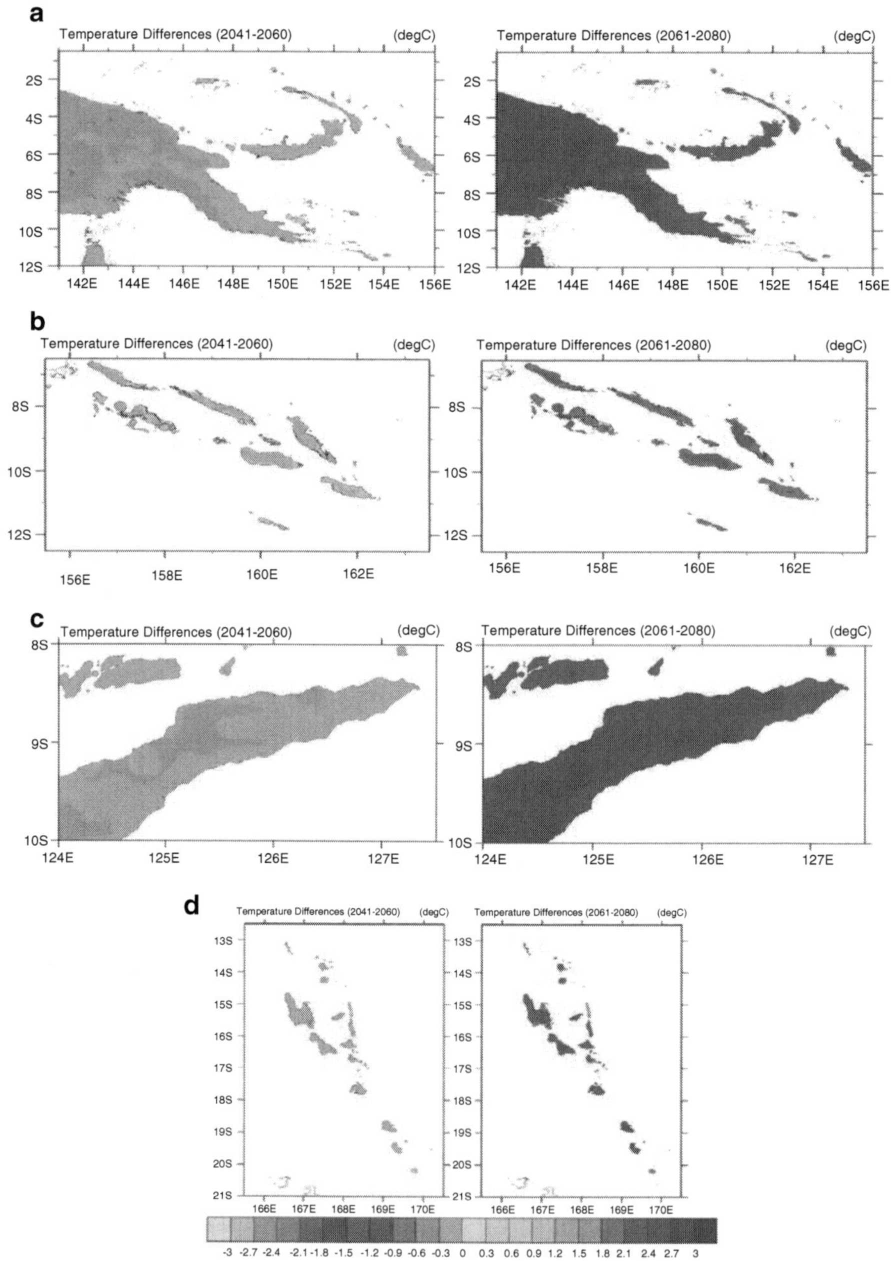

Fig. 13.1 Temperature Change under A1B Scenario: Papua New Guinea, Solomon Islands, Timor-Leste and Vanuatu using ECHAM-5 GCM. (**a**) Papua New Guinea, (**b**) Solomon Islands, (**c**) Timor-Leste, Vanuatu. *Source* ADB 2013b

Cyclone Risk

The Pacific is highly prone to natural disasters, which can inflict high levels of damage to small island states in the region with little capacity to mitigate or adapt to their effects. For example, in Samoa and Vanuatu years with damage events can cause economic damage in excess of annual GDP, while in Kiribati more than a quarter of the population is affected, on average, in each disaster year (Table 13.2).

Changes in tropical cyclone frequency for the Pacific subregion have been assessed by the Australian Bureau of Meteorology and CSIRO (2011) using 14 GCMs under the A2 (higher emission) scenario to compare the 1980–1999 period with 2080–2099. The GCMs are run with the Conformal Cubic Atmospheric Model (CCAM; McGregor and Dix 2008) at 60 km grid cell resolution, supplemented with several detection algorithms for cyclone formation, for 3 "basins" of the Pacific subregion. This analysis finds results consistent with prior global modeling studies that project a decrease in tropical cyclone frequency under climate change (e.g. Walsh et al. 2011), with the many findings indicating decreases of cyclone frequency of 30–70 %, but large variation among the models and cyclone identification algorithms (Table 13.3).

Changes in cyclone intensity were modeled by the Australian Bureau of Meteorology and CSIRO (2011), as well. Using the Maximum Potential Intensity index from the Genesis Potential Index methodology for cyclone interpretation, intensity is expected to fall based on 9 out of 14 GCMs, while the remaining 5 show very small increases. However, at the same time the analysis finds in most simulations an

Table 13.2 Key disaster statistics for countries of the Pacific from 1950–2012

	Number of disaster events	Average population affected population when event	Average economic damage in year of event	Average economic damage across all years (% of GDP)	Peak economic damage (% of GDP)
Fiji	49	6.2 %	1.5 %	0.0 %	9.0 %
Kiribati	4	26.4 %	n.d.	n.d.	n.d.
Micronesia, Fed. Sts.	8	5.2 %	0.9 %	0.0 %	6.0 %
Papua New Guinea	52	1.0 %	0.2 %	0.0 %	3.1 %
Samoa	12	19.9 %	41.6 %	0.8 %	161.7 %
Timor-Leste	8	0.3 %	n.d.	n.d.	n.d.
Tonga	14	14.6 %	7.3 %	0.1 %	28.7 %
Tuvalu	6	2.1 %	n.d.	n.d.	n.d.
Vanuatu	40	7.7 %	7.4 %	0.1 %	125.6 %

Source: EM-DAT 2009, the International Disaster Database, with adjustments for inflation

Table 13.3 Changes in expected frequencies of typhoons in Pacific subbasins

	South-east basin		South-west basin		North basin	
GCM	Most optimistic cyclone algorithm	Most pessimistic cyclone algorithm	Most optimistic cyclone algorithm	Most pessimistic cyclone algorithm	Most optimistic cyclone algorithm	Most pessimistic cyclone algorithm
CSIRO-Mk3.5	−65	10	−70	50	−80	10
ECHAM5/MPI-OM	−50	25	−80	15	−80	5
GFDL-CM2.0	−70	−20	−60	−5	−45	5
GFDL-CM2.1	−50	−10	−70	−5	−50	−5
MIROC3.2	−90	−75	−85	−80	−85	−50
UKMO-HadCM3	−55	−30	−55	10	−30	15
BCCR-BCM2.0	−90	−5	−90	10	−100	15
CGCM3.1	−20	30	−10	30	−10	−5
CNRM-CM3	−20	−35	−20	10	−30	−5
MRI-CGCM2.3.2	−25	0	−15	−5	0	0
CSIRO-Mk3.0	−50		−55		−65	
UKMO-HadGEM1	−5		5		15	
IPSL-CM4	−20		−15		−20	
ECHO-G	−35		−25		−25	
ENSEMBLE	−60	−6	−65	0	−45	−5

Source: Australian Bureau of Meteorology and CSIRO (2011)

increase in the proportion of storms with wind speeds above the current 90th percentile maximum wind speed, increasing that there may be more highly intensive cyclones.

Sector Impact Assessment

Agriculture Sector

The Pacific region has about 2.5 million hectares of land devoted to agricultural production, with more than half of the total agricultural area found in PNG, 17 % in Fiji, and 15 % in Timor-Leste (Food and Agriculture Organization (FAO 2013). The share of the agriculture sector in the total GDP of each Pacific country varies from 3 to 29 % (Table 13.1). RMI and Tuvalu devote more than 60 % of their total land area to agricultural production, while Kiribati and Tonga devote more than 40 %. Agriculture in the Pacific is dominated by production of perennial crops, such as oil palm, coconuts, bananas, and fruit. Livestock, roots and tubers, and maize are also important agricultural products. Among crops, sweet potato, yam, taro, and cassava are considered as staple foods (Wairiu et al. 2012).

Overall, global warming is expected to negatively impact crop productivity in the Pacific, although estimates vary considerably among modeling assumptions. Two sets of modeling have been performed to assess the expected effects of climate change of crop yield, both under the A1B scenario and applying DSSAT, but with different modeling nuances.

The first modeling approach (Boer and Rakhman 2011) applies RegCM3 regional downscaling of the ECHAM5 GCM under the A1B emissions scenario, with daily modeled data. Crop growth is modeled using the DSSAT model under rainfed and irrigated conditions, using the IIASA world soil database. Yields are modeled with planting dates held constant, using the optimal planting dates for yield. Carbon-dioxide fertilization effects are not incorporated in the analysis. The spatial distribution of crop production is determined using imagery from the Landsat 7 Earth Thematic Mapper Plus.

Under this approach, the largest potential yield losses are projected for sweet potato in PNG and the Solomon Islands, with losses in excess of 50 % of yield for the former by 2050 under the medium emissions scenario. For sugarcane, losses would be relatively small in 2050, but would rise in Fiji Islands by 2070 to a more substantial 7–21 %. Maize would have moderate losses of 6–14 % in Timor-Leste and Vanuatu by 2050, with a rise to 14–17 % by 2070 in the former. Results also show cassava in Fiji would also be significantly impacted with up to 37 % losses by 2070 (Table 13.4).

Table 13.4 Crop-Yield Effects (%) under the A1B Emissions Scenario Relative to a 1981–2000 Baseline

		2050		2070	
Crop Type	Country	Worst case	Best case	Worst case	Best case
Sugarcane	Fiji Islands	–9.05	–5.11	–20.55	–7.13
	Samoa	6.01	9.37	–2.30	5.69
Sweet potato	PNG	–59.16	–49.51	–68.37	–60.42
	Solomon	–14.97	–12.22	–31.02	–22.52
Maize	Timor-Leste	–10.43	–6.07	–17.03	–13.51
	Vanuatu	–13.53	–8.03	n.d.	n.d.

n.d. no data
Source: Boer and Rakhman (2011)

In parallel to this modeling work, another set of DSSAT simulations has been performed using a range of GCMs and growing condition parameters (ADB 2013c). Four GCMs were used under the A1B scenario: CNRM-CM3, CSIRO Mark 3, ECHAM 5, and MIROC 3.2 medium resolution. Monthly data were downscaled using the approach of, and were adjusted to a 100 km grid cell resolution. Monthly averages were modeled on a daily basis using a stochastic weather generator. Optimal combinations of planting dates and cultivars for each crop were selected, and yields were modeled under irrigated and rainfed conditions using soil data from the Harmonized World Soil Database (HWSD ver. 1.1), as reported by Batjes et al. (2009). The spatial distribution of crop production was determined using GLC2000 (Bartholome and Belward 2005), MODIS MCD12Q1 Land Cover 2008 L3 Global 500 m (NASA 2009), and GlobCover 2009 (ESA 2010). The crop modeling, which incorporates carbon-dioxide fertilization, generated separate scenarios for low and high fertilizer rates. In addition, one scenario includes selection of the variety in the DSSAT database that performs best in 2050.

As a result of the variation between the climate parameters of the models, there is considerable variation also in the expected effects of climate change for the same crop and location, with some runs representing yield gains, and others representing yield losses (Table 13.5). The inclusion of optimal variety choices under one scenario illustrates the current potential for adaptation, and thus is much more positive. Cassava has the highest yield losses across the countries and across scenarios, while the overall balance of yield effects is most negative in the Solomon Islands. Unexpectedly, the worst-case yield effects for sweet potato in PNG are considerably smaller than in the first analysis, which may be due to the inclusion of carbon-dioxide fertilization.

Table 13.5 Relative Changes in Crop Yields (%) under Climate Change in Year 2050 Relative to Year 2000 under the A1B Scenario

	PNG		Solomon Islands		Fiji Islands	
Crop	Worst case	Best case	Worst case	Best case	Worst case	Best case
Cassava, rainfed	–30.8	17.7	–27.8	–17.9	–36.5	–8.8
Maize, irrigated	–3.2	4.0	–9.6	0.7	–6.1	2.3
Maize, rainfed	–3.8	9.0	–16.5	–0.3	–7.0	1.0
Rice, irrigated	–8.3	12.4	–7.6	10.8	–7.1	11.7
Rice, rainfed	–7.5	11.7	–16.2	5.9	–11.0	3.5
Sugarcane, rainfed	–3.6	3.4	–12.9	0.9	–8.3	2.8
Sweet potato, rainfed	–10.9	–1.2	–15.0	1.5	–13.4	2.0
Taro, rainfed	–13.0	3.6	–18.6	–4.7	–17.5	1.1

Source: ADB 2013c

Marine and Coastal Resources

The marine and coastal environment of the Pacific region provides a significant source of food and economic security for its coastal communities and population. The region's exclusive economic zone and territorial waters that dwarf available land resources serve as home to a diverse species of marine fisheries, mangroves, and coral reefs and they contain some of the highest marine biodiversity in the world.

Fisheries are also likely to be adversely impacted by climate change, as indicated by recent modeling by Bell et al. (2011). Biological models of the marine environment have been coupled to the Coupled Model Intercomparison Project Phase 3 (CMIP3), a multi-model data set was used by the 4th Intergovernmental Panel on Climate Change (IPCC) to provide projections on the physical climate. More specifically, projected changes to the physical and chemical nature of the ocean surrounding each pacific island were incorporated into a biogeochemical modeling using the (i) Ecopath with Ecosim model (Pauly et al. 2000) to examine the effects of projected changes to on the food web and expected catches for tuna, (ii) the Pelagic Interaction Scheme for Carbon and Ecosystem Studies (PISCES, Gehlen et al. 2006) to assess the likely interaction between the phytoplankton and zooplankton in the ocean, and (iii) SEAPODYM (Spatial Ecosystem and Population Dynamics Model, Lehodey et al. 2008) to provide preliminary assessments of the vulnerability of skipjack and bigeye tuna in the tropical Pacific to climate change.

Under a high emissions scenario, catches of skipjack tuna for the western Pacific are estimated to decline by an average of more than 20 %, and for PNG by as much as 30 %. Across the entire region, total catch is projected to decrease by 7.5 % under the same scenario by 2100. For bigeye tuna, small decreases in catch (usually less than 5 %) are projected by 2035. Catches are projected to decrease by 10–30 % for many Pacific countries under the high emissions scenario in 2100.

The estimate of coral area in the Pacific in year 2000 is approximately 80 % of what would have been in the absence of thermal stress (in the pre-industrial era). The model indicates that the Pacific will experience an increase in thermal stress that would probably result in a significant decline in coral reef cover, from 88 % in the base year (1995) to 55 % in 2050 and 20 % in 2100. Coral-reef cover is further projected to be less than 1 % by 2200. Coral bleaching and decline as a result of climate change would compound effects on tourism as quantified by Tol (2011) based on climatic favorability, as well as on certain declines of oceanic fisheries catches as identified by Bell et al. (2011). In addition, the decline of coral reefs would mean the loss of areas of great importance to the sustenance of marine biodiversity.

Tourism

Tourism is a mainstay of the economy of a number of smaller countries in the Pacific. In the Cook Islands and Palau in particular, it accounts for a majority of GDP, while it is 20 % or more of GDP in Fiji, Samoa, and Vanuatu (See Table 13.1). Climate change would likely also impact tourism, which is another key economic sector of the region. An economic model was developed to appraise the potential effects of increasing temperature on the attractiveness of the Pacific as a tourism destination relative to the temperatures of countries from which tourists arrive.

Tourism, especially those in coastal areas, is one of the most important sectors in many Pacific countries. International tourism is likely to suffer from climate change as tourists tend to prefer an ideal climate for their holiday destinations. The Hamburg Tourism Model (Bigano et al. 2007) describes, at a high level of geographic disaggregation, changes in tourist behavior as a result of climate change. At the core of the model is a matrix that identifies tourism flows from one country to another. This matrix is perturbed by different scenarios of population, by changes in per capita income, and by climate change. The model also computes changes in the average length of stay and expenditures by tourists.

Modeling results indicate that the Pacific region becomes a less attractive tourism attraction and total tourism revenues are projected to fall. By the end of the century, tourist numbers are projected to be approximately one-third lower than in a business-as-usual scenario. Under all climate scenarios, the impact of climate change is to reduce tourism revenues by 27–34 % (Fig. 13.2).

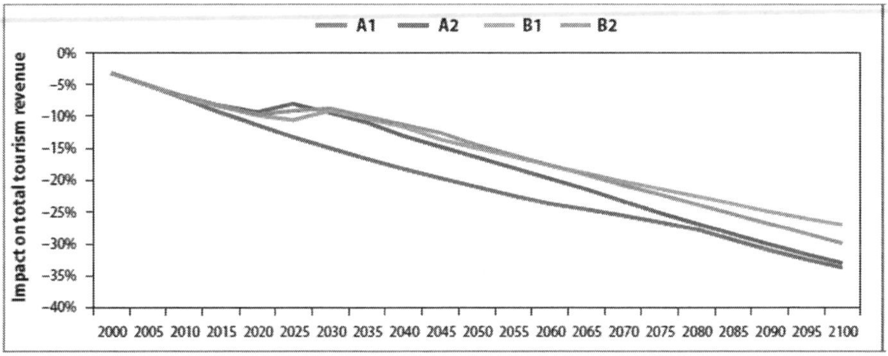

Fig. 13.2 Impact of Climate Change on Total Regional Tourism Revenue under various Emissions Scenarios. *Source*: ADB 2013b

Health

Many countries of the Pacific region currently experience poor health status that is likely to be exacerbated by climate change. Life expectancy at birth ranges from a low 55 years in Nauru to 73 in Samoa in 2005; children under 5 mortality rates range from 13 per 1,000 live births in Samoa to 74 in PNG; and access to improved sanitation range from 25 % in the FSM to 100 % in Samoa. Moreover, the incidence of communicable and non-communicable diseases is currently rising (Hanna et al. 2011). Climate change will adversely impact human health in the Pacific region through increased heat stress, higher disease transmission and risk exposure.

Under the FUND model, human health costs are valued in terms of forgone income as well as additional expenditure for treatment of illnesses. The impact assessment of climate change on human health considers the mortality and morbidity trends resulting from global mean temperature rises particularly with respect to respiratory diseases caused by thermal (heat and cold) stress and vector-borne diseases such as malaria and dengue fever.

The model uses data on temperature and precipitation and takes into account the availability of health-care services and the ability of the population to purchase medicine. Heat and cold stress is assumed to have an effect only on the elderly and among the urban population. The share of people over 65 years of age is calibrated to observed population, and is driven by per capita income. The share of the urban population from among the total population is also based on observed trends and income. In addition, the quality of housing and the availability of air conditioning are considered for thermal stress-related health disorders.

Mortality and morbidity costs together are expected to reach 0.8 % of GDP by 2100 under a high emissions scenario. Most of the estimated health costs would arise from respiratory disorders, followed by malaria, and deaths from tropical storms. By 2100, approximately 80 % of total mortality cost is projected to be

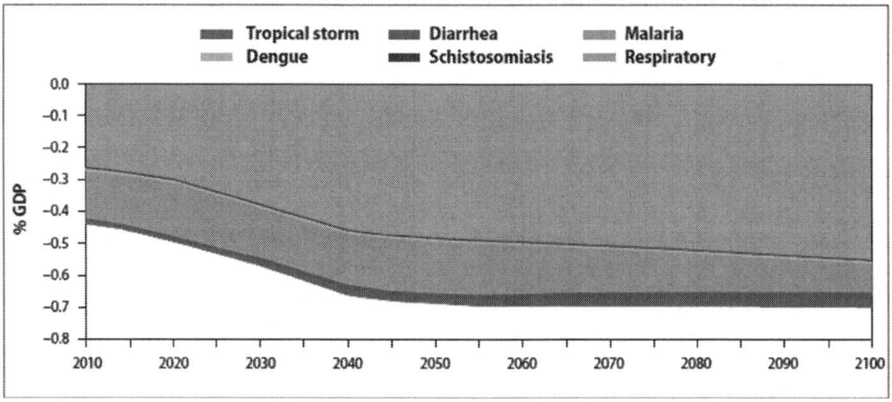

Fig. 13.3 Contributors to Mortality Cost due to Climate Change in the Pacific. *Source*: ADB 2013b

caused by respiratory disorders due to climate change, and 14 % by vector-borne diseases, particularly malaria (Figure 13.3).

Economic Impact Assessment

To supplement the climate downscaling and sectoral impact assessments, the study used two "integrated assessment models" (IAM) to examine the potential costs of climate change. PAGE09 is an IAM that values the impacts and costs of climate change for both mitigation and adaptation. It is designed to help policymakers understand the costs and benefits of action and inaction. The strength of this model is that it uses simple equations to simulate the results from more complex specialized scientific and economic models as well as illustrates the probabilistic range of outcomes and uncertainties associated with climate change. The model distinguishes eight world regions and runs ten time-periods to the year 2100 for four impact sectors (sea level, economic, non-economic, and discontinuities).

FUND 3.6 distinguishes 207 countries and runs from year 1995 to 2100 in time steps of 5 years. The model scenarios are defined by exogenous assumptions on the rates of population growth, economic growth, autonomous energy-efficiency improvements, the atmospheric concentration of carbon dioxide, global and national mean surface-air temperature, and sea-level rise. The climate module under FUND3.6 includes agriculture; forestry; sea-level rise; cardiovascular and respiratory disorders related to cold and heat stress; malaria; dengue fever; schistosomiasis; energy consumption; water resources, unmanaged ecosystems (Tol 2002a, b); diarrhea (Link and Tol 2004); and tropical and extratropical storms (Narita et al. 2009, 2010).

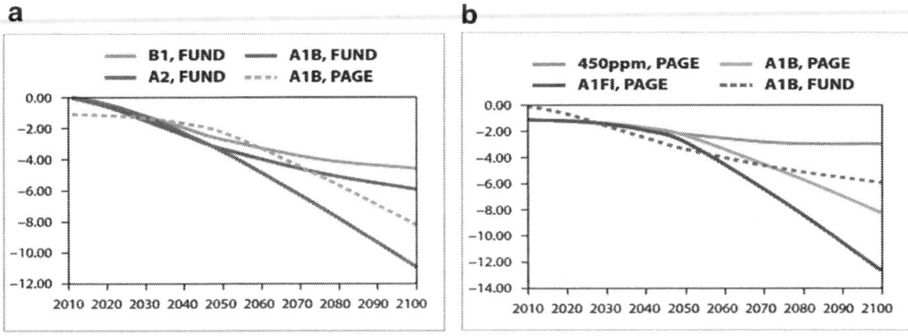

Fig. 13.4 Climate Change Cost in the Pacific (FUND 3.6 and PAGE 09 Models). (**a**) FUND 3.6 Model, (**b**) PAGE 09 Model. *Source*: ADB 2013b

Assuming the world does not deviate from its business-as-usual (BAU) scenario, the total cost of climate change in the Pacific will continue to grow over the long term. The total climate-change cost in the Pacific is estimated to reach 12.7 % of annual GDP equivalent by 2100 under the BAU scenario. Even under a low emissions scenario in which the global economy is assumed to become service-dominated, the economic loss would still reach 4.6 % of the region's annual GDP equivalent by 2100. If the atmospheric concentration of greenhouse gases were to reach 450 ppm and thus maintain global warming at approximately 2.0 °C, the economic cost would be smaller but still would reach between 2.0 and 3 % of GDP by 2100 (Fig. 13.4).

PNG will experience the most significant losses from projected climate change, reaching 15.2 % of its GDP by 2100, followed by Timor-Leste at 10.0 %, Vanuatu, 6.2 %, Solomon Islands, 4.7 %, Fiji Islands, 4.0 %, and Samoa, 3.8 % (Fig. 13.5). The negative effect on agriculture contributes the most to the total economic cost of climate change in the Pacific—approximately half of total economic cost amounting to 5.4 % of GDP in 2100 under a high emissions scenario (Fig. 13.6). Cooling cost follows second. A warmer climate will put pressure on the rapidly rising energy demand for space cooling in households and buildings around the Pacific. When income and population growth in the urban areas are considered, the cost of cooling is estimated to reach US$1,017 million or 2.8 % of the region's annual GDP equivalent by 2100. Economic impacts in the coastal areas will also be significant, and consist of three components: dryland loss, wetlands loss, and forced migration. The total impact in the coastal areas, through all three channels, is projected at US$469 million or 1.3 % of the region's annual GDP equivalent by 2100. Land-loss accounts for most of the total coastal impact. The cost of forced migration becomes significant only toward the end of the period, with rising sea level. Wetland losses remain relatively small until the end of the period.

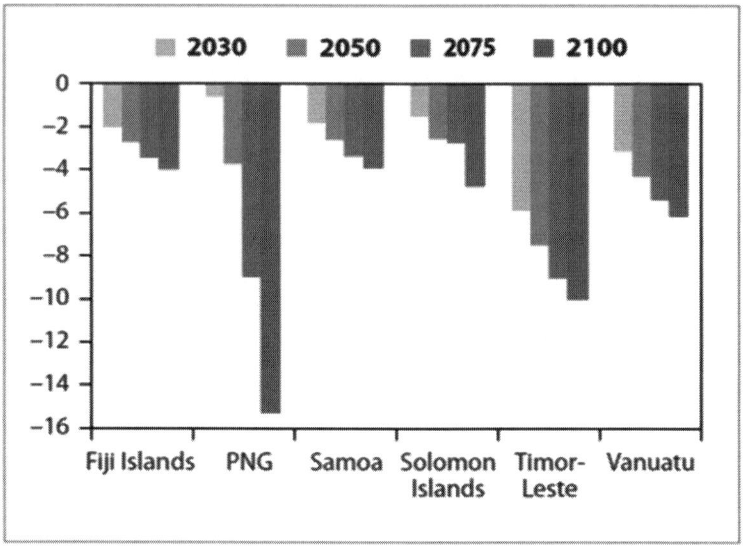

Fig. 13.5 Country-Specific Cost of Climate Change (FUND 3.6 and A2 Scenario). *Source*: ADB 2013b

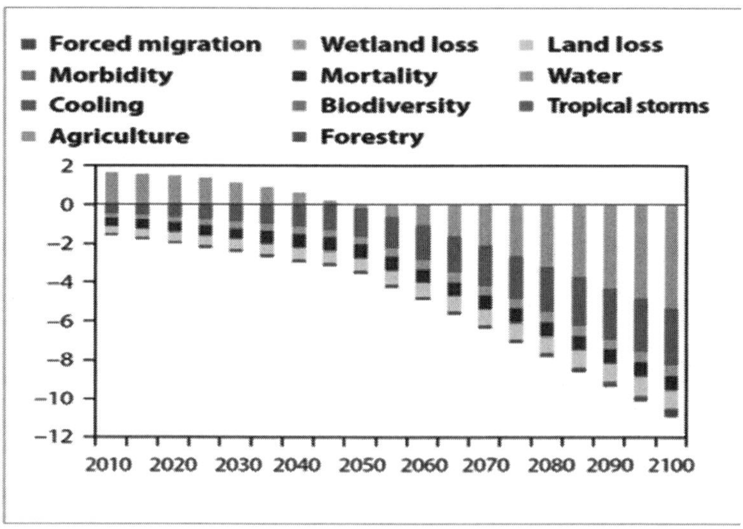

Fig. 13.6 Sectoral Composition of Climate Change Costs (FUND 3.6 and A2 Scenario). *Source*: ADB 2013b

Economics of Adaptation in the Pacific

The integrated assessment models estimate investment demands for building adaptive capacity for future climate change, as well as for climate proofing measures to establish more climate-resilient development strategies. When the adaptation targets are to contain the potential climate-change impacts until 2100, it is estimated that the Pacific region would require $447 million on average every year until 2050 (approximately 1.5 % of GDP) to prepare for the worst case (95th percentile) of climate change under the business-as-usual scenario (Table 13.6). The cost could be as high as $775 million or 2.5 % of GDP per annum. The cost of adaptation would be significantly lower under lower emissions scenarios and/or when adaptation targets to limit climate-change impacts at certain levels only up to year 2050. If the world manages to stabilize CO_2 concentration below 450 ppm, the adaptation cost is expected to be as low as $158 million or 0.5 % of GDP per annum during the same period under 2100 adaptation target, and down to annual average cost of $120 million under 2050 adaptation target.

Table 13.6 Annual average adaptation cost during 2010 2050 for the Pacific (with 2050 and 2100 Adaptation Targets)

Scenario	Adaptation target	Annual average cost	Range US$ million	Annual average cost % GDP	Range % GDP
A1FI	2100 worst case (95th percentile)	446.7	214.6–775.4	1.52	0.78–2.54
A1FI	2100 (4.5 °C, 0.70 m SLR)	284.3	131.1–483.7	0.97	0.48–1.59
A1B	2100 (4.0 °C, 0.65 m SLR)	253.1	118.9–438.4	0.86	0.44–1.43
450 ppm	2100 (2.5 °C, 0.55 m SLR)	158.3	75.2–273.2	0.54	0.27–0.89
A1FI	2050 (2.5 °C, 0.30 m SLR)	156.9	75.3–271.8	0.54	0.27–0.89
A1B	2050 (2.0 °C, 0.30 m SLR)	126.5	59.7–222.8	0.43	0.22–0.71
450 ppm	2050 (1.9 °C, 0.30 m SLR)	119.8	57.3–205.5	0.41	0.21–0.67

GDP gross domestic product, *m* meter, *ppm* parts per million, *SLR* sea-level rise
Note: The 95th percentile represents the critical point. Moving beyond, there is a low probability (at 5 % chance) that would lead to a catastrophic outcome
Source: ADB 2013b

Discussion

The analysis in this paper illustrates the potential costs of climate change. However, the estimates offered should not be considered as more than indicative, as they are all subject to a number of important limitations. Climate models are continually evolving, and there is considerable uncertainty regarding rainfall and storm projections. The temperature, rainfall and certain crop modeling results are driven by a single GCM, whereas there is still considerable variability among GCMs over the modeled timeframes. Crop growth results are contingent upon assumptions and are sensitive to rainfall, while there are many uncertainties governing fisheries migrations. Functions for climatic effects on tourism and health are reductionist, while integrated assessment models rely on difficult to verify damage functions, and may not appropriately incorporate the effects of climate damage into how economies evolve, biasing estimates downwards. Similarly, real-world autonomous adaptive responses are often not captured, biasing damage estimates upwards.

Despite these limitations, there is a consistency in the direction and the magnitude of the evidence regarding the expects of climate change on natural resource dependent economies and small island developing states, such as those that comprise much of the Pacific, which reinforces the results reported here (Brooks and Adger 2003; Hay et al. 2013; Kelman and Khan 2013). This suggests that climate change is a very real threat, which needs to be seriously addressed.

Given slow progress towards a comprehensive global agreement to limit greenhouse gas emissions, adaptation is needed to avoid large economic costs to the Pacific, and should be considered a development priority for the region. To do so, it is essential for the countries to identify low-regret strategies and responses that will build up resilience to climate change with relatively low cost, while generating benefits even in absence of climate change.

There may be a number of low-regret investment areas. Agriculture is climate sensitive, but adaptation of agronomic techniques shows potential to reduce losses. For example, genetic improvement for tolerance to abiotic stresses may help reduce current production risks (Araus and Slafer 2011). Reef protection may be particularly needed in the context of climate threats, but also is essential to manage present pressures. Improving health infrastructure and awareness can reduce risks from climate change, as well as generate long-term socio-economic benefits.

Regardless of the climate scenario and modeling approach, however, the estimated cost of climate-change adaptation is considerable. Climate change is part of the future and this analysis shows that it presents real risks, such as sea level rise and more intense rainfall, which need to be considered as future investments are made in infrastructure. This "climate proofing" entails identifying the risks posed by potential climate induced problems at project design, and addressing them during project implementation. Although climate proofing could increase upfront costs, such higher costs could be economically justified by lower total life cycle costs over the long expected period of use, as a result. Pacific countries will require substantial increases in investment, supported as appropriate with financial and technical support from the international community to take appropriate action.

Conclusions

Despite large uncertainties and limitations in the present analysis, it is clear that the effects of climate change are projected to intensify across the Pacific region in the coming decades. The annual mean temperature is expected to increase, with more volatile weather and more high intensity cyclones. The effects on key economic sectors would be primarily negative, given the region's geographic, economic, environmental, and demographic contexts. Potentially large losses are expected in agricultural production, with drops in crop productivity. Fisheries distributions will change while coral reefs will bleach, adversely affecting tourism and local economies in the Pacific. Moreover, with the increased temperature, tourist arrivals will decline and new health problems will emerge.

The overall cost of climate change in the Pacific will continue to grow over the long term. Using two IAMs, the study estimated the potential cost of climate change for the region at 12.7 % of annual GDP equivalent by 2100 under the BAU scenario, which is many times higher than the 2.2 % world GDP loss reported in the widely cited Stern (2006) review. This means that the Pacific is likely to be far more adversely affected than other world regions. The IAMs also estimate the cost of adaptation in various emissions scenarios. The estimated cost reflects the additional investment for offsetting the climate change effects, building climate-change adaptive capacity, and climate-proofing for key sectors. For example, to prepare for the worst case (95th percentile) of climate change under the BAU scenario, the region would require at $447 million on average every year until 2050 (approximately 1.5 % of GDP).

Given the uncertainties that remain, the present analysis should be considered as an initial assessment of climate risks to the Pacific, to be followed by more in-depth analysis through subsequent studies. Much more detailed understanding is needed of which risks are likely in specific locations for appropriate adaptation planning, and understanding of expected impacts should improve as climate models evolve.

Nevertheless, several important lessons arise from the study. First, climate change is not a stand-alone environmental issue, but an important development consideration, which requires development planning that integrates climate change adaptation measures building on a comprehensive policy framework that combines various sector approaches, policies, and strategies for achieving climate-resilient and sustainable development. Second, climate impacts vary by location and across sectors, and, hence, the identification of sector specific low-regret strategies and responses will be key to building

(continued)

resilience to climate change. Third, irrespective of the climate scenarios and modeling approaches applied, the estimated cost of adaptation is considerable, and early investment is crucial to guarding against the imminent threat of climate change to the region at minimal cost.

References

Anthoff D, Tol RS (2010) FUND–Climate framework for uncertainty, negotiation and distribution. University of California, Berkeley, USA

Araus JL, Slafer GA (eds) (2011) Crop stress management and global climate change, vol 2. CABI, Oxon, UK

Asian Development Bank (ADB) (2013a). Key indicators for Asia and the Pacific 2013: Asia's economic transformation-where to? How and how fast? Manila.

Asian Development Bank (ADB) (2013b). The economics of climate change in the Pacific. Manila

Asian Development Bank (ADB) (2013c). Climate change, food security and sound economic livelihoods in Pacific Islands (unpublished). Manila

Australian Bureau of Meteorology and CSIRO (2011). Climate change in the Pacific: scientific assessment and new research. Volume 1: Regional Overview. Volume 2: Country Reports. 288 pp

Bartholome E, Belward AS (2005) GLC2000: a new approach to global land cover mapping from Earth observation data. Int J Remote Sens 26:1959–1977

Batjes N, Dijkshoorn K, Van Engelen V, Fischer G, Jones A, Montanarella L, Petri M (2009) Harmonized world soil database. International Institute for Applied Systems Analysis, Laxenburg, Austria

Bell JD, Johnson JE, Hobday AJ (2011) Vulnerability of tropical pacific fisheries and aquaculture to climate change. Secretariat of the Pacific Community, Noumea, New Caledonia

Bigano A, Hamilton JM, Tol RSJ (2007) The impact of climate change on domestic and international tourism: a simulation study. Integrated Assessment J 7(1):25–49

Boer IR, Rakhman A (2011) Adaptation and impact of current and future climate variability on agriculture sector. In: Final report: The assessment of economics of climate change in the Pacific. Draft report submitted to ADB by CCROM, Bogor, Indonesia

Brooks N, Adger WN (2003) Country level risk measures of climate-related natural disasters and implications for adaptation to climate change. Tyndall Centre for Climate Change Research, Norwich

EM-DAT (2009). The OFDA/CRED international disaster database. Université Catholique de Louvain, Brussels, Belgium. Available at http://www.emdat.be/database.

European Space Agency (ESA) (2010) GlobCover 2009 land cover map, v2.3. Available at http://ionia1.esrin.esa.int/

Food and Agriculture Organization (FAO) (2013) FAOSTAT. At http://faostat.fao.org/ site/291/default.aspx (accessed 11 October 2013)

Gehlen M, Bopp L, Emprin N, Aumont O, Heinze C, Ragueneau O (2006) Reconciling surface ocean productivity, export fluxes and sediment composition in a global biogeochemical ocean model. Biogeosciences 3(3)

Hanna EG, Harley D, Xu C, McMichael AJ (2011) Overview of climate change impacts on human health in the pacific region. National Centre for Epidemiology and Population Health College of Medicine, Biology and Environment, Australian National University, Canberra, Australia

Hay JE, Forbes DL, Mimura N (2013) Understanding and managing global change in small islands. Sustain Sci 8(3):303–308

Hope C (2011) The PAGE09 integrated assessment model: a technical description. Cambridge Judge Business School Working Paper 4(11)

Intergovernmental Panel on Climate Change (IPCC) (2001) Climate change 2001: impacts, adaptation and vulnerability. In: McCarthy JJ, Canziani OF, Leary NA, Dokken DJ, White KS (Eds) Contribution of working group II to the third assessment report of the intergovernmental panel on climate change. Cambridge University Press, Cambridge, 1032 pp

Intergovernmental Panel on Climate Change (IPCC) (2007) Climate change 2007. The physical science basis. summary for policymakers. Contribution of Working Group I to the Fourth Assessment Report of the Intergovernmental Panel on Climate Change. IPCC Secretariat. Switzerland

IOC/UNESCO, IMO, FAO, UNDP (2011) A blueprint for ocean and coastal sustainability. Paris, IOC/UNESCO

Jones JW, Hoogenboom G, Porter CH, Boote KJ, Batchelor WD, Hunt LA, Wilkens PW, Singh U, Gijsman AJ, Ritchie JT (2003) The DSSAT cropping system model. Eur J Agron 18 (3):235–265

Kelman I, Khan S (2013) Progressive climate change and disasters: island perspectives. Nat Hazard 69(1):1131–1136

Lehodey P, Senina I, Murtugudde R (2008) A spatial ecosystem and populations dynamics model (SEAPODYM)–Modeling of tuna and tuna-like populations. Prog Oceanogr 78(4):304–318

Link PM, Tol RSJ (2004) Possible economic impacts of a shutdown of the thermohaline circulation: an application of FUND. Portuguese Econ J 3:99–114

McGregor JL, Dix MR (2008) An updated description of the conformal-cubic atmospheric model. In: High Resolution Numerical Modelling of the Atmosphere and Ocean. Springer, New York, pp 51–75

Narita D, Anthoff D, Tol RSJ (2009) Damage costs of climate change through intensification of tropical cyclone activities: an application of FUND. Clim Res 39:87–97

Narita D, Anthoff D, Tol RSJ (2010) Economic costs of extratropical storms under climate change: an application of FUND. J Environ Plan Manage 53:371–384

NASA Land Processes Distributed Active Archive Center (LP DAAC) (2009). MODIS MCD12Q1 Land Cover 2008 L3 Global 500m SIN Grid. USGS/Earth Resources Observation and Science (EROS) Center, Sioux Falls, South Dakota, USA.

Pal JS, Giorgi F, Bi X, Elguindi N, Solmon F, Rauscher SA, Steiner AL (2007) Regional climate modeling for the developing world: The ICTP RegCM3 and RegCNET. Bull Am Meteorol Soc 88(9):1395–1409

Pauly D, Christensen V, Walters C (2000) Ecopath, Ecosim, and Ecospace as tools for evaluating ecosystem impact of fisheries. ICES Journal of Marine Science: Journal du Conseil 57 (3):697–706

Roeckner E, Bäuml G, Bonaventura L, Brokopf R, Esch M, Giorgetta M, Hagemann S, Kirchner I, Kornblueh L, Manzini E, Rhodin A, Schlese U, Schulzweida U, Tompkins A (2003) The atmospheric general circulation model ECHAM5. Part I: Model description. Max Planck Institute for Meteorology Rep 349:127

Stern N (2006) The Stern review: the economics of climate change. HM Treasury, London

Tol RSJ (2002a) Estimates of the damage costs of climate change—Part 1: Benchmark estimates. Environ Res Econ 21(1):47–73

Tol RSJ (2002b) Estimates of the damage costs of climate change—Part II: dynamic estimates. Environ Res Econ 21(2):135–160

Tol RSJ (2011) The economic impact of climate change on pacific island nations. Unpublished report submitted to the Asian Development Bank

United Nations Economic and Social Commission for Asia and the Pacific (ESCAP) (2012) Statistical yearbook for Asia and the Pacific 2012. United Nations, Bangkok

Wairiu M, Lal M, Iese V (2012) Climate change implications for crop production in pacific islands region. ISBN: 978-953-307-887-8, InTech, DOI: 10.5772/33885

Walsh KJE, McInnes KL, McBride JL (2011) Climate change impacts on tropical cyclones and extreme sea levels in the South Pacific—A regional assessment. Global Planetary Change 80–81(2012):149–164

World Bank (2013) PovcalNet: an online poverty analysis tool (at http://iresearch.worldbank.org/PovcalNet/index.htm).

Chapter 14
The Strategic Partnerships on Climate Change in Asia-Pacific Context: Dynamics of Sino-U.S. Cooperation

Fang-Ting Cheng

Abstract It has been argued that U.S. and Chinese policies vis-à-vis the environment, are important factors in the recent stagnation of international cooperation on climate change. That notwithstanding, Sino-U.S. cooperation on matters affecting climate change, exemplified by such matters as the signature on the U.S.-China Memorandum of Understanding to Enhance Cooperation on Climate Change, Energy and the Environment, has been broadened and deepened to a considerable level in recent years. This chapter argues that the cooperative progress between the U.S. and China on climate change has come about because the two countries have shifted their positions and are now treating climate change as a strategic issue which affects their vital interests. As a result, they are more willing to cooperate strategically with each other through bilateral dialogue and programs on innovation and technology transfer. Furthermore, the chapter contends that this viewpoint has been influenced by the publication of documents which identify how climate change is likely to pose a severe challenge to a country's economic, social, or political stability.

Introduction

Efforts toward international negotiation and cooperation on tackling climate change faced great difficulty with obtaining agreement on greenhouse gases (GHGs) emissions reduction due to intense disputes between the United States and China over the responsibility. However, before the opening of the 2009 United Nations Climate Change Conference (held in Copenhagen), it could be seen that the two countries had started to strengthen bilateral dialogues, and had even initiated climate-related programs and projects through high-level efforts. Although possessing conflicting interests, the U.S. and China have shown particularly positive positions and intentions in dealing with climate issues. Since the U.S. and

F.-T. Cheng (✉)
Institute of Developing Economies, Japan External Trade Organization (IDE-JETRO), Wakaba 3-2-2, Mihama-ku, Chiba, Chiba Prefecture 261-8545, Japan
e-mail: chengft@gmail.com; Fangting_Cheng@ide.go.jp

W. Leal Filho (ed.), *Climate Change in the Asia-Pacific Region*, Climate Change Management, DOI 10.1007/978-3-319-14938-7_14

China are the biggest GHGs emitters, their actions and intentions might have influenced the international rule-making process to a significant degree. So far, analyses about the Sino-U.S. relations on climate change have mostly focused on rivalry between the two countries. This chapter, in contrast, aims to analyze the cooperative relationship/partnerships established in recent years, and to point out how and why the U.S. and China came to cooperate on issues related to climate change.

To start with, what needs to be realized here is that it is not easy for states to calculate their national absolute interests of international cooperation on climate change. This is because the uncertainty of both the causes and the effects of climate change remain strong, and this makes it difficult for states to develop cost-effective measures. Besides, climate change is an issue with cross-border[1], cross-generation[2], and hysteresis[3] characteristics, thus discouraging countries from responding to the problem immediately and actively. With that in mind, instead of conducting a cost-benefit analysis, this chapter covers a view focused on political context and policy study of addressing climate change.

Under the European initiative, demands for agreement with a multi-lateral political pact have indeed, to some extent, strengthened. However, it is hard to believe that the U.S and China, which currently together account for more than 40 % of the world's CO_2 emissions, adjusted their positions and attitudes to the problem only because of European or small islands' requests. It is obvious that the motivations for transformation in the two countries come from their own considerations. To clarity the formation of Sino-U.S. partnerships, the chapter argues that the two countries shifted their positions because they have tended to treat climate change as a strategic issue which affects their vital interests. The U.S and China

[1] "Cross-border" indicates that the cause and consequence of the environmental deterioration are not necessarily consistent geographically. For example, CFCs, notwithstanding the countries where the materials had been produced, diffused into the atmosphere beyond any borders, and caused a depletion of the ozone layer. Another example is deforestation. A large-scale deforestation in a specific region not only brings about local soil erosion, but also can affects the balance between world oxygen emissions and CO_2 absorption, and causes reduction of biodiversity. Climate change, which is very likely influenced by increasing GHG emissions, does not limit its harmful effects in developed countries. "Cross-border" environmental deteriorations are caused within specific borders, but can produce world-wide adverse effects.

[2] "Cross-generation" means that a long-term observation is needed to clarify the causal relationship between two events. In terms of climate change, although scientists have confidence in concluding that the increase in concentration of GHGs in the atmosphere has led to the warming temperature, it is difficult to assert that changing climate increases the frequency of occurrence of extreme weather events, because this assertion requires a long-term period for observing the causal relationship between global warming and dangerous weather events. Therefore, it is usually hard for policymakers to see climate change as an issue that needs to be tackled urgently.

[3] "Hysteresis" indicates that there is a time lag between a certain action and its influence. This characteristic exists in many environmental deterioration problems. In the case of climate change, it may take decades for increasing worldwide GHG emissions to cause sea level rise or extreme weather events. This time delay serves as the main reason that climate change is a cross-generation issue.

have become to hold clearer national interests through cooperation: development of mutual reciprocal and positive bilateral relations, as well as establishment of leading positions in negotiating the future international framework on climate change.

Another question that needs to be answered here is: why would nations be willing to respond to the issue? This chapter suggests that because the policy makers have recognized climate change may influence the stability of society or even security through both direct and indirect ways, thus the issue provides incentives for states to cooperate strategically. In the following section, the chapter examines environmental, social and security concerns that policy makers in both countries possess. The second part demonstrates how the U.S. and China have, in recent years, treated climate change as a strategic issue related to their environment, society and energy security.

The Natural, Social Impacts and Security Concerns of Climate Change

United States

In the United States, the adverse influence of climate change within its territories and the impact caused by extreme events outside the U.S., but indirectly affecting the country, tend to be valued more highly than regarding climate change as a task of common security for humanity (Moran 2011, p. 148). Loss of employment and an impact to economic growth and international competitiveness, which could occur in the country as a result of implementation of policies to tackle climate change, have been treated as major concerns in the U.S. Congress. Therefore, the presidential administration has had difficulties in persuading congress to adopt bills that contain U.S. acceptance of international legally-binding agreements or commitments to emission reduction targets.

However, discussions regarding the natural and social impacts of climate change have become substantial, from both theoretical and practical perspectives (Busby 2008; Barnett and Neil Adger 2007). Meanwhile, support of climate-related bills among senators has expanded in recent years compared with the time when the U.S. was considering ratification of the Kyoto Protocol. President George W. Bush's State of the Union address on January 23, 2007 marked the first time an American president recognized global climate change as "a serious challenge" (Bush 2007). Also, the U.S. National Intelligence Council issued a confidential report entitled "National Intelligence Assessment on the National Security Implications of Global Climate Change to 2030" in June 2008, and submitted it to the Congress (National Intelligence Council 2008). The report concluded that the effects the U.S. would face in 20 years would be extensive and indirect, as the adverse effects felt in other countries caused by climate change may also affect the security of the U.S.

For instance, disputes over resources such as water may cause collisions to disturb regional order; this can be regarded as a destabilizing factor in a potential conflict. In that case, the United States might request military intervention or humanitarian assistance in the conflicting area, which could cause political tension internationally. In addition, due to the deterioration of weather conditions, it is assumed that a large flow of environmental refugees would be generated, being compelled to move across borders. The report contended that the U.S.'s social stability would be threatened by the hosting of a great number of refugees from South America (ibid.).

Before his inauguration, President Barack Obama mentioned that climate change is an urgent issue concerning security that must be dealt with in a serious way (Holland 2008). Within the National Security Strategy published by the White House in May 2010, President Obama regarded climate change as a new challenge to national and global security, but one that has not been efficiently managed due to shortcomings of existing international institutions. He stated, "...we must focus American engagement on strengthening international institutions and galvanizing the collective action that can serve common interests such as combating violent extremism...and forging cooperative solutions to the threat of climate change..." (U.S. White House 2010, p. 3).

In 2010, for the first time the U.S. Defense Department Quadrennial Defense Review (QDR) included a chapter related to climate change entitled '*Crafting a Strategic Approach to Climate and Energy*,' and positioned climate change and energy as key issues that will play significant roles in shaping the future security environment (U.S. Department of Defense 2010). Also, the U.S. State Department published its first Quadrennial Diplomacy and Development Review (QDDR) in December 2010, in which climate change was considered a new challenge, noting, "New actors, good and bad, have the power to shape international affairs like never before. The challenges we face—nuclear proliferation, global pandemics, climate change, terrorism—are more complex than ever" (U.S. Department of State/USAID 2010).

Possible Impacts to U.S. Security

As pointed out by Busby, the possible impacts that climate change may present to U.S. homeland security include abrupt climate change, rising sea levels, extreme weather events and Arctic sea ice melt (Busby 2008, pp. 468–504). In particular, abrupt climate change and rising sea levels are considered a very long-term, progressive change based on scientific opinion, being hardly considered as immediate threats to U.S. homeland security. On the other hand, there are cases of extreme weather events, such as Hurricane Katrina, which struck the southeastern U.S. in 2005 and caused catastrophic destruction to the majority of New Orleans. Katrina damaged oil refineries and facilities, brought about lawlessness and more than $800 billion in losses. The hurricane killed over 1,800 people and more than 270,000 citizens were transported to shelters away from their hometown.

Even though scientists do not attribute the occurrence of Katrina to the progression of climate change, it made U.S. policy makers recognize how devastating natural disasters can be to U.S. homeland security. In fact, the occurrence of Hurricane Katrina activated discussions regarding the relationship between hurricane strength and global warming, further raising concerns in addressing climate change (Busby 2008, p. 484). As for the last concern, the shrinking of Arctic sea ice has raised disputes among neighboring countries such as Russia, Norway, Canada over right of the Northwest Passage, new sea routes, and the right of resource development. Due to this, the melting of Arctic sea ice directly links to vital interests of U.S. security (Busby 2008, pp. 489–490).

In addition to concerns for homeland security, climate change may also present security problems for the U.S. by affecting its national interests overseas. Busby listed examples such as damaged overseas assets or U.S. military bases, loss of facilities, flow of environmental refugees and regional instability due to the occurrence of armed conflict, the emergence of failed states and large humanitarian disasters (Busby 2008, p. 498). Other researchers suggest that depletion of water resources, crop failure and a strengthened feeling of despair caused by variations in the amount of precipitation may induce direct military conflicts (Hendrix and Glaser 2007). These arguments share the common view that it is not only conflicts or regional instability originate in environmental factors, but also the occurrence of natural disasters which relate to climate change can influence U.S. national interests of extraterritorial security.

China

After the enforcement of the Kyoto Protocol, the Chinese government has released several official documents to clarify its position and policies in regard to tackling climate change, as well as its concern toward the aggravation of the problem and the impact on China. Although the Chinese government has been emphasizing climate change as a matter of social and economic development by asking developed countries to take responsibilities in mitigating climate change, it recognizes climate change as a threat to its natural environment, agricultural development and resource protection (Chinese State Council 2007).[4] In the 2007 national report, the government saw climate change as a serious challenge to its modality of current social and economic development, energy structure, ability of independent development of energy technology, protection and use of forest resources, agricultural adaptation of

[4] See Chap. 3 in Chinese State Council (2007): "Influence and Challenge of Climate Change on China."

climate, protection and development of water resources, as well as its ability to respond to threats due to sea level rises (ibid.).

China has shown concern with adopting climate change policies due to the expected adverse influence on its economic and social development programs (ibid.). China asks for the right to take domestic development as the first priority while insisting industrialized countries should take responsibilities (Smith and Lennon 2008). However, as a result of its principle of economic growth, China faces domestic problems of environmental degradation and huge economic and social costs. Observed changes in the atmosphere within China are consistent with a global scale, thus making appropriate domestic measures urgently necessary. China's position on climate change evolved into not denying that climate change, especially global warming, has caused adverse effects in various areas in China, such as desertification, drought, and floods (Chinese State Council 2007, pp. 4–5). Through the national performance of implementing climate policy, it has been witnessed that the Chinese government is responding to the issue more actively, although there are still challenges for the authorities to strike a balance between environmental protection and the high level of economic growth and political stability.

In addition to environmental concerns, a reduction of energy would cause serious consequences to the maintenance of environmental conservation and sustainability owing to the rapid growth in population.[5] In 1993, China became a net importer of petroleum, and oil imports had exceeded 100 million tons by 2003. Domestic oil production in China has been ranked the best in the world but only meets half of its domestic demand (*People's Daily* 2010). Meanwhile, China became a net importer of natural gas starting in 2008, and in 2009 there was a gap of 4 billion square meters of natural gas between domestic production and consumption (International Energy Agency 2009). In the face of the circumstances of relying on foreign oil and natural gas, expanding the use of renewable and clean energy serves as an important tool to complement domestic demand for fossil fuels, and would be a rational choice for China.

Examples of Serious Environmental Degradation in China

There are two concerns to Chinese territory caused by climate change. Firstly, a flooded homeland, a decreasing coastal territory and the erosion of coastal areas will all result from sea level rise, creating huge economic losses in developed areas. The submergence of islands may change the maritime border of China and

[5] In China, the increase in population has slowed down in recent decades, with the rate of population growth from 1990 to 2000 at 1.0 %, while the rate from 2004 to 2030 is estimated to be 0.4 %. However, the annual economic growth rate between 1990 and 2002 was 9.7 % on average. The annual economic growth rate is forecast to be 5.0 % from 2004 to 2030.

neighboring countries, which could significantly relate to its national interest of security (Zhang 2010, pp. 61–81).

The second concern of climate change is a decline in the quality of soil. This can be illustrated with the expanding desertification in Xinjiang, Inner Mongolia, Tibet, Qinghai, Gansu, Hebei, Shaanxi and Ningxia. From the 1950s to the end of the 1990s, the desertification in northern China has shown a tendency toward expansion. The current area of desertification is 1.2–1.3 times larger than it was between the mid-50s and mid-70s (Zhang 2010, pp. 83). Also, the process of desertification in Xinjiang, Inner Mongolia, Qinghai, and Gansu has become "obvious" (China State Forestry Administration 2005).[6] The desertification in these provinces has reached 318,600 km^2, accounting for 3.32 % of the total area of the territory (ibid.). Change in climate has a particularly decisive impact regarding the desertification, owing to the small population in the dry region. Desertification does not only result in the disappearance of residential and agricultural land, but also increases the frequency and strength of sandstorms.

A typical example is the problem of desertification within Wuwei City, Minqin County, Gansu Province. This is accepted as an example of the adverse influence that progressing climate change causes toward soil degradation within China. In Minqin County, located downstream of the Shiyang River Basin, the single water resource in the county, the precipitation has been increasing because of rising temperatures. Nevertheless, evaporation reaches up to 24 times the precipitation amount, reducing water levels in the upstream area, and significantly decreasing the amount of available water resources for human activities in the downstream and midstream areas. Therefore, there are not only ecological degradations in the Shiyang River Basin to consider, making the subject of desertification in the Minqin area more serious than ever. The lack of water resources and the deterioration of life quality has forced 26,500 residents—who became so called environmental refugees—to relocate within the last decade (Dai et al. 2008, p. 322).

Since climate change can present impacts on natural environment, society and even security in both U.S. and China, policy makers tend to value the importance of combining different kinds of approaches in dealing with various problems related to climate change. Two dimensions of approaches, stated as countermeasures and preventive measures in this chapter, have been addressed to a significant degree under the U.S. and China bilateral cooperation. These will be explained in the following section in order to provide the framework to understand the strategic partnerships formed between the two countries.

[6] "Obvious" means that the quality of the soil is at the stage between desertification and non-desertification due to water shortage and the excessive use of land. This type of land, even though not yet sandy, is clearly showing tendencies of desertification (China State Forestry Administration 2005).

Countermeasures and Preventive Measures to Tackle Climate Change

The complexity of solving climate change comes from the particularity of the problem itself, instead of simple conflicting national interests among developed and developing countries. For the issue of climate change, which is caused by both anthropogenic and natural factors,[7] there have been difficulties in reaching a comprehensive global agreement with effective solutions. In searching for solutions, stakeholders consider that it is necessary to make use of dynamic policy instruments.

Since climate change is occurring and partially is due to spontaneous atmospheric variation in weather, as shown in Table 14.1, such that governments can respond via exogenous approaches, say *countermeasures*, by introducing or accepting tools or means such as the construction of breakwaters or strengthening facilities of water conservancy in order to reduce the damages brought by natural disaster. Input of financial resources, existing technologies or equipment is considered necessary since part of environmental degradation such as sea level rise and rise in temperature are estimated to be inevitable.[8]

The another approach to address climate change emphasizes to mitigate an thropogenic GHG emissions since the industrial revolution have been leading to the rising concentration of GHGs in the atmosphere and ongoing global warming/ climate change. By implementing endogenous approaches, stated as *preventive measures* here, people can contribute to the prevention of climate change exacerbation by reducing GHG emissions which originate from human activities. This

Table 14.1 Countermeasures and preventive measures to tackle climate change

	Approaches	
	Counter measures	Preventive measures
Objectives	· Responses to inevitability of environmental degradation	· Prevent environmental degradation from occurring
	· Reducing damages brought by natural disasters	· Managing potential risk of serious environmental destruction
Measures	· Input or acceptance of tools or means	· Innovations or conversions of system and technology

Source: author

[7] See IPCC (2007) Table SPM.2: "Recent trends, assessment of human influence on the trend and projections for extreme weather events for which there is an observed late-twentieth century trend."

[8] See IPCC (2007) Table SPM.3: "Projected global average surface warming and sea level rise at the end of the twenty-first century."

effort may not eliminate climate change itself, but to some extent could contribute to the mitigation of the problem.

For the purpose of reducing GHG emissions, preventative measures include developing innovations to, or conversion of, economic and social systems as well as environmental and resource-utilization technologies (see Table 14.1). The innovation of systems and technology will result in the creation of processes, measures and devices in economic development and advanced technology. Conversion refers to implementation of innovated systems and technologies, under which the consumption of resources would be more efficient, energy-saving and environmentally friendly. The concepts of innovation and conversion of systems and technology in coping with climate change can be illustrated by that of ecological modernization, which focuses on the development and creation of new environmental technologies that provoke new economic interests and employment (Wurzel and Connelly 2010).

Owing to the nature of the climate change, countermeasures, which are close to *adaptation*, aiming to lessen the vulnerability and respond to the adverse effects, have been receiving attention and are making concrete progress during the process of the post-Kyoto negotiation.[9] Also, negotiations with a focus on preventive measures, that is, *mitigation*, such as the pledged targets of GHGs emission reduction and technology transfer between countries with a different degree of development, have remained its importance in the UN processes.

Although there have been strong confrontations among the Annex 1 and non-Annex 1 countries, they have agreed to cooperate on reducing or slowing the rise of GHG emissions.[10] If we take the U.S. and China, for example, even though disagreements between the two countries have been identified as the reasons for stalemate of negotiation under the United Nations Framework Convention on Climate Change (UNFCCC), Sino-U.S. cooperation on climate change in recent years has been expanded and intensified substantially (Han et al. 2009).[11] To analyze the bilateral cooperation, which can be understood under the framework shown as Table 14.1, climate policies in each country and the establishment of bilateral strategic partnerships will be discussed in details below.

---

[9] For enhancing adaptation actions, Annex 1 Parties (developed countries) have committed to offer financial support for developing countries, especially those vulnerable to climate change, under the Green Climate Fund, which was adopted at the Sixteenth Conference of Parties (COP16) of UNFCCC in Cancun.

[10] Parties have agreed to recognize that "deep cuts in global greenhouse gas emissions are required ... to hold the increase in global average temperature below 2 °C above preindustrial levels." Also, they recognize "the need to consider ... strengthening the long-term global goal on the basis of the best available scientific knowledge, including in relation to a global average temperature rise of 1.5 °C." See UNFCCC (2011).

[11] Right after the closing of COP 15, UK climate secretary Ed Miliband accused China of trying to "hijack" the climate deal by pointing out, "We did not get an agreement on 50 % reductions in global emissions by 2050 or on 80 % reductions by developed countries. Both were vetoed by China, despite the support of a coalition of developed and the vast majority of developing countries" (Miliband 2009).

The U.S. and China Strategic Partnerships on Climate Change

In this section, a focus will be given to the analysis of Sino-U.S. strategic cooperation on climate change by overlooking the frameworks and projects agreed/carried out in recent years. Since the chapter also aims to provide an analytical viewpoint for future case studies, details of implementation of those frameworks and projects are omitted here at this moment.

The Sino-U.S. confrontation on climate change came to the surface at the thirteenth Conference of Parties (COP13) of the UNFCCC. Held in Bali, the U.S. and some developing countries, especially emerging economies, had a dispute over the burden-sharing between developed and developing parties under the post-Kyoto regime. The Bali Road Map was adopted at COP13, structuring the rough image of the future framework.[12] However, there was no huge progress from Bali to Copenhagen. The Copenhagen Accord, which the parties as a whole only agreed to "take note of," was agreed upon compromise within a few countries against the backdrop of political chaos.[13] Although the adoption of the Cancun Agreement in 2010 restored confidence in the UN process, pivotal compromises need to be made among major parties to set the structure of the new framework.

However, the stagnation of UN negotiations does not indicate that there is no progress toward Sino-U.S. bilateral cooperation. Before the Copenhagen conference, the U.S. and China had prepared several documents, particularly those related to energy and the prospect of cooperation, in order to adjust each position under the Bali Action Plan (see Table 14.2). Looking into the bilateral cooperative relationship on climate change shows that Sino-U.S. cooperation is associated firmly with energy efficiency and technological development. In June 2008, the fourth Strategic Economic Dialogue (SED) was held, at which the two governments adopted a platform called the Ten-Year Framework for Cooperation on Energy and Environment (TYF) (see Table 14.2). Under the TYF, China and the U.S. have been cooperating through projects in the fields of clean air and water, clean and efficient transportation, conservation of wetlands and nature reserves, efficient and secure electricity, and energy efficiency. In addition, the two countries signed EcoPartnerships in order to achieve the stipulations of the TYF, while showing their willingness of expanding existing partnerships.

In July 2009, the "Memorandum of Understanding to Enhance Cooperation on Climate Change, Energy and Environment" (MOU) was released, strengthening the

---

[12] The Bali Road Map (Decision 1/CP.13) was agreed to negotiate a post-Kyoto framework by establishing the Ad Hoc Working Group on Long-term Cooperation Action under the Convention (AWG-LCA). The AWG-LCA was aimed to complete negotiations within 2 years on four main topics: A Shared Vision, Mitigation, Adaptation, and Finance and Technology Transfer.

[13] The Copenhagen Accord was accepted only in a closed-door meeting among about 30 parties. The parties as a whole only agreed to "take note of" the official decision, though, because of strong opposition from some developing countries.

Table 14.2 Sino-U.S. cooperation on climate change

Year	Cooperation and measures	Level
June 2008 SED 4	**Initiating 10-year energy and environment cooperation framework (TYF) and eco-partnerships** · Five goals of cooperation were set for the first phase: (1) Energy saving of electric power systems and logistics, (2) Improvement of efficiency of transportation, (3) To deal with water pollution, (4) To deal with air pollution, (5) Protection of forests and wetlands.	Vice-President
July 2009 S&ED I	**Joint press statement of the first strategic and economic dialogue**	Minister
July 2009 S&ED I	**Memorandum of understanding to enhance cooperation on climate change, energy and environment** · First official document of cooperation on climate change. To build and strengthen 10 fields of the cooperative relationship such as energy-saving, renewable energy, clean coal, and Carbon Capture and Storage (CCS), etc. · A commitment to strengthen the eco-partnership, and regard U. S.-China cooperation as a platform for dialogue to construct climate change policy dialogue and cooperative framework. In addition, an agreement to implement UNFCCC thoroughly, effectively and continually.	Minister
November 2009	**U.S. China joint statement** · (1) To emphasize the mutual roles of enhancing sustainable development, and agreement to take mitigation actions according to each domestic situation; (2) To promote successful international negotiations on climate change, notably to agree upon the GHGs emission reduction target of developed countries and National Appropriate Mitigation Actions (NAMAs) of developing countries; (3) Establishment of a clean energy research center; (4) Establishment of The Electric Vehicles Initiative; (5) To launch a new energy saving action plan; (6) To launch a new renewable energy partnership; (7) To develop a large scale CCS project and promote the use of clean coal; (8) To launch a new Shale Gas Initiative, and (9) To launch The Energy Cooperation Program (ECP).	President
November 2009	**Protocol for Cooperation on a Clean Energy Research Center** · Each invests 150 million USD, approximately half of the whole budget. The establishment of set research subjects such as the energy efficiency of buildings, clean energy including CCS, clean vehicles as a priority, the supporting the launch of The Electric Vehicles Initiative, and to introduce cooperative related CCS on a large scale.	Minister
November 2009	**Memorandum of cooperation to build capacity to address climate change** · The Establishment of The Renewable Energy Partnership, for promoting cooperation between government and private sectors. To Utilize private resources and technology to further promote the wide spread use of clean energy technologies. To support the continuation of The Energy Policy Dialogue.	Minister

(continued)

Table 14.2 (continued)

Year	Cooperation and measures	Level
January 2011	**Sino-U.S. joint statement** ・ Agreement to continue negotiations on climate change in order to realize the energy security of each country. Welcome and implement The Cancun Agreement, while supporting the UN framework as well as selecting South Africa as the host of COP17.	President
May-11 S&ED III	**Improvement of (TYF) and EcoPartnerships** ・ Signature of the six new eco-partnerships under the TYF.	Minister
April 2013	**Establishment of the U.S.-China bilateral working group on climate change (CCWG)**	Vice-President
July-13 S&ED V	**Sino-U.S. joint statement** ・Agreed to work together with other countries to use the expertise and institutions of the Montreal Protocol to phase down the consumption and production of hydro fluorocarbons (HFCs), among other forms of multilateral cooperation.	President

References: Zhang 2010, pp. 237–239;Committee on U.S.-China Cooperation on Electricity from Renewable Resources et al. 2010, pp. 205–216
Source: author

TYF. The central topics for actions included "(1) Energy conservation and energy efficiency; (2) Renewable energy; (3) Cleaner uses of coal, and carbon capture and storage (CCS); (4) Sustainable transportation, including electric vehicles;(5) Modernization of the electrical grid; (6) Joint research and development of clean energy technologies; (7) Clean air; (8) Clean water; (9) Natural resource conservation, e.g., protection of wetlands and nature reserves; and (10) Combating climate change and promoting low-carbon economic growth."[14] In addition, in order to facilitate policy cooperation on climate change, the two countries agreed to cooperate by establishing The Climate Change Policy Dialogue (see Table 14.2).

Followed by the Sino-U.S. Joint Statement announced by President Obama and President Hu Jintao on November 17, 2009, additional progress was made. Firstly, the U.S., which had raised opposition to the definition of the principle of common but differentiated responsibility and capability in December 2007, officially accepted that principle in tackling climate change. Secondly, the two leaders expressed their willingness to take actions to reduce GHG emissions. In return for China's cooperative stance, the two countries declared to promote cooperation on provision and acceptance of financial assistance and technology transfer for adaptation and mitigation based on equality and mutual reciprocity.

With the MOU, the U.S. and China agreed to formulate a new action plan for the improvement of energy efficiency under the TYF. Moreover, two documents, a

[14] "Appendix: A Timeline of U.S.-Chinese Cooperation on Clean Energy and Climate Change," *The Power of Renewables: Opportunities and Challenges for China and the United States*, 2010, p. 212.

"Memorandum of Cooperation to Build Capacity to Address Climate Change"[15] and a "Protocol for Cooperation on a Clean Energy Research Center"[16] were signed, with the purpose of achieving mutual goals in addressing climate change. As agreed in the Protocol, both China and the U.S. will, before 2015, finance a Clean Energy Research Center with a minimum of US$150 million in order to provide convenience and to facilitate the opportunity for interaction between engineers and researchers of both countries. Topics such as energy efficiency of buildings, clean energy (including CCS), clean vehicles and clean coal technology are given precedence. Furthermore, the two countries agreed to introduce projects of CCS on a large scale, to facilitate the quick actions of technical development, and the use, spread and transfer of the technology.

Climate and Energy Cooperation under the Strategic and Economic Dialogue

Starting in 2009, China and the U.S. have been holding a periodic U.S.-China Strategic and Economic Dialogue (S&ED). In the second S&ED, held in May 2010, the two countries announced an action plan based on the TYF, and signed an Implementation Plan for EcoPartnerships. Furthermore, the two parties agreed to hold the first U.S. and China Energy Efficiency Forum, Electric Vehicles Forum, the first Renewable Energy Forum, Energy Policy Dialogue, Oil and Gas Industry Forum, and to start the Clean Energy Research Center Working Group, as well as the Renewable Energy Partnership[17] (also see Table 14.2). Both the U.S. and China consider energy efficiency and technology cooperation as common interests; for this reason, they are trying to standardize industrial specifications and regulations in technology between each other.

It is clear that cooperation related to environmental protection and energy covers a great proportion of the strategic partnership in the S&ED, while including both counter measures and preventive measures (see Table 14.3). One of the reasons is that the Obama administration tends to expand the rights and interests of his country by integrating the standards and regulations with China on technology such as clean energy and automobiles through Sino-U.S. cooperation (Sasaki 2011, p. 10).

[15] The full title is "Memorandum of Cooperation between the National Development and Reform Commission of the People's Republic of China and the Environmental Protection Agency of the United States of America to Build Capacity to Address Climate Change."

[16] The full title is "Protocol between the Department of Energy of the United States of America and the Ministry of Science and Technology and the National Energy Administration of the People's Republic of China for Cooperation on a Clean Energy Research Center."

[17] The partnership aims to promote the construction of a collaborative relationship between the two countries' public and private sectors. It also accelerates the application of clean energy through technologies and competitiveness that private companies possess in order to strengthen energy security and address climate change.

Table 14.3 Agreed measures under U.S.-China strategic partnerships

Approaches under the U.S.-China strategic partnerships	
Counter measures	Preventive measures
· Carbon capture and storage (CCS)	· Energy conservation and energy efficiency
· Clean water	· Renewable energy
· Clean air	· Clean energy technologies
· Natural resource conservation	· Electric vehicles
· Capacity building	· Emission/consumption reductions of HFCs
· Accurate and reliable observation of GHGs	

Measures agreed between the U.S. and China are included but not limited to those of that listed in the table
Source: author

Hillary Rodham Clinton, the Secretary of State, delivered a speech before attending the fifth East Asia Summit to emphasize the common strategic interests between the two countries in international agreements by stating "the two countries share the responsibility of constructing an obvious strategy in addressing climate change," when speaking regarding issues with China (Clinton 2010).

Chinese President Hu Jintao visited the U.S. in January 2011, and the subject of climate change was also taken up in a Joint Statement at the Sino-U.S. Summit. The two countries appreciated the cooperative stance and approaches that had been initiated, and agreed to support the Clean Energy Research Center, Renewable Energy Partnership, Joint Statement of Cooperation on Energy Security, TYF, The Cancun Agreement, and to continue to negotiate under the UN process (Ministry of Foreign Affairs of the People's Republic of China 2011). Later, the third S&ED was held in May 2011. Climate change, environmental protection and energy security were again central topics. At the third S&ED, the two countries signed six new EcoPartnerships under the TYF.[18]

In addition to new EcoPartnerships, the U.S. and China confirmed that the two countries share common goals on climate change, energy and environmental protection. For example, based on the Memorandum of Cooperation to Build Capacity to Address Climate Change, one of the urgent highlighted tasks is to strengthen the capacity of understanding regarding the exact amount of GHG emissions that China produces. The U.S. National Oceanic and Atmospheric Administration and the Chinese Meteorological Administration agreed to strengthen common research on developing the ability for accurate and reliable observation and understanding the nature of GHGs (U.S. Department of State 2011).

Recent partnerships are focused on the establishment of the U.S.-China Bilateral Working Group on Climate Change (CCWG) and joint effort to phase down the emission and consumption of hydro fluorocarbons (HFCs) (see Table 14.2). This is the first trial for both countries to cooperate on emission cut of GHGs. On the other

[18] Remarks at U.S.-China EcoPartnerships Signing Ceremony, U.S. Department of State, Washington D.C., May 10, 2011.

hand, the CCWG aims to catalyze cooperative efforts by playing a role to facilitate enhanced policy dialogue among stakeholders. So far, the working group has issued its first report to S&ED 5 in July, 2013, with recommendation of new action initiatives that should be launched for the next step in bilateral cooperation (Report of the CCWG to the Strategic and Economic Dialogue 2013).

Compared to Sino-Japan and Sino-Europe cooperation on environmental issues, the Sino-U.S. environmental cooperation is at a relatively early stage because the U.S. had not been providing official assistance to China in the field of environmental protection and energy. However, since climate change, as a complicated issue, can be partially mitigated from improvement of energy efficiency and systematic innovation, further cooperation between the U.S. and China is assumed to be rational. Although there seems to be deadlock in the international negotiation, Sino-U.S. environmental cooperation is not simply focusing on strengthening the exchange of information or energy-related technologies, but also aiming to build a strategic foundation for a breakthrough under the UNFCCC. The two countries have promoted the cooperation on climate change and energy through discussion amongst top political leaders in each country at a rapid pace, enhancing environmental cooperation from strategic positions.

As analyzed in the section, Sino-U.S. cooperation on climate change has been boosted to a significant degree by opening dialogues and carrying out various projects in recent years. However, these efforts have not been clearly reviewed in this chapter due to the reason that many of the partnerships/projects are currently in progress. Therefore, it might be necessary to further investigate Sino-U.S. cooperation through case studies, such as projects under TYF, operation of the Clean Energy Research Center, etc., in order to specifically indicate problems or limitations that the strategic partnerships have encountered through recent cooperation.

Conclusion

It is apparent that the cooperative progress between the U.S. and China on climate change has been accelerated in recent years. Both countries have become to treat climate change as a severe challenge to their environment, society, economy and energy security. The reasons can be considered as follows.

First of all, China has been experiencing rapid economic growth at the expense of environmental destruction, low energy efficiency and a huge outlay of GHG emissions. Secondly, as for the U.S. economy, it has problems of unemployment and stagnating economic growth. With this background, dealing with climate change provides a good opportunity for a breakthrough for China and the U.S. to move away from excessive dependence on fossil energy. And such activity will have a positive effect on the future economic growth of both countries, as they will be deeply involved with the conversion of energy structures and consumption. As a result, the current climate policies of both countries actually promote a long-term development strategy, instead of only a simple ambition to mitigate the climate change itself.

(continued)

In the analysis, it is understood that the central task of Sino-U.S. cooperation regarding climate change focuses on the enhancement of energy efficiency and the development of clean technology. Under the S&ED and related partnerships, China and the U.S. have been emphasizing mutual efforts of promoting innovation and technology transfer. Also, along with the enhancement of adaptive countermeasures, such as research and development of the CCS, the U.S. and China demonstrate the importance of preventive measures for mitigating the occurrence of climate change by supporting mutual cooperation on renewable/clean energy, electric vehicles, and so on. The cooperation has presented opportunities regarding the achievement of sustainable development without causing substantial contradiction to each other. And this is why climate change can be brought up as an appropriate subject in building a strategic collaborative relationship through mutual confidence and understanding between the two countries.

Meanwhile, after Copenhagen conference, the international negotiations for agreeing on post-2020 frame work, have made certain progress substantially in mitigation and adaptation (e.g. Parties pledge their GHGs emission reduction targets or actions and receive evaluation, that is, the acceptance of "pledge and review" approach), finance (e.g. the launch of the Green Climate Fund) and technology transfer (e.g. the constitution of the Climate Technology Centre and Network). It might be interesting to explore whether the Sino-U.S. strategic cooperative partnerships and dialogues have positively influenced the architecture of future climate regime under process of UNFCCC negotiations.

Also, it is crucial to understand the nature of Sino-U.S. strategic partnerships by continuously observing the progress. By doing so, we may come to realize if the Sino-U.S. partnerships on climate change have properly established and maintained in an authentic strategic manner, or are nothing more than empty promises or ad hoc solutions for climate deadlock. In any case, the U.S. and China both plays a significant role in policy-making of forging post-2020 climate agreement, it deserves continued and careful attention and observation.

References

Barnett J, Neil Adger W (2007) Climate change, human security and violent conflict. Polit Geogr 26(6):639–655

Busby J (2008) Who cares about the weather? Climate change and U.S. national security. Secur Stud 17(3):468–504

Bush GW (2007) State of the union addresses of the American presidents. Jan 23, 2007

China State Forestry Administration (2005) Desertification in China and its status report. June 2005

Chinese State Council (2007) China's national program against climate change. June 4, 2007

Clinton HR (2010) America's engagement in the Asia-Pacific. Speech of the U.S. Secretary of State, October 28, 2010

Committee on U.S.-China Cooperation on Electricity from Renewable Resources, Policy and Global Affairs, National Research Council, Chinese Academy of Sciences and Chinese Academy of Engineering (2010) The power of renewables: opportunities and challenges for China and the United States. The National Academies Press, Washington, DC

Han G, Hallding K, Carson M, Thai H, Nilsson M (2009) China-U.S. relations and domestic politics on the road to Copenhagen 1: making virtue of necessity. Stockholm Environment Institute, Stockholm

Hendrix CS, Glaser SM (2007) Trends and triggers: climate change and civil conflict in Sub-Saharan Africa. Polit Geogr 26(6):695–715

Holland S (2008) Obama says climate change a matter of national security. Reuters, December 9, 2008

Intergovernmental Panel on Climate Change (IPCC) (2007) Fourth assessment report (AR4), summary for policymakers (SPM)

International Energy Agency (2009) CO_2 emissions from fuel combustion, 2009th edn. OECD/IEA, Paris

Mao DS, Yu CG, Ming Z (2008) Research on anti-desertification and control in Gansu Minqin Oasis. Arid Zone Res 25(3):322

Miliband (ed) (2009) The road from Copenhagen. The Guardian, December 20, 2009

Ministry of Foreign Affairs of the People's Republic of China (2009) Sino-U.S. joint statement. November 17, 2009

Ministry of Foreign Affairs of the People's Republic of China (2011) Sino-U.S. joint statement. January 20, 2011

Moran D (ed) (2011) Climate change and national security: a country-level analysis. Georgetown University Press, Washington, DC

National Intelligence Council (2008) National intelligence assessment on the national security implications of global climate change to 2030. Archived from the original on 2013-06-14 http://www.dni.gov/testimonies/20080625_testimony.pdf

People's Daily (2010) China's oil production ranks 4th in world, only meets half domestic demand. People's Daily, February 9, 2010

Sasaki T (2011) Features of the Obama administration's economic strategy towards China. Int Trade Invest Q 83:10

Smith J, Lennon ATJ (2008) Setting the negotiation table: the race to replace Kyoto by 2012. In: Campbell KM (ed) Climatic cataclysm: the foreign policy and national security implications of climate change. Brookings Institution Press, Washington, DC, pp 205–207

U.S. Department of Defense (2010) The quadrennial defense review 2010. February 2010

U.S. Department of State (2011) U.S.-China strategic and economic dialogue 2011: outcomes of the strategic track. May 10, 2011

U.S. Department of State (2013) Report of the U.S.-China climate change working group to the strategic and economic dialogue. July 10, 2013

U.S. Department of State and U.S. Agency of International Development (USAID) (2010) The quadrennial diplomacy and development review. December 2010

United Nations Framework Convention on Climate Change (UNFCCC) (2011) UN decision 1/CP.16. The Cancun agreements: outcome of the work of the ad hoc working group on long-term cooperative action under the convention, FCCC/CP/2010/7/Add.1., March 15, 2011
U.S. White House (2010) National security strategy 2010. May 2010
Wurzel R, Connelly J (eds) (2010) The European Union as a leader in international climate change politics. Routledge, New York
Zhang H-B (2010) Climate change and China's national security. Shi shi chu ban she, Beijing

Chapter 15
Strengthening Climate Change Adaptation in Nepal: Needs and Perspectives

Bimal Raj Regmi, Cassandra Star, Apar Paudyal, and Ram Chandra Karki

Abstract Nepal is one of the countries most vulnerable to the impacts of climate change, due to its geographical fragility and socio-political circumstances. The effects of climate change are observed across regions, society and villages, mostly impacting the wellbeing of poor and marginalized households. This research explores the existing challenges of managing climate change adaptation in Nepal. The research is based on a case study of three different locations of Nepal, representing different landscapes. The findings show that the impact of climate change differs based on the socio-economic characteristics of households and communities. Poor and marginalized households seem to be more affected by the impacts of climate change than the rich and well off, because of their resource limitations. The findings also show that responses to and management of climate change at the local level are constrained by limited information and knowledge on climate change, inadequate access to technology and services and other governance challenges. The findings imply that local, national, regional and international collaboration is needed, to address the knowledge gap and issues related to financing and technology transfer in Nepal.

Introduction

The scientific evidence indicates that human activities are mostly responsible for changing the climate (IPCC 2007). The nature of climate change risk and vulnerability varies across regions and countries, over time, and depends on unique socio-

B.R. Regmi (✉) • C. Star
Faculty of Social and Behavioural Sciences, Flinders University, GPO Box 2100, Adelaide, SA 5001, Australia
e-mail: regm0003@flinders.edu.au; Cassandra.star@flinders.edu.au

A. Paudyal
Practical Action—Nepal, Narayan Gopal Sadak, Maharajgunj House no. 2677, Kathmandu, PO Box 15135, Nepal
e-mail: apar.paudyal@practicalaction.org.np

R.C. Karki
Department of Hydrology and Meteorology, Babarmahal, Kathmandu, GPO Box 406, Nepal
e-mail: rammertro@hotmail.com

W. Leal Filho (ed.), *Climate Change in the Asia-Pacific Region*, Climate Change Management, DOI 10.1007/978-3-319-14938-7_15

economic, biophysical and other conditions. Among the many Least Developed Countries (LDCs), Nepal is one of the most vulnerable to the impacts of climate change due to its fragile ecosystems and socio-political circumstances (NCVST 2009). Data from the climate risk index ranking of 1993–2012 indicates that Nepal is among the 20 most vulnerable countries (Sönke and Eckstein 2013). The effects of climate change are already being observed across society and villages in Nepal, mostly impacting the well being of poor and marginalized households. Several studies show that poor, women and marginalized households living in rural areas of Nepal are suffering from the negative impacts of climate change (Jones and Boyd 2011; Regmi and Bhandari 2012).

The vulnerability context of countries like Nepal emphasizes the need for effective adaptive measures to deal with climate change risks and impacts. Adaptation has emerged as a central area in climate change research, in country-level planning, and in implementation of climate change strategies (IPCC 2013). Nepal has in recent years made significant progress in devising policies and conceiving programmes on climate change adaptation (Helvitas 2011). However, the implementation of policies and plans remain largely uncertain due to lack of a clear strategy on how to implement them, and governance issues (Pant and Gautam 2013).

The literature shows that lack of information, knowledge management and technology transfer are major hurdles to promoting adaptation and climate resilience programmes in Nepal. Several studies have indicated the prevalence of a gap in knowledge and information on the impacts of climate change in Nepal. Bolch et al. (2012) argue that climate change projections in Nepal are speculative and incomplete due to poor understanding of climate change processes, combined with the diversity of climatic conditions and the extreme topography. In this case, undertaking multiple location-specific studies by government and Nepali stakeholders would enrich the knowledge and understanding of climate change in Nepal.

Therefore, this chapter is timely as it provides a multi-dimensional analysis of the climate change context in Nepal by outlining different factors that shape the climate change vulnerability and adaptive capacity of communities. It also adds value to the on-going debate on integrating climate change adaptation in development, by providing evidence of the associated links between the socio-economic conditions of households and their degree of exposure and vulnerability to climate change. In addition, it contributes to filling an important information gap by providing empirical case studies on the existing opportunities and challenges of managing climate change adaptation in Nepal.

This chapter specifically examines the existing state of climate change practices and the future need to strengthen climate change adaptation responses in Nepal. By providing an analysis of the opportunities and constraints of implementing mainstreaming initiatives, it contributes to an enriched understanding of how climate change adaptation responses should be designed and practiced. This research will therefore potentially assist decision makers to develop effective policy and governance mechanisms for strengthening climate change adaptation at the local level.

Methodology

Approach

The study was conducted in Nepal using a comparative case study methodological approach, in order to generate specific evidence on climate change in different geographical locations of Nepal. It also sought to enrich understanding in general of the complex issue of climate change. Case studies emphasize detailed contextual analysis in specific locations or conditions and their relationships. They also allow an empirical examination of the theoretical hypothesis of the research (Yin 1994).

Research Sites and Participants

The study involved in-depth research in six different geographical locations in Nepal. The districts selected cover three major ecological belts: high hill, mid hills and terai, representing different altitudinal zones of Nepal. Three major development regions of Nepal were included, namely the central, western and mid-western regions, in order to capture different vulnerability contexts. Within the development region, one district was selected based on the National Adaptation Programme of Action (NAPA) vulnerability criteria (refer to Table 15.1). Likewise, in each district, one Village Development Committee (VDC) was selected. The selection of the VDCs was carried out based on the available socio-economic and disaster data. The VDCs selected for this research are among the most vulnerable in the districts because they faced severe climate change impacts and had low coping capacity.

The participants in the research included policy makers, practitioners and communities in the selected research locations. A total of 17 policy makers, 28 practitioners and 288 households were identified to participate. Purposive sampling was used in order to capture the experiences of policy makers and practitioners in

Table 15.1 Research sites and selection criteria

VDCs selected	Districts	Ecological region	Major climate change disasters reported in the districts	Categorization of vulnerability by national adaptation programme of action (NAPA)
Ramche and Bhorle	Rasuwa	High hill	Landslide	Moderate
Dhungegadi and Bangesaal	Pyuthan	Mid hill	Drought	Low
Betahani and Holiya	Banke	Terai	Flooding	Very low

Table 15.2 Showing the features of the research stations that was used for analysing climate data

Station name	District	Index no.	Type of station	Lat (degree/min)	Long (degree/min)	Elevation (m)	Remarks
Nepalgunj	Banke	0416	Climatology	28/04	81/37	144	Plains
Bijuwartar	Pyuthan	0505	Precipitation	28/06	82/52	823	Mid hills
Ghorai	Dang	0515	Synoptic	28/03	82/30	634	Mid hills
Dhunche	Rasuwa	1055	Climatology	28/06	85/18	1,982	Mountains

climate change. Of the total 288 households, there were 96 households for each district. Stratified simple random sampling was employed to best represent the socio-economic diversity of households and communities. These sampling strategies helped the researchers to best identify different categories of households for the research process.

Research Methods and Analysis

The study employed mixed approaches to data gathering. Participatory data gathering methods such as interviews and focus group discussions were used. Climate trends were analyzed using data collected between 1981 and 2010 from four reference stations located in the terai plains, mid hills and mountain areas of Nepal (Table 15.2). For all reference station data, Mann-Kendall (non-parametric test) and the least square fitting statistical analyses were utilized to test the significance of trend in time series. Trends were also identified using the least square fitting method and Sen's slope method. These tests were applied because they are simple, robust and can cope with missing values and values below a detection limit (Gilbert 1987).

The presence of a statistically significant trend was evaluated using the Z value of normal distribution. A positive (negative) value of Z indicates an upward (downward) trend. In this analysis, if the value of Z was greater than 1.96 or less than -1.96, then it was considered a significant trend at the 5 % level. In addition to this non-parametric test, the simple least square fitting method was also utilized to identify trends and their significance based on R2 values.

Limitations of the Study

The analysis undertaken in this chapter has some limitations to be taken into account when reviewing the findings. The first limitation is related to the availability of climate data; a limitation often encountered with data holdings in the developing world (IPCC 2013). In analyzing the trends, data gaps were noted.

For the Nepalgunj station, data was missing for only 1 or 2 years over the entire period. As there were some gaps in data on temperature at the Bijuwartar station in Pyuthan, data was taken at a proxy station from nearby Ghorai (Dang) for both the temperature and rainfall analyses. There were some data on continuous temperature and rainfall missing at the Dhunche and Ghoari stations respectively and this impacted the data analysis. The second limitation concerns the focus of the study. The study was conducted in Nepal and focused on a few districts and VDCs of Nepal.

Findings

Temperature and Rainfall Trend

The results showed a decreasing trend in average maximum and minimum temperature in mountain areas compared to the mid hills and terai plains. This result contrasts with a general observation of an increasing trend in temperature with a vertical rise in topography. This may be essentially due to gaps in continuous data; primarily the lack of temperature data on the Rasuwa district. However increasing trends in mean temperature were observed from the terai plains to the mid hills.

In Nepalgung station, the temperature rise trend was highest in the Pre-Monsoon season (0.04 °C/year) and lowest with a slightly negative trend during the winter season (−0.01 °C/year). The trends during the monsoon and post-monsoon seasons were 0.002 and 0.004 °C/year respectively. At the Ghorai station, where data were available from 1989 to 2010, the mean annual maximum temperature showed a rising trend of 0.06 °C/year. The rising trend was highest (0.12 °C/year) during the winter season and lowest (0.04 °C/year) during the monsoon season. In Dhunche, there was a sharp decline in mean annual maximum and minimum temperatures by 0.02 and 0.05 °C/year (Table 15.3).

Table 15.3 Temperature change trend by stations

Temperature parameters	Temperature change (°C/year)		
	Nepalgunj	Ghorai	Dhunche
Mean annual maximum temp	0.01	0.06	−0.02
Mean pre-monsoon maximum temp	0.04	0.05	−0.02
Mean monsoon m maximum temp	0.002	0.04	0.11
mean post-monsoon maximum temp	0.004	0.05	−0.04
Mean winter season maximum temp	−0.01	0.12	0.018
Mean annual minimum temp	0.02	0.01	−0.05
Mean pre-monsoon minimum temp	0.05	0.04	−0.1
Mean monsoon minimum temp	0.01	−0.03	−0.04
mean post-monsoon minimum temp	0.04	0.02	0.02
Mean winter minimum temp	0.03	0.01	−0.04

For all stations, the temperature trend was identified using both Sen's Slope method and the least square fitting method, and the significance of trends was tested with the Mann Kendall test as well as the R2 value in least square fit. The trend value obtained from both the least square fitting and Sen's slope methods did not differ greatly. Temperature trends were not consistently significant. In Nepalgunj, the rising trend in mean annual minimum air temperature was significant, but the trend for the mean annual maximum air temperature was non-significant. While in Ghorai, the significance of the temperature increase was reversed for two temperature parameters. At the Dhunche station there were non-significant trends for all the temperature parameters across all seasons (Table 15.4). The variation might be due to the limited availability of data from this station over the entire period.

In terms of rainfall, the results indicate a high variability in rainfall trends between the terai plains, mid hills and mountains. There was a sharp decline in rainfall in the terai plains compared to other parts of Nepal. A decrease in winter and post-monsoon season rainfall was observed, which concurs with the responses of communities. The significance of the temperature trend was different from season to season and from maximum and minimum temperature parameters.

The annual rainfall total shows a large inter-annual variability, with the value ranging from 1,000 to 2,000 mm. At Nepalgunj station, the annual rainfall total and rainfall total for all seasons showed a decreasing trend. The decreasing trend in total annual rainfall was about 14.7 mm/year. In the mid hills, Ghorai station had a slight increase in total annual rainfall (5 mm/year), however in Bijuwartar there was a decreasing trend of 9.6 mm/year. In comparison to Ghorai, there was a decreasing trend for all seasonal rainfalls in Bijuwartar. However, in Dhunche, the total annual rainfall displayed an increasing trend (1.6 mm/year), and by season the highest increasing trend was observed around the monsoon (20.9 mm/year). However the pre-monsoon, post-monsoon and winter rainfalls showed a decreasing trend by 4.31 mm/year, 6 mm/year and 3.5 mm/year respectively (refer to Table 15.5).

Rainfall trends were analyzed using both Sen's Slope method and the least square fitting method, and the significance of trends were tested with the Mann Kendall test as well as the R2 value in least square fit. The findings show that during the winter and post-monsoon seasons, all four stations showed a decreasing trend. This indicates a decrease in winter and post-monsoon season rainfalls over large areas of Nepal. Relatively, the significance of the rainfall trend is very low (Table 15.6).

Analysis of temperatures and rainfalls at all the three stations clearly shows that climate change is occurring in the study area. There was a sharp increase in temperature and frequent variability in rainfall observed at all three sites. In comparison with the rainfall data, there were more significant trends in temperature parameters. The perception of communities also shows that more than 98 % of respondents across the three regions had experienced increasing temperature and a large variability in rainfall patterns.

Table 15.4 Analysis of station temperature trends using Mann-Kendall with Sen's slope method

Station name	Time period	Season	Max air temperature Z value	Sen's slope (change per year)	Significance of trend	Min air temperature Z value	Sen's slope (change per year)	Significance of trend
Nepalgunj	1981–2010	Monsoon	0.04	0.003	No	1.87	0.01	No
		Post-monsoon	0.45	0.006	No	2.70	0.04	Yes
		Winter	−0.38	−0.008	No	1.8	0.03	No
		Pre-monsoon	2.02	0.05	Yes	2.69	0.05	Yes
		Annual	1.25	0.01	No	2.21	0.02	Yes
Ghorai	1981–2010	Monsoon	1.63	0.05	No	−1.54	−0.02	No
		Post-monsoon	2.35	0.05	Yes	0.51	0.02	No
		Winter	2.99	0.1	Yes	0.58	0.02	No
		Pre-monsoon	0.67	0.05	No	0.51	0.03	No
		Annual	2.72	0.06	Yes	0.42	0.01	No
Dhunche	1981–2010	Monsoon	2.1	0.09	Yes	−0.85	−0.04	No
		Post-monsoon	−0.4	−0.04	No	0.7	0.02	No
		Pre-monsoon	−0.27	−0.01	No	−0.96	−0.12	No
		Winter	0.18	0.02	No	0.7	0.04	No
		Annual	−0.17	−0.01	No	−1.17	−0.06	No

Table 15.5 Rainfall change trend by station

Rainfall parameters	Precipitation (mm/year)			
	Nepalgunj	Ghorai	Bijuwartar	Dhunche
Total annual rainfall	−14.787	5.0094	−9.6523	1.66
Total winter rainfall	−1.5011	−1.1973	−0.2	−3.5
Total pre-monsoon rainfall	−0.6591	2.7996	−0.87	−4.31
Total monsoon rainfall	−11.594	3.6438	−5.3	20.94
Total post-monsoon	−1.0736	−0.0323	−1.12	−6.07

Impact of Climate Change on Different Livelihood Resources

The evidence shows that the sectors most impacted by climate change in the regions are agriculture and water. The losses of major agriculture crops and production decline have been major issues at the study sites. The farmers in the Pyuthan, Rasuwa and Banke districts mentioned losing crops of the traditional varieties of rice due to rainfall variability. According to the farmers, these local varieties need continuous rainfall for 4 months in order to mature and yield. Farmers at all the three sites also reported a declining yield of barley, wheat and legumes (see Table 15.7). They indicated that the seedling period had changed, affecting the production yield. This supports the research finding of Devkota et al. (2013) that seeding rice before and after the optimal seeding dates reduces crop yield and yield stability significantly because of spikelet sterility induced by both high and low temperatures.

Climate change had also visibly impacted forests and biodiversity. A shifting of vegetation patterns and crops is evident at higher altitudes in Nepal. The field evidence from the Rasuwa district indicates that a succession of the vegetation had been observed in the Thulo Syabru area. People indicated that rhododendron and juniper are now appearing at higher altitudes than their normal distribution range. Similarly, communities are experiencing changes in the plant behavior and flowering season of rhododendron. A recent study carried out in the Western Himalayan eco region by Joshi et al. (2012) showed that with increasing temperature, a shift to higher altitudes in all forest types was observed. A research study in India reveals that the suitable bioclimatic envelope for rhododendron has shrunk considerably under the envisaged climate change scenario (Kumar 2012).

Climate change is also a major threat to water resources and the hydrological cycle. The global circulation climate model also suggests that climate change will impact total flows, seasonal runoff, high- and low-flow conditions, and surface-groundwater interactions (Manandhar et al. 2012). Climate change was also impacting water resources in the study areas. Communities at the research sites perceived that depletion of water resources had impacted them directly. Villagers in the Rasuwa and Pyuthan districts expressed that the drying up of springs and rivers was a major concern for them, reporting that almost 50 % of the local springs and water sources had dried up in their villages. Likewise, communities in the Banke

Table 15.6 Analysis of station rainfall trends using Mann-Kendall with Sen's slope method

Season	Pre-monsoon			Post-monsoon			Winter			Monsoon		
Station name	Z	Sen's slope	Significance	Z	Sen's slope	Significance	Z	Sen's slope	Significance	Z	Sen's slope	Significance
Nepalgunj	−0.52	−0.74	No	−1.31	−1.13	No	−1.69	−1.28	No	−1.84	−11.98	No
Bijuwartar	−0.89	−1.3	No	−1.18	−0.88	No	−0.27	−0.31	No	−0.99	−4.69	No
Dhunche	−0.1	−0.42	No	−2.29	−4.64	Yes	−2.03	−3.47	Yes	2.14	22.2	Yes
Ghorai	0.9	1.8	No	−0.05	−0.17	No	−0.98	−0.91	No	0.66	7.40	No

Table 15.7 Impacts of climate change on agriculture

District	Impact on agriculture	
	Cereals (loss of species)	Legumes (declining yield)
Rasuwa	Local landraces/varieties of rice—(Marsi, Bhangari, RatoDhan, Kalokathe, Bunglange, Sikrimarshi)	Bhatmas (soybean), Bodi (beans), Rice bean, Gahat (horsegram), Kheseri and Mas (blackgram)
Pyuthan	Local rice varieties (Simtharo, Marsi, Hasi)	Wheat, Lentil, Peas
Banke	Local rice varieties (Anadi, Dhunmuniya, Masuli, Shyamjira, Tilki, Barma, Sungapankhi, Deruwa, Rudhan)	Wheat, Lentil, Chickpea, Mustard

district reported excessive water and flooding in the rainy season and a massive decline in the water table during the winter.

The evidence presented in this section clearly indicates that climate change is already impacting upon major livelihood sectors such as the agriculture, biodiversity and water sectors. Other areas may also be impacted by climate change impacts, such as health and migration. However, due to lack of access to data, the total picture of climate change impacts is difficult to ascertain.

Socio-economic Context

This section of the paper provides a brief summary of the socio economic contexts of the study sites, arguing that vulnerability at the local level is high because climate change is having a greater impact relative to the existing coping and adaptive capacity.

The findings show that the lack of access to education for women leaves them more vulnerable to climate change impacts compared to men. The mean education ration in the study VDCs is 51.7 %, which is lower than the national average (CBS 2011). In comparison with males, the females have a much lower educational status. The household survey shows that only 22.89 % of the female populations within the six study VDCs are literate. There is variation within VDCs; in the Ramche VDC only 5.4 % of females are literate, whereas the Bangesaal VDC of Pyuthan district has the highest female literacy rate of 35.1 % (refer to Table 15.8). These data show that the women in the study districts are less privileged than men in terms of access to education and other support services.

The majority of the households in the study areas are dependent on climate-sensitive sectors such as agriculture. The survey data show that 79 % of the total population in the study VDCs is dependent on agriculture for their livelihood. The dependency is highest in the Bhorle VDC, where 93 % of the population is dependent on agriculture. Only 67 % of the population in the Holiya VDC of the Banke district is dependent on agriculture, as there are alternative livelihood sources in this district compared to Rasuwa. It should be noted that the average

Table 15.8 Socio-economic features of the study sites

VDCs	Education (%)		Food sufficiency (month/year)	Land size (ha)	Income per annum (rupees)	Agriculture dependency (%)
	Male	Female				
Bhorle	46.10	24.40	5.00	0.35	50,145.00	93.00
Ramche	14.30	5.40	5.50	0.32	44,816.00	82.00
Dhugegadi	57.00	43.00	9.00	0.73	120,000.00	70.00
Bangesaal	64.90	35.10	7.00	0.71	132,000.00	74.00
Bethani	33.52	24.60	5.40	0.67	74,699.00	88.00
Holiya	26.01	4.84	5.80	1.26	78,600.00	67.00

Table 15.9 Human development index of the study areas

District	HDI	GDI	HPI	HEI	Below poverty line (%)
Rasuwa	0.39	0.376	54.5	0.439	46
Pyuthan	0.45	47.9	0.41	0.01	38
Banke	0.46	34.4	0.579	0.01	40

Source: DDC-Banke (2004), DDC-Pyuthan (2004), DDC-Rasuwa (2004), UNDP (2004)

household size in the study area is 6.5, which is higher than the national average. The findings indicate that the large household size in the study areas has placed enormous pressure on the limited land for food supply and production.

Access to land is one of the major issues for food security. The average land size is 0.67 ha in the study areas. The Rasuwa district has only 0.33 ha whereas the Banke district has an average of 0.97 ha. The average food sufficiency status was 6.28 months indicating that a household can have sufficient food for only 6 months. In the Rasuwa and Banke districts, 20–30 % of the population has food sufficiency, while the rest of the population has to either consume less or seek alternatives. Comparatively, households in the Rasuwa and Banke districts have a lower annual average income (refer to Table 15.8). Likewise, according to the national census data, 15.7 % of Nepali households do not have adequate food to eat (CBS 2011).

An assessment of the socio-economic conditions of the populations at each of the study sites reveals that most of the households live below the poverty line and are food insecure. Based on UNDP (2004) data with regard to the Human Development Index (HDI), Gender Development Index (GDI), Human Poverty Index (HDI) and Human Empowerment Index (HEI), the Rasuwa district can be considered the most vulnerable. Rasuwa has a HDI of 0.39, which is lower than the national average of 0.463. Further, 46 % of the population in Rasuwa is living below the poverty line, i.e., under one dollar of income per day. The HDI and poverty ratios in the other study districts are also very low (Table 15.9).

The findings show that the poor and women suffer most from the impact of climate change. The majority of the respondents (92 %) with a low income perceive that they are experiencing significant impacts of climate change on their water resources compared to respondents with a better income. During focus group

Table 15.10 Impact of floods on the well being of different categories of respondents

Impact of loss	Non-poor (household %)	Poor (household %)	Total (household %)
Male	10.5	3.3	5.0
Female	15.8	19.7	18.8
Children	21.1	18.0	18.8
All	52.6	59.0	57.5
Total	100.0	100.0	100.0

Source: Interview with households, Banke

Table 15.11 Impact of climate change on women households

	Year	
Time needed to collect water (min/day)	2009 (%)	2012 (%)
Within house or no time needed	19.1	10.0
Less than 15 min	50.1	27.1
15–30 min	18.0	20.6
More than 30 min	22.8	42.2

Source: VDC-Dhungegadi (2009)

discussions, the participants shared that more than 70 % of the victims of flooding and landslides were from poor households. For example, ward number 5 of the Bangesaal VDC reported that the outbreak of water-borne diseases (cholera and diarrhea) in 2008/2009 took the life of two people belonging to poor families. Similarly, the household survey of the Banke district indicated that children and females are equally vulnerable to the effects of floods. Moreover, poor females are more vulnerable than non-poor females (Table 15.10).

Women's vulnerability to climate change and variability is strongly influenced by their socioeconomic status and gender roles (Figueiredo and Perkins 2012). A comparison of the data generated from the focus group discussion and secondary sources in the Pyuthan district indicate that between 2009 and 2012, there was a reduction in local water availability. In recent years, 42 % of the women have had to travel more than 30 min to collect water compared to only 23 % in 2009 (Table 15.11). Similarly, according to the majority of the female interview respondents in the Banke district, their workload has also been increased with intensive agriculture practices required due to the impact of climate extremes like drought and land degradation.

Capacity to Cope and Adapt

This section discusses the existing capacity of communities to cope with and adapt to climate change in the study areas. The field information shows that there are limited traditional coping practices available to deal with climate change at the

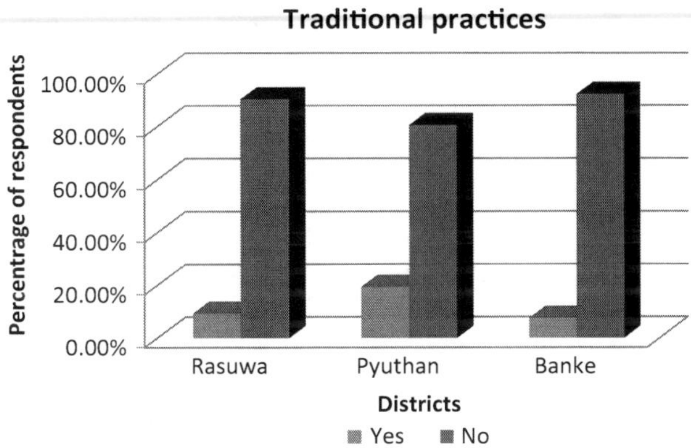

Fig. 15.1 Household perception on the use of coping practices

local level. The household survey indicates that more than two thirds of the respondents (above 80 %) in all the districts lack sufficient traditional practices to deal with climate change. This is because they have limited access to knowledge and technology on climate change adaptation. Some households have adopted very few local practices. For example, at the time of the study, in the Banke district less than 5 % of non-poor households had initiated construction of physical infrastructure such as drainage canals or water storage tanks around their house/land. Likewise, only around 10 % of households had opted to change their crop patterns to include mixed crops, or changed their cropping system or fertilizer (Fig. 15.1).

It was found that most of the traditional practices adopted by the communities are ineffective to deal with climate disasters. The household survey of all three districts revealed that more than two thirds of respondents argued that their practices are ineffective and inadequate to mitigate climate change impacts, often due to the scale and magnitude of the impacts. For example the use of vegetative check dams proved ineffective during massive flooding in the Holiya and Bethani VDCs. Almost 89.1 % of the respondents in the Bangesaal VDC of Pyuthan district perceived that the existing adaptation options were ineffective because they lacked information, knowledge and technology on dealing with climate change risks and impacts (refer to Fig. 15.2).

The findings reveal that the external support from the government and non-government service providers on climate change was limited. More than 90 % of respondents in the Rasuwa district indicated that they had not received any support from the government to deal with climate change, and more than 50 % of respondents in the Pyuthan and Banke districts perceived that there was no support. For example, around 32 % of the households in the Banke district perceived that the government was doing nothing to mitigate the vulnerability of the flood impacts, whereas around 27 % of households responded that they did not know anything about the governments' initiatives on climate change.

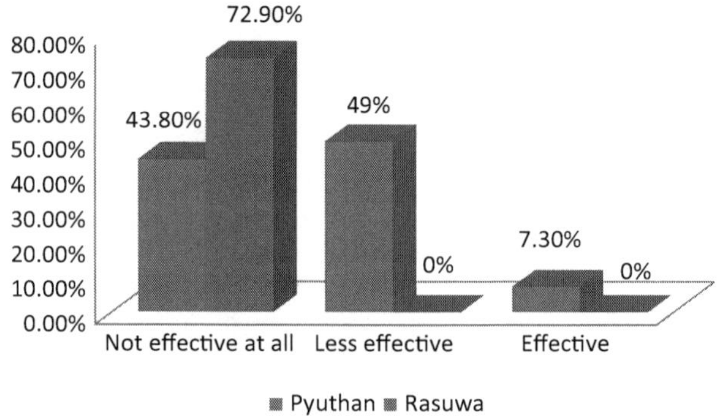

Fig. 15.2 Household perception on effectiveness of current traditional practices

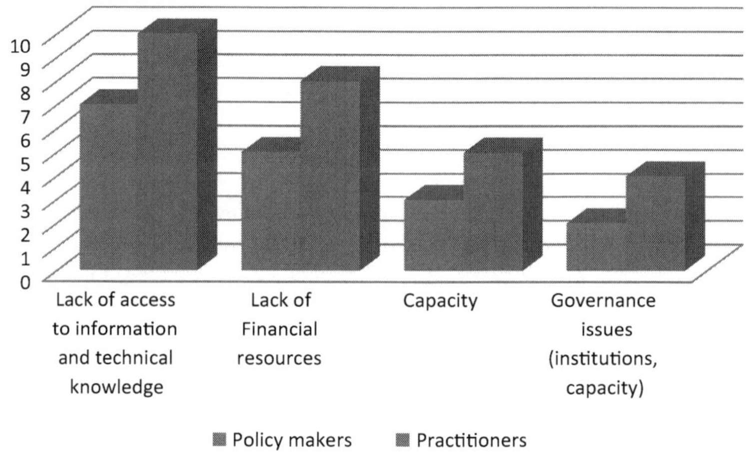

Fig. 15.3 Perception of respondents on challenges in promoting climate change adaptation

Challenges to Implementing an Adaptation Programme

There are gaps in information, knowledge and technology on climate change at both local and national levels. The interviews with communities in the Rasuwa and Pyuthan districts revealed that more than 80 % of the respondents were not aware of climate change and related issues. Likewise, at the local level the government and NGO practitioners also revealed that they had limited knowledge on climate change and that technology transfer is not occurring. Most of the policy makers and practitioners revealed a lack of access to knowledge and technology as one of the major barriers preventing them from taking affirmative action on climate change. The policy makers and practitioners also identified that a lack of financial resources and capacity, as well as governance issues, are challenging progress to promote climate change adaptation in Nepal (see Fig. 15.3).

Another challenge in promoting climate change adaptation at the local level concerns governance and institutions. According to the majority of the practitioner and community respondents, the local government is weak, has insufficient capacity and is under resourced. Government respondents reported a lack of information and knowledge on climate change as a major obstacle in decision making. In addition, the political instability and lack of a local election in the last 15 years has made the local government weak and fragile. The local government officials in an interview revealed that the current capacity and resources of the local government are insufficient to support additional services for climate change adaptation.

Discussion

The findings of this study indicate that climate change will make the achievement of development goals in Nepal more complicated by adding an extra burden to the poor, marginalized and vulnerable households. The data clearly show that communities in all three geographical locations studied have a high degree of exposure and sensitivity to climate change and a low adaptive capacity to deal with climate extremes. It was found that the poorest and most excluded were the most vulnerable to the impacts of climate change, both because of their high dependency on climate-sensitive resources and their lack of access to material, social, political and economic resources.

One of the major constraints to climate change adaptation for the districts identified in this research is a lack of information and data on climate change. There were constraints observed in this study in accessing meteorological data which affected the analysis. In addition, there are issues with the availability and access of districts to scientific and meteorological data related to climatic variables. This can be attributed in part to the relatively short length of records, about 30 years (Shrestha and Aryal 2011). Similarly, the precipitation and temperature data, used earlier, are constrained by locality since they were taken at a different altitude, aspects and geographical location from the study area. Although local information is useful in understanding local contexts and climate change impacts (Berkes and Jolly 2002; Byg and Salick 2009), it is limited by lack of access to technology and specific knowledge required to deal with climate extremes.

The findings in this research support earlier findings which indicate that unsustainable development has resulted in poor and inequitable distribution of adaptive capacity (Yohe et al. 2006). Traditional responses and the existing local governance structure alone are not sufficient to address contextual vulnerability. Existing resources and practices often fail to address the uncertainties and scale of climate change impacts. Yamin et al. (2005, p. 9) also argue that the complexities of climate change necessitate addressing the *structural* causes of vulnerability that cannot be dealt with in a piecemeal, project-by-project fashion.

The findings in this paper demonstrate that, to deal with climate extremes, local responses must be supported with technology transfer and knowledge and skills.

The findings support the notion that effective adaptation measures in development-deficient situations require 'transformational' approaches that necessitate adaptive co-management and joined-up actions. Co-management can be considered a knowledge partnership between organizations, from local to international (Berkes 2009). In addition, joint working or a 'joined-up' government strategy is useful in bringing together a number of public, private and voluntary sector bodies to work across organizational boundaries towards a common goal (Kavanagh and Richards 2001; Ling 2002; Wilkins 2002).

The most suitable adaptation response strategy specific to Nepal and other LDCs, as argued earlier, is to adopt an adaptive co-management approach where the government and stakeholders identify a common local and national-level mainstreaming strategy for knowledge management, resource mobilization and institutional development. The evidence from this research argues that a governance structure that is inclusive and owned by multiple stakeholders has the potential to overcome institutional, technological and financial barriers.

Conclusion

This study aimed to explore the existing challenges of managing climate change in Nepal.

The findings show that climate change has emerged as an additional burden to the existing problems of socio-economic underdevelopment, inequality and lack of access to sufficient services at the study sites. The weak socio-cultural context necessitates that interventions consider adaptation and development together in order to address both poverty and vulnerability. It also implies that stand-alone climate change adaptation interventions are not sufficient for addressing the root causes of vulnerability, because a fragmented approach will further marginalize households.

This chapter has identified several challenges for implementing climate change adaptation responses in Nepal and other LDCs. The gaps in information, knowledge, technology and financial resources are already undermining the local and national-level responses on climate change adaptation. In particular, the findings showed that adaptation responses at the local level are currently ineffective because of limitations within local knowledge and practices to deal with climate extremes as well as inadequacies in government and NGO services and support for the local communities.

The findings clearly demonstrate that, in order to strengthen climate change adaptation in Nepal, an integrated interventions and technological support might be required at different levels, which could include the individual and collective responsibility of the international community, national-level government, development agencies and the local communities. Specifically, there is a need to place a greater emphasis on information and knowledge management that can help poor and vulnerable households to

(continued)

access information and technology to adapt to the adverse impacts of climate change. This could happen if local, national, regional and international actors and agencies collaborate with each other in devising policies and establishing knowledge and technology sharing mechanisms and systems. This highlights an urgent need to identify tools and methods to practically facilitate these collaborations in Nepal and other LDCs.

References

Berkes F (2009) Evolution of co-management: role of knowledge generation, bridging organizations and social learning. J Environ Manag 90(5):1692–1702

Berkes F, Jolly D (2002) Adapting to climate change: social-ecological resilience in a Canadian western Arctic community. Conserv Ecol 5(2):18

Bolch T, Kulkarni A, Kääb A, Huggel C, Paul F, Cogley JG, Frey H, Kargel JS, Fujita K, Scheel M (2012) The state and fate of Himalayan glaciers. Science 336(6079):310–314

Byg A, Salick J (2009) Local perspectives on a global phenomenon—climate change in Eastern Tibetan villages. Glob Environ Chang 19(2):156–166

CBS (2011) Preliminary findings of the national census 2011. Nepal Centre Bureau of Statistics, Kathmandu

Intergovernmental Panel on Climate Change-IPCC (2013) Working Group I contribution to the IPCC fifth assessment report climate change 2013: the physical science basis summary for policymakers. IPCC WGI AR5, Intergovernmental Panel on Climate Change, Sweden

DDC-Banke (2004) District profile Banke. District Development Committee (DDC), Nepalgunj

DDC-Pyuthan (2004) District profile Khalanga. District Development Committee (DDC), Pyuthan

DDC-Rasuwa (2004) District profile Rasuwa. District Development Committee (DDC), Dhunche, Rasuwa

Devkota K, Manschadi A, Devkota M, LaLamers JPA, Ruzibaev E, Egamberdiev O, Amiri E, Vlek PLB (2013) Simulating the impact of climate change on rice phenology and grain yield in irrigated drylands of Central Asia. J Appl Meteorol Climatol 52(9):2033–2059

Figueiredo P, Perkins PE (2012) Women and water management in times of climate change: participatory and inclusive processes. J Clean Prod 60:1–7

Gilbert RO (1987) Statistical methods for environmental pollution monitoring. Wiley, New York, NY

Helvitas (2011) Nepal's climate change policies and plans: local communities' perspective. Nepal Helvitas-Swiss Intercorporation, Kathmandu

Intergovernmental Panel on Climate Change-IPCC (2007) Climate change 2007: impacts, adaptation and vulnerability. Contribution of the Working Group II to the fourth assessment report. Intergovernmental Panel on Climate Change, Sweden

Jones L, Boyd E (2011) Exploring social barriers to adaptation: insights from Western Nepal. Glob Environ Chang 21(1):1262–1274

Joshi P, Rawat A, Narula S, Sinha V (2012) Assessing impact of climate change on forest cover type shifts in Western Himalayan Eco-region. J For Res 23(1):75–80

Kavanagh D, Richards D (2001) Departmentalism and joined-up government. Parliam Aff 54 (1):1–18

Kumar P (2012) Assessment of impact of climate change on Rhododendrons in Sikkim Himalayas using Maxent modelling: limitations and challenges. Biodivers Conserv 21(5):1251–1266

Ling T (2002) Delivering joined–up government in the UK: dimensions, issues and problems. Public Adm 80(4):615–642

Manandhar S, Pandey VP, Ishidaira H, Kazama F (2012) Perturbation study of climate change impacts in a snow-fed river basin. Hydrol Process 27(24):3461–3474

NCVST (2009) Vulnerability through the eyes of vulnerable: climate change induced uncertainties and Nepal's development predicaments. Institute of Social and Economic Transition (ISET), Kathmandu

Pant D, Gautam K (2013) Policy provisions and local response on climate change adaptation in Nepal. Regional climate change adaptation knowledge platform for Asia. Stockholm Environment Institute, Bangkok

Regmi B, Bhandari D (2012) Climate change governance and funding dilemma in Nepal. TMC Acad J 7(1):40–55

Shrestha AB, Aryal R (2011) Climate change in Nepal and its impact on Himalayan glaciers. Reg Environ Chang 11(1):65–77

Sönke K, Eckstein D (2013) Global climate risk index 2014: who suffers most from extreme weather events? Weather-related loss events in 2012 and 1993 to 2012. Germanwatch, Berlin

UNDP (2004) Human development report (HDR) Nepal. United Nations Development Programme, Kathmandu

Village Development Committee (VFCC)-Dhugegadi (2009) Local adaptation plan of action (LAPA). Village Forest Coordination Committee (VFCC)-Dhugegadi, Pyuthan

Wilkins P (2002) Accountability and joined-up government. Aust J Public Adm 61(1):114–119

Yamin F, Rahman A, Huq S (2005) Vulnerability, adaptation and climate disasters: a conceptual overview. IDS Bull 36(4):1

Yin RK (1994) Case study research: design and methods. Sage, Los Angeles

Yohe G, Lasco R, Ahmad QK, Cohen S, Janetos T, Perez R, Ebi K, Lankao PR, Malone E, Malone T (2006) Perspectives on climate change and sustainability. Change 25(48):49

Chapter 16
Climate Change Policy and Sustainable Energy Development in Fiji: Implications to Pacific Island States

Anirudh Singh and Atesh Gosai

Abstract Fiji's National Climate Change Policy (FNCCP), established in 2012, encourages the government to take up climate change mitigation initiatives, as part of Fiji's contribution to global efforts to reduce greenhouse gas emissions.

Energy use is one of the largest contributors to global warming, and the substitution of fossil fuel with renewable energy, coupled with energy efficiency measures, provide the most effective way for mitigating climate change. It would therefore be expected that the country's energy development plans would bear strong synergies with the objectives of the National Climate Change Policy.

This chapter presents a brief overview of Fiji's National Climate Change Policy, followed by a review of the recent energy developments in the country and its new National Energy Policy (NEP). It is found that there is a clear mismatch between the NEP and the NCCP on the issue of energy considerations with regard to climate change mitigation. Indeed there is no mention in the NEP of the existence of the nation's National Climate Change Policy and its commitments to the global climate change mitigation efforts through energy efficiency and the use of renewable energy.

Plausible reasons for the absence of a linkage between the two policies are put forward, and possible methods to redress the problem are suggested.

Introduction

Most PICs have established National Climate Change Policies (NCCPs) to fight the adverse effects of climate change. At the same time, these nations have also developed National Energy Policies (NEPs) to oversee the development of their energy needs.

It is well known that the primary means of reducing greenhouse gases (GHGs) and thus mitigating climate change is through the reduction of the use of fossil fuels

A. Singh (✉) • A. Gosai
School of Engineering and Physics, Faculty of Science, Technology and Environment, The University of the South Pacific, Laucala Campus, Suva, Fiji
e-mail: anirudh.singh@usp.ac.fj; gosai.atesh@gmail.com

W. Leal Filho (ed.), *Climate Change in the Asia-Pacific Region*, Climate Change Management, DOI 10.1007/978-3-319-14938-7_16

and the more efficient use of energy. As both these processes relate to energy use, their implementation will be governed by the nation's NEP. One thus expects a linkage between the respective NCCPs and NEPs of all nations. It is interesting to discover how this occurs in the case of the Pacific Island Countries (PICs).

The PICs are faced by similar energy challenges, arising, amongst other things, from their lack of fossil fuel resources, their remoteness and the absence of economically favourable fuel supply chains. These island nations have developed policies at both the regional and national levels with a view to improving their energy status (Singh et al. 2013). These policies have been driven primarily by the need to ensure sustainable energy for all, including those in remote and rural areas of the country.

Fiji, with a population of 850,000 and a land area of 18,000 km^2 shared amongst its 300 islands (consisting of the two main islands and the outer islands), is one of the more developed PICs. But its varied geology of volcanic islands interspersed with coral atolls means that the country is perennially faced with energy challenges that include the full range of energy issues encountered by the other PICs. These include

1. Impracticality of large grid supply due to the small size of an island population
2. Absence of indigenous fuel resources
3. Fuel supply issues due to the remoteness of an outer island
4. Lack of science and technology infrastructure to develop and produce energy technologies (including renewable energy technologies) locally.

Fiji thus provides a representative case study for the issues facing the sustainable energy development of the PICs.

This chapter begins with an overview of Fiji's National Climate Change Policy (FNCCP 2012), and considers its implications on the country's national energy development plans. A review is presented of Fiji's current status of energy development, and the chapter next examines the current National Energy Policy of the country. Relationships between the two policies are then critically examined especially with respect to the global climate change mitigation efforts.

The chapter finds that while Fiji's two national policies are excellent documents on their own rights, there is little linkage between the two. Indeed it is found that the need to provide the country with access to affordable and reliable modern energy services, and not climate change mitigation, provides the driving force for the development of the NEP. There appears to be complete oblivion amongst its drafts-persons of the existence of the National Climate Change Policy and especially its recourse to methods of energy use (including renewable energy and energy efficiency) for climate change mitigation. Possible reasons for this dichotomy of purpose between the two policies are considered, and plausible methods for the avoidance of such policy mismatch are explored.

Fiji's National Climate Change Policy

Climate change poses a barrier to the sustainable development of Fiji through its impact on the country's biodiversity, particularly with respect to its coastal and marine eco-systems. To deal with this issue, the Fiji government in 2007 adopted the *Fiji National Climate Change Policy Framework (FNCCPF)* which defined the country's position on the climate change issue. A review of the framework was initiated in 2011, and the outcome of the process was the development of the *Fiji National Climate Change Policy (FNCCP)* (2012). This policy provides "a platform for (climate change) coordination amongst sectors", and "direction on national positions and priorities regarding climate change adaptation and mitigation". At the same time it recognizes the need for an "interdisciplinary and multi-sectoral approach emphasized in Agenda 21 of the 1992 United Nations Conference on Environment and Development held in Rio de Janeiro (FNCCP 2012, *Forward*). At the heart of the policy are its eight objectives and the list of strategies for the achievement of these objectives. Table 16.1 lists the eight objectives, together with the most important strategies for each.

An inspection of the table reveals that the policy is built around the key themes of adaptation and mitigation with subsidiary policies to facilitate these two important objectives. Examination of the FNCCP document also reveals that the strategies recommended to achieve the second of these objectives have renewable energy, energy efficiency and energy use as their main basis. These strategies are important enough to be considered in detail.

The list of strategies recommended by the FNCCP for the achievement of climate change mitigation is:

1. Develop joint programmes and cooperation agreements between relevant sectors to reduce and avoid greenhouse gas emissions.
2. Develop and implement national, industrial, commercial and household energy efficiency programmes...
3. Assess and utilize appropriate renewable energy sources...
4. Support the implementation of the REDD-Plus policy...
5. Access international financing instruments to support renewable energy, energy efficiency, waste management and carbon trading initiatives.
6. Control and reduce emissions from existing private and public vehicles.
7. Control the ages of imported and second-hand vehicles and introduce alternative fuel powered vehicles.
8. Develop activities and infrastructure that promote the reduction and avoidance of fossil fuel consumption.
9. Support the enforcement of legislation on open burning in residential and commercial locations as stated by the Environmental Management Act (2005).
10. Formalize collaboration arrangements and commitments of members of committees working in the area of climate change mitigation...
11. Establish a national monitoring and evaluation system to calculate GHG emissions and assess Fiji's mitigation efforts.

Table 16.1 The objectives of the Fiji national climate change policy, and examples of strategies

Objective	Statement of objective	Example of strategy
1. Mainstreaming	Integrate climate change issues in all national and sector policy and planning processes	Incorporate climate change into national plans and budgets in line with the climate change policy framework
2. Data collection, storage and sharing	Collect, manage and use accurate and scientifically sound climate change-related data and information	Establish, within the climate change unit, a clearing house mechanism for climate change data and information (research and projections on climate variability, climate trends, etc.) to foster data accuracy and efficient information sharing
3. Awareness raising	Increase awareness and understanding of climate change-related issues across all sectors and at all levels in Fiji	Conduct awareness-raising workshops and sessions for policy makers, decision makers and local and national planners on climate change issues
4. Education and training	Integrate climate change in school curricula, tertiary courses, and vocational, non-formal education and training programmes	Review and update the current primary and secondary school curricula, and the tertiary and vocational education courses to ensure the inclusion of local accurate and current climate change information, and to encourage student research around the issue of climate change
5. Adaption	Reduce the vulnerability and enhance the resilience of Fiji's communities to the impacts of climate change and disasters	Integrate disaster-related risk reduction and climate change adaptation strategies and actions into national sectoral planning to streamline responses
6. Mitigation	Reduce Fiji's greenhouse gas emissions and implement initiatives to increase the sequestration and storage of greenhouse gases	Develop joint programmes and cooperation agreements between relevant sectors to reduce and avoid greenhouse gas (GHG) emissions
7. Financing	Ensure sustainable financing for climate change efforts	Ensure that national budgeting processes include the assignment of funds for climate change mitigation and adaptation research, planning and programme implementation
8. International and Pacific region participation	Effectively participate in, and contribute to, international and Pacific region climate change negotiations, discussions, commitments and outcomes	Strengthen international negotiation skills of Fiji delegation members and improve their understanding of international policies relating to climate change

Source: Republic of Fiji national climate change policy

It is seen that six out of these eleven strategies are directly related to renewable energy, energy efficiency or energy use. This clearly brings out the importance of energy considerations to one of the two main themes that the FNCCP addresses. It is thus of interest to consider the energy mix used in Fiji currently, and in particular to examine the extent of its renewable energy development and the attention given to energy efficiency. This is the subject of the following section.

The Current Energy Status of Fiji

Energy Mix for Power Generation in Fiji

In Fiji indigenous energy sources are limited to biomass, hydro, wind and solar, with some (yet to be fully explored) geothermal potential. Hydro has been the main source of energy for the national power supply system in Fiji since the commissioning of the Monasavu Hydro Scheme in 1983. The Fiji Electricity Authority (FEA), the government-owned power utility is responsible for electricity supply nationally where it is financially and economically viable.

The FEA operates four grid systems in Fiji: one on VitiLevu, two on the island of Vanua Levu (Labasa and Savusavu) and one on Ovalau. The VitiLevu grid, known as the VitiLevu Interconnected System (VLIS) is the largest grid system in Fiji and accounts for the distribution of over 94 % of FEA's generation (Dornan and Jotzo 2011). It accounts for almost all the energy generated by the renewable technologies by FEA. The other three small grids use oil-based generators to generate electricity. On the second largest Island, Vanua Levu, there is no extensive electricity grid system due to the high cost of diesel generation and sparsely clustered population over the entire region.

According to the 2007 census, the FEA provides grid connected power to nearly 75 % of the Fiji population (Fiji Census 2007). The company operates with a vision of "Energizing Our Nation" and aims to provide clean and affordable energy solutions to Fiji with at least 90 % of energy generated through renewables by 2015 (FEA 2013). The electricity demand in Fiji has been growing at 5 % per annum over the last 5 years and is expected to increase significantly in the coming years. The peak load including all the grid system is estimated at 138 MW. FEA has a total installed capacity of about 215 MW. In 2012 about 63.6 % of the electricity was generated from hydro and 33 % from the diesel generators (FEA 2012). The remainder is provided by the Independent Power Producers (IPPs) that contribute about 2–3 % of Fiji's electricity, while wind and solar provide around 1 % of FEA power. The electricity mix (see Fig. 16.1) shows that hydro and thermal generation was approximately equal in 2010. With the recent addition of 40 MW hydro capacities at Nadarivatu, the renewable contribution has become the lead source of electricity generation.

A large proportion of mineral fuel imported in Fiji is used for transportation, electricity generation and for home economics. The price of mineral fuel imports

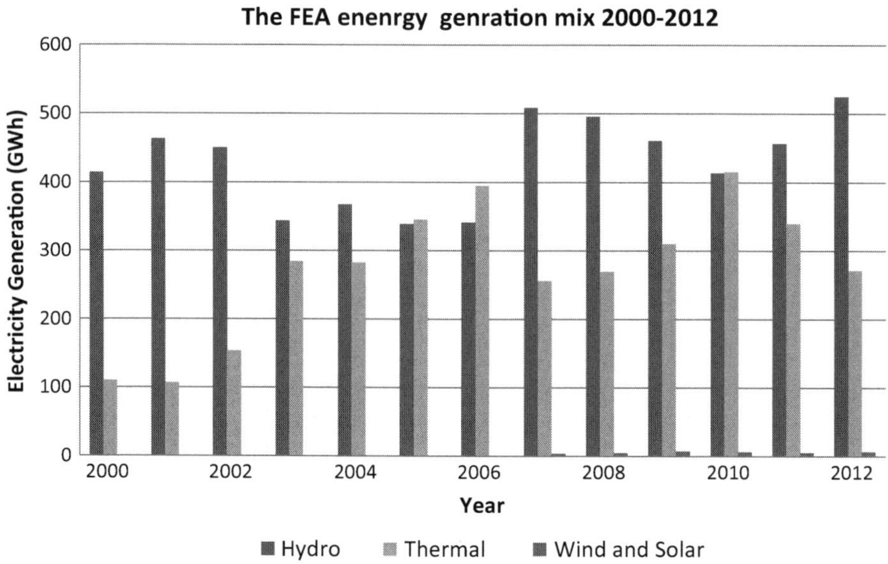

Fig. 16.1 The Fiji electricity authority (FEA) electricity generation mix, 2000–2012 (FEA 2012)

has increased in value from about FJD784 million in 2005 to about little over FJD1.2 billion in 2012 (Fiji Islands Bureau of Statistics 2012). The heavy dependence of the industry on the imported fuel highlights the vulnerability of the economy to the rising fuel prices. The amounts of monies spend on buying mineral fuel (Fig. 16.2) shows an exponential increase since 2000.

The amount of mineral fuel consumed by each sector (Fig. 16.3) shows that transport sector is the largest consumer. The National Energy Security report (2010) forecast to 2020 shows a constant demand.

Renewable Energy Development in Fiji

Hydropower

In past years, hydro resource has supplied a significant proportion of electricity to the national grid and is considered to be the cheapest source of electricity available to FEA. Currently it accounts for nearly 64 % of Fiji's electricity demand, producing 525 GW in 2012, and providing a total generation capacity of 130.2 MW. Most of the hydro proven resources are already tapped. However, there is still about 170 MW promising hydro resources that remain to be investigated (Fiji Department of Energy 2010) which together would generate approximately 300–400 GWh of electricity yearly. The FEA has commenced feasibility studies to determine the viability of developing hydro projects in the upper Navua River area, Wailoa

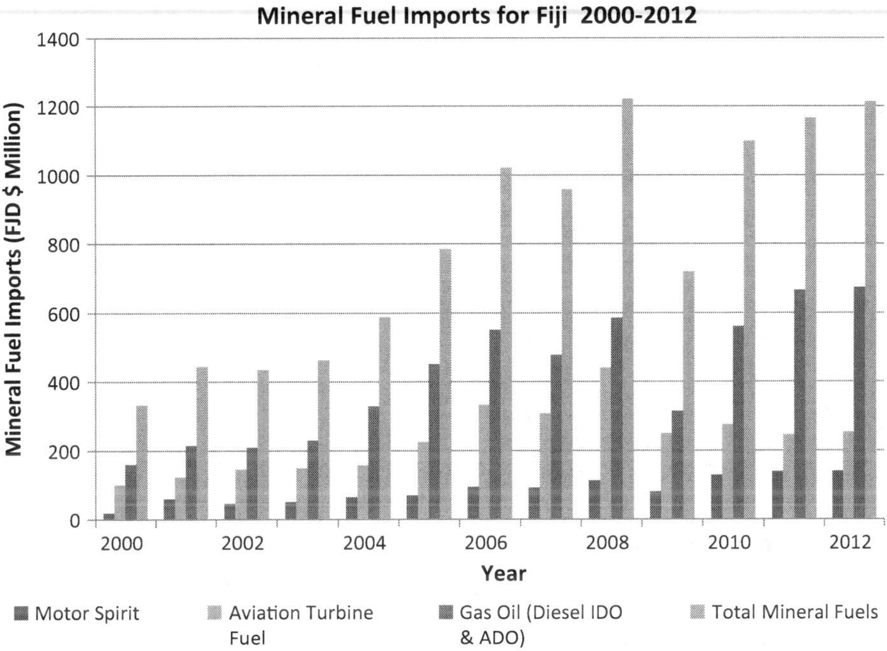

Fig. 16.2 Cost of mineral fuel imported to Fiji between 2000 and 2012 (*data source*: Fiji Islands Bureau of Statistics 2012)

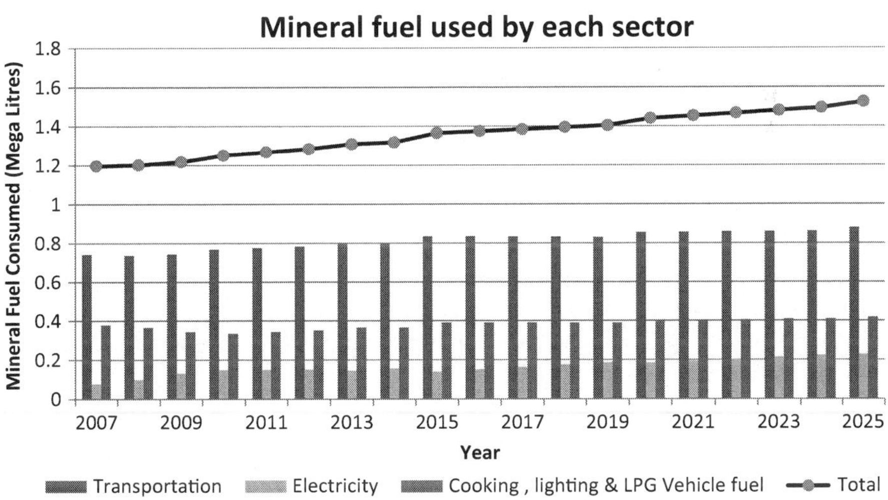

Fig. 16.3 Mineral fuel consumed (2007) and the estimated demand by sector (Snowy Mountains Engineering Corporation 2010)

Table 16.2 Hydro schemes in Fiji in 2014

Project/Scheme	Location	Year commissioned	Capacity (MW)
Monasavu	Wailoa, VitiLevu	1983	80
Wainikasou Hydro	VitiLevu	2004	6.4
Nagado	Nadi, VitiLevu	2006	3
Nadarivatu	VitiLevu	2012	40
Wainikeu	Savusavu, Vanua Levu	1992	0.8
Bukuya	Ba, VitiLevu	1989	0.1
Buca	Cakaudrove, Vanua Levu	2011	0.03
Muana, Cakaudrove	Vanua Levu	1999	0.03
Koro	Kadavu	1994	0.02
Nasoqo	VitiLevu	1984	0.004
Vatukarasa	VitiLevu	1993	0.003

downstream and at Waibutasavu (FEA 2012). The Fiji Department of Energy (FDoE) is currently engaged in performing feasibility studies around Fiji to identify potential hydro sites to provide electricity for communities not served by the FEA grid. The Fiji Department of Energy has surveyed around 118 hydro sites around the country since 2009 (Fiji Department of Energy 2009). The total installed micro-hydro capacity possible is estimated to be around 990 kW.

Table 16.2 summarizes Fiji's hydro based schemes, and shows that Monasavu is still the largest hydro installed capacity. The most recent addition of Nadarivatu (40 MW) has contributed substantially to meeting the demand, thus reducing the reliance on diesel generation.

Following a FJD10 million Chinese government grant, a hydro plant is expected to be implemented soon at Somosomo in Taveuni, the third largest Island in Fiji. It will be a 700 kW, run-of-the river system with installation of 2 by 350 kW turbines. The electricity will be provided to the populated areas on the West coast via a 10 km grid system. Environment Impact Assessment at the site was completed by a local company (Corerega Environmental). The consultants to the project, Rawalai (2013) reported that the construction work on the hydro project was supposed to start by end of July, 2013. However, due to delays in the completion of the project formalities, the construction work was delayed. According to Katirewa (2014, Interviewed by Atesh Gosai, Personal Communication, 10 March) the projects road access was completed in March 2014 and the construction work is expected to begin in the second quarter of 2014.

Wind

According to the wind monitoring data revealed by the FDoE, the average wind speed at their monitoring sites varies between 4 and 6.5 ms^{-1} at 40 m above ground level (a.g.l) (Fiji Department of Energy 2011). This shows that Fiji has potential for power generation from wind. Kumar and Prasad (2010) also found that the annual average wind speed is 6.5 ms^{-1} at 50 m a.g.l in Fiji. Wind monitoring stations are setup by the FDoE at various locations (Gamu, Korotogo, Vunatovau, Waibogi, Kavukavu, Tamuka, Kadavu, Vunisea, Vadravadra and Nacamaki) around Fiji to find potential sites for wind power generation. The average wind speed at these sites range between 4.7 and 7.0 ms^{-1} at heights between 10 and 48 m a.g.l. Currently the wind monitoring at two other sites are being carried out by the FDoE and these are Dakuilomaloma village in Lau and Nabouwalu Government Station in Bua at 30 m a.g.l (Fiji Department of Energy 2013). In 2012 the FDoE installed two 34 m monitoring stations, one at Matacawalevu in Yasawa and another one at Nauouo in Ovalau (Fiji Department of Energy 2012).

According to Katirewa (2014, Interviewed by Atesh Gosai, Personal Communication, 10 March) by end of 2014 there will be 15 of 50 m NRG wind monitoring systems installed around Fiji. Due to some issues with land leasing there is a delay in the installation. The equipment has already arrived in the country and it is currently kept at Clay Energy's warehouse.

Presently, there is only one wind farm in Fiji. The Butoni wind farm is a 10 MW grid connected system operated by FEA. It has a total of 37 Vergnet 275 kW wind turbines installed at a hub height of 55 m. When installed, the Wind Farm was expected to produce 11.5 GWh per annum. A survey carried out on the wind farm in 2009 showed that the wind farm is not producing as expected due to insufficient wind resource around Butoni hills (Karan 2009). However FEA reported that in 2009 a total of 7.2 GWh of electricity was generated by Butoni wind farm (FEA 2011), which is the maximum energy generated since the establishment of the wind farm.

Solar PV

Solar resource is abundantly available in Fiji, being more intensely available in the western part and the Northern part of Fiji. Grid-connected solar-based generation has been very limited to date in Fiji despite the country's high solar insolation rates, which range from 4.5 to 5.47 $kWh/m^2/day$ (Fiji Department of Energy 2006). The Fiji Meteorological Service has six pyranometers stations in VitiLevu (Nadi, Vaturu, Monasavu, Nacocolevu, Koronivia, Laucala Bay); one in the Lau Group (Vanuabalavu); and, two in Vanua Levu (Dreketi and Seaqaqa). There is also an on-going data recording at the Nabouwalu Hybrid Power Station (since 1996) where the annual daily average insolation is 4.5 $kWh/m^2/day$ (Fiji Department of Energy 2011). The University of the South Pacific has also installed pyranometers at three of its wind monitoring sites (Rakiraki, Saqani and Laucala Bay) in Fiji. In addition, the FDoE is also engaged in monitoring solar resources at remote sites around Fiji to find the potential sites to produce electricity.

The earliest grid-connected electricity generation system in Fiji was the 10 kW system installed in Navutu, Lautoka in 1997. In 2011 two new grid connected PV systems were implemented: a 10 kW PV system at Tuvu in Lautoka (by an NGO called FRIENDS Fiji) (DIREKT 2012) and a 45 kW system at USP (lower campus at Laucala Bay). The latter system was funded by The Korea International Cooperation Agency (KOICA). In 2012 two more grid connected systems were installed, at Port Denarau Marina in Nadi and at the Turtle Island Resort in the Mamanuca Group. The system at Port Denarau Marina is 122 kW and the latter system, the largest in Fiji is 260 kW. The Port Denarau marina is the largest solar powered Marina in the world (Lal et al. 2013; Raturi 2013). Additionally Turtle Island Resort is the world's first 100 % renewable energy resort.

On a different scale, the FDoE under its rural electrification scheme has been installing the solar home systems (SHS) in the rural communities that do not have access to grid connected supply. Installation under this scheme started in 2008 in Vanua Levu, with installation of 209 projects. Since 2008, the demand for the SHS increased rapidly, prompting more installation to be carried out. Currently 3,680 SHS are installed all over Fiji (Valemei 2014). According to Valemei, this year (2014) there will be an installation of another 1,200 SHS in all rural villages and maritime islands. The number of requests received has been increasing every year and the FDoE is on schedule to tender another new contract in May 2014 for the supply and installation of another 1,000 or more SHS projects.

Biomass and Biofuel

Fiji has substantial potential for biomass industry because of its forests, bagasse from sugarcane industry and hog fuel from the timber industry. Presently a significant amount of Fiji's total energy needs are met by biomass (Singh 2009). Firewood is the main source of energy used for cooking mainly in the rural areas. Bagasse from the sugarcane industry and the hog fuel from the timber industry are currently two forms of biomass used to generate electricity at a commercial scale. The latter is used by the logging industry for process heat and as well as electricity generation. The FSC uses the bagasse for electricity generation and to process steam for sugar production. A significant potential for development of biofuel industry from indigenous resource such as coconut is also present in the country. Coconut oil and biofuel extracted from coconut can be used as a substitute for diesel fuel. Research on biofuel extraction using Jatropha fruit is in progress.

Fiji currently has two grid connected biomass fed generation systems. FSC uses bagasse to generate electricity using its 31 MW plant. Secondly the Tropik Wood Industries Limited (TWIL) uses the waste stream from the timber industry to produce electricity from its old 3 MW and the newer 9.3 MW plant.

Biogas

The Department of Energy has been involved in the installation of Biogas projects all over Fiji. Currently there are 20 sites where the FDoE has completed the installation of bio-digesters. The largest installed biogas system is 20 m^3 in size. According to Laliqavoka (Interviewed by Atesh Gosai, Personal Communication,

10 March 2014, Personal Communication) all the current installed projects are used for cooking only however, now the FDoE is planning to bring models that can also be used for electricity generation.

Geothermal Resources

Fiji is located near the "Pacific Ring of Fire" where volcanic eruptions and earthquakes are caused by Pacific plate movements and volcanic activities. Chances are more than likely that adequate geothermal resources exist in Fiji, especially in the northern and eastern parts of Fiji. Although there is currently no electricity generated from geothermal in Fiji, there is evidence of geothermal resources in both Vanua Levu and VitiLevu. Studies conducted by Japanese consultants have shown that a total of 20 MW geothermal potential exists around Fiji. This would involve 5 MW power stations installed at the two large deposits in VitiLevu (Tavua and Busa) and at Savusavu in Vanua Levu (Dornan and Jotzo 2011). Feasibility study conducted by the FDoE also indicates that there is potential for steam generation in Labasa, with an estimated temperature of 125 °C at 500 m below ground. The estimated temperature at Savusavu is around 160 °C, however deep drilling is necessary to verify the available resource data (REEEP 2010).

The above review reveals that renewable energy features well in Fiji's current energy mix. To achieve the aims of the National Climate Change Policy however, one requires a considered approach in the selection of both renewable energy and energy efficiency measures that optimize the prospects of carbon emissions reductions. This calls for a coordinated programme of action that controls how energy development takes place, and uses the optimum reduction in carbon emissions as a measure of the success of this programme.

Fiji has a (recently reviewed) National Energy Policy (NEP) that aims to provide for its future energy needs. Does this policy provide the programme of action required by the NCCP? To answer the question, one must take a close look at the new NEP itself.

The Fiji National Energy Policy

Fiji adopted its first National Energy Policy in 2006. This policy has recently undergone a thorough review to take into account the numerous developments that have taken place. Chief amongst these has been the *United Nations Sustainable Energy for all by 2030 (SE4ALL by 2030)* programme.

The revised document, Fiji National Energy Policy 2013 (FNEP 2013), whose final draft was completed in November 2013, was prepared by the *National Energy Policy Review Advisory Committee*, consisting of the Department of Energy (chair), the Department of Transport and the Ministry of Transport, Works and Public Utilities, the Ministry of Strategic Planning, National Development and Statistics, the Climate Change Unit of the Ministry of Foreign Affairs, the Reserve Bank of Fiji, the Deutsche Gesellschaft fur InternationaleZusammenarbeit (GIZ), and the United Nations Development Programme (UNDP). The final draft of the FNEP (6 November 2013) was to have been presented for Cabinet approval at the time of writing.

The policy has a vision for a **resource efficient, cost effective and environmentally-sustainable energy sector for Fiji**, and the following three objectives:

1. To provide all Fijians with access to affordable and reliable modern energy services.
2. To establish environmentally sound and sustainable systems for energy production, procurement, transportation, distribution and end-use.
3. To increase the efficient use of energy and the use of indigenous energy sources to reduce the financial burden of energy imports on Fiji.

The document begins by presenting an overview of the energy sector of Fiji under the area headings of **grid-based power supply**, **rural electrification**, **renewable energy**, **transport**, **petroleum and biofuels** and **energy efficiency**. It then sets targets for the energy sector in line with the three objectives of the United Nations SE4ALL by 2030 policy. These are

1. access to modern energy services,
2. improving energy efficiency
3. improving the share of renewable energy

The policies are then stated in terms of each of the key areas of the energy sector mentioned above. An abbreviated list of the policy statements are given in Table 16.3.

The FNEP provides for the implementation of the policy by suggesting the appropriate institutional arrangements for the coordination, planning and policy development, stakeholder consultation, regulation and reporting, monitoring and evaluation of the project.

Comparing the FNCCP and FNEP

To determine how far the Fiji National Climate Change and National Energy Policies are aligned, it is best to begin with some pertinent observations relating to the first of these policies. We note that the FNCCP contains certain overarching features that must apply to other sectors to ensure its objectives are facilitated.

Firstly the preface of the NCCP points out that it is a "guideline for (all) sectors to ensure that the impacts of climate change are considered in their (respective) planning and implementation programmes". Also the second of its goals is to "promote integration of climate change issues in national planning, budgeting and implementation processes".

Table 16.3 The statements (abbreviated) of the new Fiji national energy policy

S. no	Energy policy	Policy statement
1.	Grid-based power supply	*Priority policies* 1.1 increase private sector investment in large-scale electricity generation... 1.2 Increase private sector investment in small-scale grid-connected renewable energy generation by establishing economically justified feed-in tariffs... 1.3 Strengthen transparency and effectiveness of the regulation of the electricity industry *Remaining policies* 1.4 Improve the efficiency and effectiveness of management of the Fiji Electricity Authority (FEA) grid
2.	Rural electrification	*Priority policies* 2.1 develop a national electrification master plan, showing how each un-electrified area of Fiji will be served 2.2 establish a dedicated electrification fund... *Remaining policies* 2.3 The FEA will be responsible for electrification by grid extension in accordance with the national electrification master plan 2.4 improve the effectiveness and sustainability of the existing management models for off-grid rural electrification including Renewable Energy Service Companies... 2.5 establish a framework for encouraging off-grid rural electrification projects by non-government providers including community-based organizations
3.	Renewable energy	*Priority policies* 3.1 Maintain a comprehensive assessment of Fiji's renewable energy resources, including hydro, wind, biomass, solar and geothermal resources 3.2 Make all data on renewable energy resources available to the public and prospective investors through a single national repository at the Department of Energy 3.3 Conduct further investigations into geothermal energy resources with a view to identifying a pilot project for development *Remaining policies* 3.4 Research and promote new renewable energy technologies including assessing their technical and economic viability... 3.5 Promote and improve guidelines and technical standards for renewable energy technologies... Work with industry associations, civil society, non-government organizations and communities to identify and remove barriers to the uptake of renewable energy technologies in power generation and transport
4.	Transport	*Priority policies* 4.1 Promote the fuel efficiency of imported motor vehicles in order to reduce petroleum consumption 4.2 Investigate the potential and cost-effectiveness of energy efficiency and renewable energy solutions for sea vessels, including biofuels, solar and sail-assisted sea transport... 4.3 Support the development and implementation of the

(continued)

Table 16.3 (continued)

S. no	Energy policy	Policy statement
		Department of Transport's land and marine transport policies that encourage a shift towards more energy efficient forms of land and sea transport... *Remaining policies* 4.4 Explore the costs and benefits of introducing mandatory fuel efficiency standards based on those applied internationally 4.5 Explore the potential for the introduction of hybrid and electric cars 4.6 Promote the fuel efficiency of the existing motor vehicle fleets... 4.7 Support voluntary actions of businesses, industry and private users, including the aviation industry and commercial land transport, to implement energy efficiency measures 4.8 Promote public transportation, including buses and railways 4.9 Promote cycling and establishment of cycle paths in urban areas... Encourage town planning and work with local councils to encourage consideration of energy aspects...
5.	Petroleum and biofuels	*Priority policies* 5.1 reduce the cost of imported petroleum products by negotiating directly with fuel suppliers... 5.2 improve the transparency of petroleum supply, including collecting data on fuel imports, re-exports... 5.3 continue research to explore the potential for increased production and use of biofuels *Remaining policies* 5.4 encourage the use of relatively environmentally-friendly petroleum products such as LPG... 5.5 keep fuel standards up to date and ensure they are adhered to including in rural and remote areas... 5.6 prepare and maintain contingency plans for fuel supply to ensure there is adequate supply during business as usual and during emergency times... 5.7 continue to apply national biodiesel and ethanol fuel standards in a way that supports the uptake of these commodities
6.	Energy efficiency	*Priority policies* 6.1 continue to increase public education and awareness of energy efficiency by providing targeted information to end-users... 6.2 extend the current system of energy labelling and minimum energy performance standards to all widely imported electrical appliances 6.3 develop and implement an energy information database *Remaining policies* 6.4 monitor and improve the existing customs and tax incentives that are in place to encourage the use of energy efficient appliances and equipment 6.5 update the codes and standards for buildings and industry 6.6 promote energy efficiency in the public sector as a platform for demonstrating the feasibility of energy efficiency projects

(continued)

Table 16.3 (continued)

S. no	Energy policy	Policy statement
		6.7 Strengthen the enabling environment for the energy service companies to undertake and finance public and private sector energy efficiency projects 6.8 support voluntary efforts by the business community to improve energy efficiency including public recognition of best performers...

These provisions ensure that **the NCCP has an oversight on the development of the policies of all other sectors (including energy).** The question that arises is *whether the NCCP was successful in having this oversight in the case of the NEP, which was developed one year afterwards.*

Secondly, the preface of the NCCP encourages relevant sectors to "take up climate change mitigation as part of Fiji's commitment to the global efforts to reduce greenhouse gas emissions". Also, the fourth goal of the policy is to "guide sectors to develop appropriate climate change adaptation and mitigation strategies".

The first of the mitigation strategies it recommends is to "develop joint programmes between relevant sectors to reduce and avoid greenhouse gas emissions". As the main device for GHG reduction are the replacement of fossil fuels by renewable energy, and the practice of energy efficiency, **there should be a joint programme between the energy and other sectors to achieve this GHG reduction aim**. *Does such a joint programme exist?*

When one examines the National Energy Policy (NEP), it becomes clear that its main objectives are to provide energy access to all Fijians, "establish systems for energy production, procurement, transportation, distribution and use", and to increase renewable energy share and the practice of energy efficiency. Thus the main thrust of the policy is the provision of energy on a more equitable basis to the people of Fiji and at less cost to the country as a whole.

How are these objectives related to the overarching requirements of the NCCP? On close examination, one finds that there are few (if any) provisions within the NEP that are intentionally provided to meet the objectives of the NCCP. There is a mention of concern for the environment in the second of its objectives, but this is only in passing. There is no doubt that both renewable energy and energy efficiency will achieve the aims of mitigation, but the motivation for these provisions within the NEP is concerned more with providing energy access and reducing costs rather than reducing carbon emissions. There are no requirements that they are implemented in accordance with the objectives of the NCCP, which require an overall reduction the GHG inventory of the country at any time. As the energy demand of the country grows with time for instance, there will be a general increase in emissions, and there is no assurance that the simple provisions contained in the present NEP will ensure that the net inventory is actually reduced or at least maintained at previous levels.

The above observations should reveal that there has been little, if any, coordinated development of the NEP in light of the overarching mitigation requirements as laid down by the NCCP.

Possible Solutions to the Policy Mismatch

Solutions to the policy mismatch noted above may be achievable if the following points are noted:

1. The initiative to coordinate the development of national sectoral plans to tackle emissions and climate change impacts must come from the NCCP, as it already has a mitigation strategy specifically devoted to such a coordinated effort.
2. Success would be expedited if a mutual objective for both policies could be identified. This may not lie within the set of policies already propounded.
3. An appropriate mutual objective is provided by sustainable development of the country.
4. Energy provides a common denominator for the NCCP, NEP and sustainable development.
5. To remove the policy mismatch, the two policies should be ideally reviewed together, preferably by a government department that has the required authority.

Conclusions

The above review of Fiji's National Climate Change Policy (NCCP) and the National Energy Policy (NEP) reveals that the objectives of the NEP bear little relation to those of the NCCP. This is true even though an overarching aim of the NCCP is to guide the policy development of all sectors (including energy) to ensure that they are responsive to the need for climate change mitigation.

The fourth goal of the NCCP is to guide all sectors to develop climate change adaptation and mitigation strategies. While the NEP does in fact propose the replacement of imported fossil fuel by renewable energy and the employment of energy efficiency measures, the reason for the choice of these energy strategies, however, resides firmly in their ability to reduce the country's reliance on expensive fossil fuel imports. The selection of these climate change mitigation measures as NEP strategies is therefore purely coincidental, and there is no assurance that the implementation of these measures by the NEP will necessarily lead to greenhouse gas reductions in all situations.

Overall, it is clear that little attempt has been made to coordinate the development of the NEP with the objectives of the NCCP. To ensure that the development of policies are integrated in their aims and coordinated in

(continued)

their strategies requires a more holistic development pathway than the piece-meal development of individual sector policies. This can be achieved, for instance, through the establishment of a policy master-plan that sets the essential requirements for the development of all sector policies of the country.

In the present situation, the policy mismatch that has occurred between the NCCP and the NEP can be ameliorated, and the objectives of the two policies better integrated, if a concurrent review is undertaken of both policies by a section of the government that has the mandate for such action.

References

DIREKT (2012) Transnational recommendation report. The Hamburg University of Applied Sciences, Hamburg. Available at http://www.direkt-project.eu

Dornan M, Jotzo F (2011) Electricity generation in Fiji: assessing the impact of renewable technologies on cost and financial risk. In: Australian agricultural and resource economics society 55th annual national conference, Melbourne, 8–11 February. The Australian National University, Melbourne

FEA (2011) Annual report 2011. Fiji Electricity Authority, Suva

FEA (2012) Annual report 2012. Fiji Electricity Authority, Suva

FEA (2013) FEA energising our nation. Available at http://www.fea.com.fj/pages.cfm/who-we-are/our-mission-vision.html. Accessed 10 July 2013

Fiji Census (2007) Available at http://www.statsfiji.gov.fj/index.php/2007-census-of-population

Fiji Department of Energy (2006) National energy policy. Ministry of Energy and Mineral Resources, Fiji Islands

Fiji Department of Energy (2009) Fiji's hydro potential report, vol 2. Fiji Department of Energy, Suva

Fiji Department of Energy (2010) National energy security situation report. Snowy Mountains Engineering Corporation, Suva

Fiji Department of Energy (2011) Fiji renewable energy report. Asian and Pacific Centre for Transfer of Technology of the United Nations—Economic and Social Commission for Asia and the Pacific, New Delhi

Fiji Department of Energy (2012) Department of Energy. Watts News. Ministry of Works, Transport and Public Utilities, Suva

Fiji Department of Energy (2013) Department of Energy. Available at http://www.fdoe.gov.fj/index.php/renewable-energy/accounts. Accessed 12 Aug 2013

Fiji Islands Bureau of Statistics (2012) Key statistics. FIBS, Suva

FNCCP (2012) Fiji national climate change policy. Available at http://www.spc.int/lrd/component/docman/cat_view/137-all/133-climate-change/209-climate-change-general/315-pacific?limit=5&limitstart=0&order=date&dir=DESC

FNEP (2013) Fiji national energy policy. Available at the Fiji Department of Energy www.fdoe.gov.fj

Karan M (2009) $34 m failure: FEA admits lack of study in wind farm project. Fiji Times, 25 August. Available from www.fijitimes.com/story.aspx?id=127996. Accessed 20 May 2012

Kumar A, Prasad S (2010) Examining wind quality and wind power prospects on Fiji Islands. Renew Energy 35:536–540

Lal R, Pratibha J, Stolz E (2013) Denarau Marina powered with solar. Fiji Sun, 23 February. Available from http://www.fijisun.com.fj/2013/02/23/denarau-marina-powered-with-solar/. Accessed 6 March 2014

Raturi A (2013) A fully renewable power sector for the PICs: opportunities and challenges case studies from Tokelau and Fiji. Available from http://www.icsu.org/latin-america-caribbean/news-centre/AtulRaturiICSUpresentation.pdf. Accessed 6 March 2014

Rawalai L (2013) Work on $20 m project to start. The Fiji Times, 12 July. Available from http://www.fijitimes.com/story.aspx?id=239469. Accessed 5 March 2014

REEEP (2010) Policy DB details: Fiji Islands (Internet). Available from http://www.reeep.org/index.php?id=9353&text=policy&special=viewitem&cid=1. Accessed 2 Feb 2012

Singh A (2009) The sustainable development of Fiji's energy infrastructure: a status report. Pac Econ Bull 24(2):141–154

Singh A, Fifita S, Mario R, Bijay P (2013) Sustainable energy development in the Pacific: the evolution of energy frameworks and national policies. In: Leal Filho W, Mannke F, Mohee R, Schulte V, Surroop D (eds) Climate-smart technologies—integrating renewable energy and energy efficiency in mitigation and adaptation responses. Springer, Berlin, pp 129–140

Snowy Mountains Engineering Corporation (2010) National energy security situation report. Prepared for Fiji Department of Energy

Valemei J (2014) E-mail to Atesh Gosai, 10 March 2014

Chapter 17
Building Resilience to Climate Change Impacts and Socioeconomic Attributes of Rural Households in Solomon Islands

Michael Otoara Ha'apio and Ricardo Gonzalez

Abstract The source of livelihood varies amongst the urban centers and rural areas in the Solomon Islands. Most of the people within the communities rely on subsistence activities, agriculture, forestry and marine resources for survival. This research aimed to perform a descriptive analysis of the socioeconomic attributes of rural households that participate in the Coral Triangle initiative (CTI) and Mangrove rehabilitation project (MRP) in selected rural villages of Solomon Islands. Household surveys were conducted in order to raise information on the socioeconomic attributes of participant households. The analysis revealed that households from Sairaghi (***project site 1***), and Oibola (***project site 3***), rely mainly on marine resources for their income; whilst in Naro (***project site 2***) they rely mostly on agriculture. Consumable items were the main household expenses, followed by education. It was identified that the villagers also begun to invest in a series of new business ventures that could potentially prepare them better to cope with impacts and risks from climate change in the future. We found that beneficiaries' expectations on potential benefits from project outcomes were high. CTI and MPR projects are vital for the communities as most villages are settled along the coastal lines, however ownership of conservation initiatives must be taken by villagers in order to assure of their support and a higher sustainability of the projects. In spite the immediate project impacts that slow the economic activities at local level, villagers perceive that benefit obtained from restoration would assist them to be resilient

M.O. Ha'apio (✉)
Pacific Centre for Environment and Sustainable Development, The University of the South Pacific, P O Box 1168, Suva, Fiji

Idea Connection Pacific (ICP), Business Consultancy Services, Panatina Valley, Honiara, Solomon Islands
e-mail: mhaapio@gmail.com

R. Gonzalez
Faculty of Business and Economics, School of Economics, The University of the South Pacific, P O Box 1168, Suva, Fiji

Department of Forest Sciences, Faculty of Agricultural and Forest Sciences, Universidad de La Frontera, Temuco, Chile
e-mail: ricardo.gonzalez@usp.ac.fj

W. Leal Filho (ed.), *Climate Change in the Asia-Pacific Region*, Climate Change Management, DOI 10.1007/978-3-319-14938-7_17

against the impact of climate change, given the fact that they will be able to obtain such project benefits in the near future.

Introduction

Climate change is one of the major challenges for Pacific Islands Countries and Territories (PICT), including Solomon Islands. It threatens not only the people's livelihood and living standards, but also the viability of isolated communities and rural households. The country has half a million population from which 85 % comprises rural households that rely on subsistence activities from agriculture, forestry and marine resources (Albert et al. 2010a, b). Although local and international remittances at little but increasing rates also contribute as alternate sources of income. Like any PICTs, Solomon Islands has been identified as one of the vulnerable countries in the region to the adverse impacts of climate change. The high vulnerability lays on many factors like excessive dependence on foreign aid, remoteness, the fact that the majority of the population lives within 1.5 km from the coastline and, also the higher poverty level that according to Dyoulgerov et al. (2011) reached 22.4 % in 2008.

The need to analyze government strategies to mitigate climate change and adapt to its impacts in Solomon Islands has motivated this study. The Solomon Islands (SI) government, international donor partners and agencies have spent more than USD4.8 million (MECDM 2010) in the last 5 years, in initiatives aiming to integrate Climate Change (CC) in developing, planning and budgeting, capacity building and implementation of adaptation practices to increase resilience to climate change. The donor aid partners include the US Agency for International Development (USAID), Global Environment Facility (GEF) and, Asian Development Bank (ADB) have assisted in this process. The cost to address climate change impacts to developing countries including Solomon Islands is a burden and increasing annually (Huhtala and Ambrosi 2010).

In light of these increasing costs and with deep concern on the risks that climate change poses, PICTs has adopted the "Pacific Islands Framework for Action on Climate Change 2006–2015" through the Pacific Islands Forum (PIF) in 2005. Such a concern was confirmed by PIF[1] leaders at the Apia meeting in 2011 (Slade 2012) and a Climate Change Communiqué at the Majuro meeting in 2012.

The regional concern translates from deep seated climate challenges at country level. For example, Sore (2010) pointed out that climate change is one of the important impediments for small islands developing (SID) countries, like Solomon

[1] The PIF is an organization compounded by 16 independent and self-governing states in the Pacific Region with a vision to seek a region that is respected for quality of its governance, sustainable development of its resources, for the full observation of democratic values and its defense and promotion of human rights.

Islands, to meet their development goals. Thus, the low-lying coastal islands, small population, susceptibility to natural disasters, remoteness to markets, vulnerability to external shocks, and excessive dependence on international trade and foreign aid are factors that make Solomon Islands more vulnerable to impacts from climate change. Climate change becomes a cross-cutting issue that impacts all development sectors in the country, making Solomon Islands particularly vulnerable (Wickham et al. 2012). The Government of Solomon Islands (SI) had included climate change into national development strategies and decision making, where all sectors of the economy are expected to be negatively affected by climate changes with increasing adaptation costs (Dyoulgerov et al. 2011).

Sanderson and Islam (2007) explained that, in broad terms, the two major policy options for climate change are mitigation and adaptation. They define mitigation as the act of reducing greenhouse gas (GHG) emissions with the goal of slowing or preventing climate change, whereas adaptation is the act of reducing vulnerability to the effects of climate change. The mitigation actions are evaluated in terms of cost effectiveness whereas adaptation measures must be evaluated in terms of benefits. An example of adaptation practice is the Coral Triangle Initiative (CTI), which is a regional initiative covering six countries: Philippines, Malaysia, Indonesia, Timor-Leste, Papua New Guinea, and Solomon Islands in Southeast Asia and Pacific Triangle. The Solomon Islands component was implemented in rural communities on 2008, with the main objective to conserve corals reefs with emphasis in economic development and building resilience. The general idea between conservation and building resilience to climate change is that when the participants of the CTI conserve their resources (marine area and coral reefs) they also increase their capacity and potential to build reliance on sources that will prevent them from over depending on those resources (coral reefs) for their survival. For example, in the western province introduction of MPA in the surrounding reefs increase the number of home stay at the site because of the increase of tourist coming to dive and do research at this site. Besides conservation increase the value of fish stock in the area which the respondents could fish to earn money and in turn assist them to invest in businesses that will diversify their dependency on marine resources. Furthermore, it ensures that the coral reefs grow and in long term provide barriers against high sea waves coming directly into the villages.

Prior to the introduction of CTI in the country there were also some organizations such as the Ministry of Climate Change, Disaster Management and Meteorology (MECDM), Ministry of Fisheries and Marine Resources (MFMR) and a few NGOs that have participated in programs aimed at conserving the environment. The Nature Conservancy, World Fish and WWF have participated in marine conservation across the country (van Berkering et al. 2007). Despite these initiatives the coral reefs are still under over-use pressure from humans. For example, there has been a growing demand for coral in Solomon Islands for the international aquariums and curio trade in the recent years, as well as for betel nut lime (made from live coral) in the local markets which further intensifies stress on the reefs (Albert et al. 2012). These activities had resulted in unprecedented removal of various coral types, and localized destruction of reef's ecology and habitats, with major ecological impacts on other reef dependent species, like fish and invertebrates. The reef degradation can affect the resilience of the whole ecosystem, and its ability to recover from both natural and anthropogenic disturbances. The degraded coral reef system may also lead to negative socio-economic effects on communities that depend on them.

The CTI has aimed at rehabilitating and conserving the coral reefs and marine ecologies. The country is in the process of adopting the program on Reducing Emissions from Deforestation and Forest Degradation (REDD+). This is a United Nation funded initiative that started in 2011 and, to date is yet to be implemented effectively. REDD+ goes beyond deforestation and forest degradation to include the role of conservation, sustainable management and enhancement of forests, including also, the rehabilitation of mangroves. It is expected that REDD+ will complement the conservation efforts of CTI.

The CTI initiative has three major objectives: (1) Ensure food security; (2) build community resilience and, (3) halting the decline in ecosystem productivity. By the introduction of more effective management of coastal and marine resources, it was expected to build resilience to the impacts of climate change and human-induced activities and, also increase the ecosystem productivity. The CTI is relevant because of the fact that it addresses adaptation to climate change impacts and risks by targeting economic development goals in rural communities. Beside the two CTI case studies we also looked at one Mangrove Rehabilitation Project (MRP), with similar objectives as the CTI. The MPR is a private initiative supported by the community members of Oibola site, Malaita Province. Unlike the CTI, no aid donor has financially supported the MPR in this community. The community leaders and members have decided to take the leading role in this initiative to replant their mangrove at their own costs. Replanting mangroves has played a significant role in conserving habitats and reservoirs of biodiversity, home of many species some as yet undiscovered at this site. Mangroves provide a buffer from hurricanes; protect inland areas from surge and, plants help to hold the soil preventing from soil erosion.

Over the past decades, previous to CTI and MPR initiatives, villagers at these selected sites had adopted some of the most destructive fishing methods in order to catch enough to support their families, producing the destruction of corals. The lack of land for farming at either commercial or self-consumption scale has put even more pressure marine resources as source of income.

The introduction of MRP and CTI aimed at restoration of marine habitat. Villagers were benefited from these projects by the introduction of Marine Protected Areas (MPAs). MPAs are demarcated sea areas in which villagers were not allowed to fish throughout the year or mutually agreed period of time (Dudley 2008). The villagers had to travel outside these demarcated zones to fish and find other marine resources for their domestic and commercial needs. By doing this over the years, fish and marine resources at these MPA sites are expected to replenish and increase in size and volume. Villagers benefited by fishing in surrounding sites as the spillover from the MPAs increased into the allocated fishing sites. Villagers were able to increase catch and income as this process in the period 2009–2012.

These economic benefits assisted villagers to improve their livelihood and provide them the opportunity to build resilience to the devastating impacts of climate change at these sites.

In addition, through conservation of the marine resources (corals and mangroves) at these sites with adopting of sustainable methods of harvesting, this has enabled the villagers to depend on these resources for their livelihood amidst the increasing adverse impact of climate change. It has provided them with the opportunity to build resilience initiatives with the income earned and adapt to the communities standards of living. For example with the income earned from the conservation of the resources has helped them to pay for their children's school fees and invest in new business activities that will help them to be resilient with the impact of climate change at these sites. We have also discovered during the trip that more villagers, particularly from Oibola have promoted education as one of the strategies to counter the impact of climate change. Study also revealed that conservation of marine resources is also a means of poverty eradication (van Berkering et al. 2007). Furthermore, the way how the level of economic development in rural areas of Solomon Islands is related to the effects and risks from climate change may give us a light on how building resilience to climate change. Only after constructing a base line on the socio-economic status of rural communities we may be able to explore the relationship among such status and their resilience to cope with climate change risks.

The objective of this study therefore is to *describe the socioeconomic situation* of households in three communities hosting CTI and the MRP. The evaluation aims to identify sources of income; differentiate activities for subsistence from commercial ones, the level of integration to local markets and their resiliency to climate change. The study also explores difficulties and challenges households face in building their own resiliency and how these challenges are related to their socioeconomic situation.

Study Sites

The study covers two CTI and one MRP project sites in the country. The first CTI project was at Sairaghi, Western Province: the second, at Naro, Guadalcanal Province and, the MRP project, at Oibola, Malaita Province. These were chosen because of the vulnerability of these communities to the impacts of climate change and also, because of their wider geographical, economic and social distribution.

Sairaghi Project

The objective of Sairaghi project was to restore the coral reefs as consequence, a higher level of villager's income was expected to improve their livelihood,

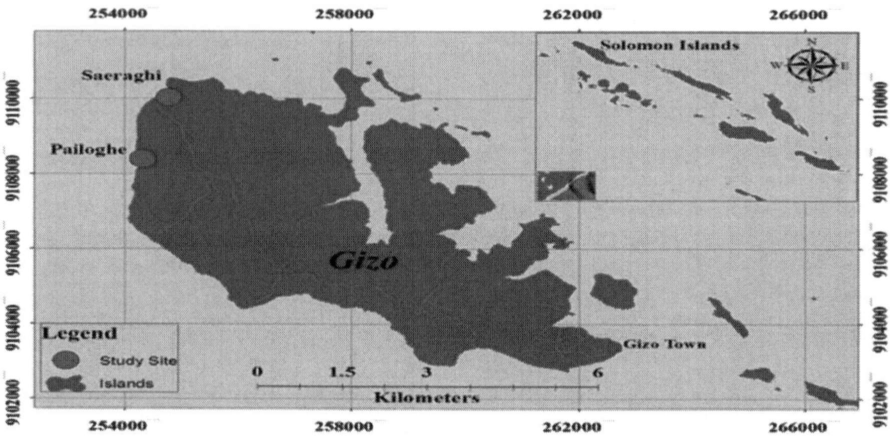

Fig. 17.1 Location map of Sairaghi project

making them more resilient to impacts from climate change. This CTI project is located along the western coast of Gizo Island and has a total of 640 people living within the project area in two villages, Sairaghi and Pailoghe (Fig. 17.1). The island hosts the administrative headquarter of Western Province. This province has a population of 76,649 people (Gagahe 2011) and 24,214 households. The main livelihood sources are fishing and agriculture. Tourism shows an increasing dynamism in the area.

Naro Project

The Naro CTI project is located along the western coast of Guadalcanal Province, with a site population of 514 people (Fig. 17.2). One of the objectives of this CTI project is to relief the dependency of the villagers from marine resources by establishing MPAs and promoting farming as alternate source of income. This action was expected to ensure the replenishment of marine resources and benefit future generations (Lal and Holland 2010). The population of Guadalcanal province is 93,613 people (Gagahe 2011) and 17,163 households. Naro villagers rely for their livelihood mainly on fishing and agriculture. From this location, they travel less than 2 h to Honiara Central Market to sell their products. This relative better access to markets allows them to obtain better prices than villagers from the other two sites: Sairaghi and Oibola.

The decision to establish the CTI at this site was a collective one made by villagers in order to conserve the coral reefs and to raise funds through the introduction of MPA to assist with the completion of the community church

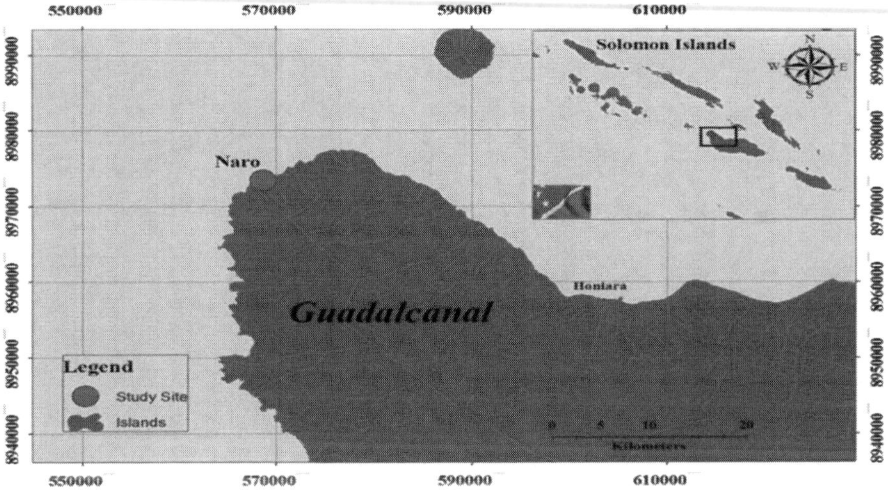

Fig. 17.2 Location map of Naro project

building. The villagers expressed that their church building was incomplete for more than 20 years. With the CTI and MPA the fish in the area were able to replenish and thus they were able to catch and earn enough money to complete the building. This immediate benefit has caused the villagers to extend this initiative to conserve their marine their resource to meet both their short term and long term livelihood benefits.

Oibola Project

The Oibola MRP project covers four small villages with a 469 people and 87 households. They are located on Malaita Province which has a population of 137,596 people (Gagahe 2011) and it is about 15 minutes' drive away from Auki, the provincial headquarter. The objective of implementing the MPR was to ensure the rehabilitation of mangroves in the area in order to provide habitat for fish and other marine resources. The primary sources of income at this site are marine resources, particularly shell money, necklace production and fishing within the reefs in the surrounding lagoon. Shell money production although traditionally regarded as the main source of income is under severe pressure due to depletion of the shells in the surrounding reef and lagoon (Fig. 17.3).

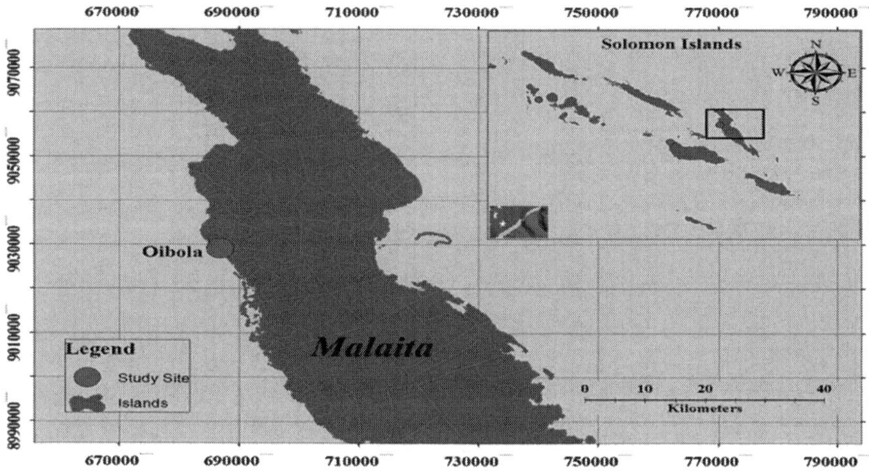

Fig. 17.3 Location map of Oibola project

Materials and Methods

In order to evaluate the potential benefit of these projects on the villagers' livelihood and how they build resilience against the impact of climate changes, we primarily used household surveys and semi-structured interview questionnaires (on our focus group) and expert opinions to gather the relevant information necessary for compiling this paper.

Household Surveys The first instrument focused on rising socioeconomic attributes and perception level of the benefits from CTI and MRP at household level. The survey consisted of 14 sections and 177 questions.

With an aim to arrange this work in a representative manner, 110 survey questionnaires were designed and distributed at these sites.

The other variables that we also have investigated in our survey included "the sources and level of income", "the number of family members", "how do they judge their farm operations in the previous years", "the types of businesses in which participants were involved in" and "income source". We requested the participants to respond considering their responses against the benefits obtained from the CTI and MPR, like the level of income.

Semi-structured Interview Questionnaires The second instrument was used to elaborate and verify the household heads respondents' views on the socio-economic attributes and the third instrument was used to engage expert opinions from the government and climate change implementing agencies' officials on their opinion on how effective the CTI and MRP projects were implemented at these sites. These engagements are critical to ascertain the level of benefits these implementing agents perceived that the people gained from these projects particularly CTI since it was funded by the government and aid donors. The questionnaire consisted of four sections with 50 open questions.

There were 30 interviewees representing the focus and expert groups selected and interviewed purposively from the study areas. The survey questionnaires were targeted mainly towards the participated household heads and individuals both male and female (age 25–72). Besides the main socio-economic factors we also investigate the "reasons why there was a general shift in agriculture to fisheries" and "why villagers were attracted to go to urban centers". Furthermore, we also examined "the reasons why the respondents at home spend less time at farms than other business ventures". With the expert and focus groups we have interviewed seven officials from the government (ministry of environment, climate change and ministry of fisheries), five from the NGOs and further five from the former NGO and project employees. From the project sites we have further interviewed four groups from each site, the group interviewed consisted of "youth's representatives, mothers' unions, village elders, and government workers such as teachers and nurses".

In engaging the perception of the respondents we designed and provided them with alternate levels of benefits (for example, rank from 1 to 5) or possible scenarios ("satisfied with the level of your income" or "not satisfied with your income") and requested them to tick which number or statement that suited them according to their judgment. We then grouped those rankings and compiled them in tables as presented in the results and discussions section for our analysis discussions.

In the process of designing and formulating the measurement instruments, we have adopted a livelihood sustainable development approach (Kelman and West 2009). This had assisted us in critically analyzing the responses of the participants against the satisfaction and their perceptions towards the projects implementations and how useful the projects are in assisting them adapt to the increasing adverse impact of climate change.

Results and Discussions

A total of 110 household surveys were conducted at all three project sites, 50 in Sairaghi, 30 in Naro and, 30 in Oibola, between May and June 2013. In addition there were also 30 focus groups and expert opinions interviews were made at the sites including the capital city, Honiara. Majority of respondents from the three sites were male. Thus, in Sairaghi they were 80 %, in Naro 60 % and in Oibola 63 %. The Table 17.1 shows the basic socio-economic characteristics of the respondents from the three project sites.

Table 17.1 The social characteristics of villagers in the study area (three sites)

Gender	Sairaghi	Naro	Oibola
Male	80 %	60 %	63 %
Female	20 %	40 %	37 %
Total	100 %	100 %	100 %
Marital status			
Single	4 %	2 %	2 %
Married	37 %	23 %	20 %
Widower	1 %	3 %	1 %
Divorced	2 %	0 %	3 %
Widow	1 %	0 %	1 %
Total	45 %	27 %	27 %
Educational level			
No formal education	2 %	1 %	0 %
Primary school	12 %	12 %	7 %
High school without grad	8 %	5 %	5
High school general	17 %	8 %	15 %
Tertiary	6 %	2 %	1 %
Total	45 %	27 %	27 %
Does the family member live with the family?			
Yes	35 %	15 %	19 %
No	10 %	12 %	8 %
Total	45 %	27 %	27 %
Years stay away from home			
No stay away from family	35 %	15 %	19 %
Between 1 and 12 months	1 %	0 %	1 %
More than 12 months but less than 24 months	0 %	3 %	3 %
More than 24 months but less than 36 months	8 %	4 %	1 %
More than 36 months but less than 48 months	1 %	5 %	1 %
More than 60 months	0 %	1 %	2 %
	45 %	27 %	27 %
Average monthly income	SBD1,423	SBD1,936	SBD1,787
Average monthly spending	SBD973	SBD1,233	SBD1,103
Main sources of income			
Fishing and marine products	40 %	30 %	60 %
None farm products and others	36 %	13 %	13 %
Own farm	8 %	43 %	7 %
Wages	16 %	13 %	20 %
Time spent on farms in 1 year (2012)			
0 months	22 %	47 %	47 %
3 months	48 %	0 %	50 %
6 months	24 %	7 %	0 %
9 months	2 %	20 %	3 %
More than 11 months	4 %	27 %	0 %

(continued)

Table 17.1 (continued)

Gender	Sairaghi	Naro	Oibola
Types of businesses owned by family members			
No formal business	26 %	10 %	13 %
Fishing and marine products	40 %	30 %	60 %
Handicraft and shell money	36 %	13 %	13 %
Agro-businesses	8 %	43 %	7 %
Commerce and trading	12 %	17 %	10 %
Others	10 %	10 %	20 %

Sairaghi, Western Province

The age of the respondents ranged from 33 to 70 years, with a mean of 43 years. From respondents, 8 % were chiefs, 62 % were household heads, 20 % were house wives, 6 % were relatives and 4 % were none relatives.

In terms of education, 4 % of the respondents from this project site did not attend formal education. From those who obtained formal education, 42 % have reached primary level and 54 % secondary school level. Of this 54 %, 36 % did not complete their education. In terms of *income generation*, 60 % of the Sairaghi respondents had casual employment as the main source of income, 32 % had regular jobs, mainly from nearby township of Gizo and 8 % responded to have no form of consistent income. It was also confirmed that majority of these respondents who confirmed casual employment as their main sources of income depend on the marine resources for their income and livelihood.

Naro, Guadalcanal Province

The age of the respondents from this site spreads from 25 to 72 years with an age range of 47 years. The survey respondents were structured as follow: 3 % chiefs, 37 % household heads, 23 % house wives, 7 % relatives and 10 % others. Compare with the other two sites, this site has experienced good support of the community members especially the youths and the church leaders.

The analysis showed that 67 % of respondents from this site have completed their education while 33 % have either abandoned (23 %) or incomplete (10 %) studies, both at primary and secondary levels. This was the highest percentage of respondents with no formal education at all the three project sites.

Oibola, Malaita Province

The age of respondents was ranged from 25 to 68 years, with a mean of 43 years. From the age distribution and interviews we know that many young people have left the village to seek for better opportunities at urban centers to find employment and support their families.

There were 7 % of chiefs and village elders, 50 % of household heads, 27 % of house wives, 13 % relatives and 3 % of non relatives who have responded to the survey.

In terms of education, we noted that 63 % of the respondents have completed their education while 23 % did not and, 13 % had already abandoned their education endeavors. Although there were schools at the surrounding villages, the rate of respondents with incomplete education background was relatively high.

Discussion

In terms of household income, a previous study which was conducted on a similar project at Arnavon, Choiseul province in 2007 for the economic impact of MPAs on the surrounding communities had revealed that the average household income of respondents was SBD795 (USD104) per month. Compared to these three sites, the average monthly household income ranged from SBD1,450 (USD194.00) to SBD1,787.00 (USD134) (Table 17.2). This was about USD90.00–USD238.00 more than the average income of the residence of the Arnavon.

Respondents from Naro community have a higher average income than Oibola and Sairaghi because of their proximity to the capital city and thus the central market (von Thunen's rule). This is reflected in their relatively higher cash flow compared to Oibola and Sairaghi sites. On the other hand, they also recorded a relatively higher level of expenses as compared to the two other sites. This high monthly spending also indicated their access to cash and the weak saving attitude this community has.

Further analysis showed these project sites were depended on similar sources of income structures. This was because there was no huge difference in the income base and sources at these selected sites. Because of this we identified and narrowed down the sources of income to four main categories. From those categories, it was

Table 17.2 Average monthly income and spending for the three project sites

Project site	Ave: income (p.m.)	Ave: spending (p.m.)
Sairaghi	SBD1,423.00	SBD973.00
Naro community	SBD1,936.00	SBD1,233.00
Oibola community	SBD1,787.00	SBD1,103.00

Table 17.3 Sources of income for the project sites

Source of income	Sairaghi (%)	Naro (%)	Oibola (%)
Fishing and marine product	40	30	60
None farm products and others	36	3	13
Own farm	8	43	7
Wages	16	13	20

evident that Sairaghi and Oibola community sites were depended on marine ecology for their main source of income compared to Naro project site.

Oibola respondents have depended largely from the marine resources compared to both Naro and Sairaghi (See Table 17.3). This was mainly because of the scarcity of farm land due to salt water intrusion on traditional farming area. The Sairaghi respondents though dependent on marine resources, they also have enough land for farming while Naro project site has abundance of farm land both for commercial and domestic farming. With regards to paid wages sources, it varied from 13 to 20 % at the project sites.

The analysis shows that majority of the respondents were not able to afford the basic family needs with their current level of income, 46 % of the respondents confirmed that their income was not enough to meet bare needs in life; these basic needs were including school fees for their children and private medical cost of family members. A further 38 % of all the respondents rated their income as sufficient but only to meet necessary expenses such as education and basic health cost, 14 % confirmed that their income was sufficient but not enough to buy decent things in life. The remaining 3 % confirmed that they could meet their children's school fees, private health costs with some expensive goods with their income without much restriction.

In terms of commercial activities, overall 37 % of the respondents believed that there was no difference in their operations from 2012 compared to 2011. The year 2011 was used as bench mark for comparison of 2012, the most recent year of farming operation before this study. From the study there were 35 % who confirmed their operations as "worse" in the same period. Furthermore; 16 % of the respondents have compared their operation as "worst" while 2 % couldn't compare any farm operations with the previous year. The majority of respondents confirmed that the profitability of their business operations was deteriorating instead of positively improving. This was reflective of the poor market produce by Sairaghi and no farming operation with the Oibola site. Although Naro site which generally has good farming land also seen most of its villagers finding jobs in non-farm sectors such as taxi services etc. in the main city, Honiara. In terms of profitability at each project site (Table 17.4) more than 58–80 % have confirmed that their profitability either was "worse", "worst" or "no commercial farming" to compare in the last year. It was important that operations at these projects sites were profitable in order to attract villagers to continue operating at these provincial sites.

This is important because in order to retain the villagers with their farming business, there must be some form of motivations such as improvement or growth

Table 17.4 Profitability of farms in 2012

Descriptions of profitability	Sairaghi (%)	Naro (%)	Oibola (%)
Better	14	7	7
Similar	20	43	13
Worse	42	2	33
Worst	8	20	14
No commercial farming	8	10	33

in their profitability to keep them in operation. The cost of transportation, access to market, deteriorating quality of farm land and adverse impact of the changing climate were some of the main contributing factors that have contributed to the decline in the villagers' profitability.

Because of the declining farming profitability at these project sites it was evident that respondents also spent less time in their farms. Looking at the result, 22–47 % of respondents spent between "0 and 1 month" in their farm operations. Oibola and Naro sites have recorded the highest respondents under this category. The reason for this was that at Naro, a good number of their respondents have confirmed to find employment and other opportunities besides farming in the area. For example, some of its respondents worked as bus and taxi drivers in the transport sector at this site. Some of the villagers reside in the area but have employment in the city thus they spend only weekends on their farms. At Oibola there were less time spent in farms because mostly there was no farm around the site. Over all from the result (Table 17.5) it was evident that 70 % of Sairaghi spent less than 6 months during the past 12 month on their farms. Oibola project site has recorded 97 % for the same while Naro project site has recorded the highest percentage of respondents to spend more than 6 months (54 %) in their farms.

Further analysis showed that there were several reasons (Table 17.6) why villagers at these sites were abandoning their farms and spending more time on other livelihood activities. From results, it was evidenced that 59 % of the overall respondents revealed that farm lands were becoming low in quality, 11 % each of the respondents have confirmed that there weren't enough capital to do commercial farming and the other 11 % revealed that the land area allocated for farming was too far and costly to cultivate. In addition, 5 % revealed that there wasn't any road access to the suitable farm land; a further 5 % confirmed there were conflicts over ownership of the land while 4 % revealed that their next farming plot were in a protected area. There were 2 % who have confirmed that agriculture was no longer profitable while 4 % still believed that other reasons also contributed towards general abandoning of the agricultural activity within the area. From this finding it was highlighted that "poor land quality" and "lack of capital and resources" were the main reasons for the mass desertion of agricultural activities. Long distant from markets and production was also another contributing factor to these obstacles. To revert this growing trend the responsible authorities should try to address these factors at these respected sites.

Table 17.5 Time spent in the farm 2012

Periods (times) in months	Sairaghi (%)	Naro (%)	Oibola (%)
0 months	22	47	47
3 months	48	0	50
6 Months	24	7	0
9 months	2	20	3
More than 11 months	4	27	0

Table 17.6 Reasons why agricultural production is declining at the three sites (combine)

Descriptions (Reasons)	Percentage (%)
Agriculture not profitable	2
Not enough capital and other resources to farm	11
Land is low quality	59
There is no road access to farm land	5
Too far-farm operation (distance)	11
Conflict over ownership	5
Land located on protected area	5
Other factors	4

It was further evidenced that a reasonable number of villagers at these sites have migrated to other parts of the country, especially to urban centers in search of better employment and other opportunities. From the result (Table 17.3) it showed that generally 69 % of the respondents remained with their families throughout the year while 31 % left their homes. Those who left were mostly the strong men and youths that should cultivate the farm land and participate in other livelihood activities in the villages. The analysis also showed that the length of time these villagers left their respective communities varied. For example the result (Table 17.9) has showed that between 53 and 78 % have remained at their village site during the period. It was evident that 30 % of Oibola respondents left home between 12 to more than 48 months. Naro site has reported 47 % and Sairaghi 22 % for same. The general finding from the result showed that between 12 and 17 % have left these sites 2 or 3 years. This was the longest duration, whereby the highest percent of respondents had left these sites. This analysis revealed that despite the increasing number of villagers leaving these sites, there were still others who remained in the villages and have adapted to the village environment even when faced with the adverse impact of climate change at their selective villages (Table 17.7).

The results have shown that there were several reasons why the villagers have left the project sites to mostly urban centers during the study period. The result (Table 17.8) has shown that between 53 and 78 % of respondents remained at these sites during the study period. Furthermore between 16 and 20 % respondents left for employment, between 1 and 10 % left for business related reasons while a small

Table 17.7 Duration of time members stayed away from home, 2008–2012

Duration in months	Sairaghi (%)	Naro (%)	Oibola (%)
0 months	78	53	70
Between 0 and 12 months	2	3	0
Between 12 and 24 months	2	13	3
Between 24 and 36 months	12	17	13
Between 36 and 48 months	4	13	7
Between 48 months–	2	0	7

Table 17.8 Reasons to stay away from home

Purpose stay away from home	Sairaghi (%)	Naro (%)	Oibola (%)
No stay away from home	78	53	70
Work	16	20	17
Business	1	7	10
Combinations	3	3	3
Others	2	0	0

percentage between 2 and 3 % left for combination of business, work and other reasons. It was evident that a larger portion of respondents at these sites have left the villages mainly for work related reasons as a result of the financial hardship these people faced at these sites.

From the individual project site analysis, we noted that Naro site has reported the highest respondents with 20 % who left to find employment. This was possible because of the proximity of the capital city which is less than 2 h drive away by truck. Oibola respondents who faced with the difficulty of finding alternate sources of income to their depleting marine resources has recorded the second highest respondents (17 %) with Sairaghi at 16 % for the same reason. It was also evident from the result that between 7 and 10 % respondents from Naro and Sairaghi have left their respective sites for business purposes. This has reflected the real situation at Sairaghi site as most of its villagers only reside around its locality. Furthermore the distance between Sairaghi and Honiara, the capital city of the country is quite far and would be costly for one to leave this site in search of other opportunities.

During their absence from villages, respondents have participated in various occupational activities, mainly at urban centers. They usually sent money earned from these various occupational activities to their families as part of the local remittances to support them met some of their basic livelihoods needs. With those who remained at the villages, overall result showed that they have also ventured in various livelihood activities to help them cope with the financial difficulties and assist them in building resilience to impact of climate change. The analysis therefore revealed that overall; the highest percentage of respondents depended on marines resources as their main source of income at the three sites.

Table 17.9 Types of businesses the respondents were participated in around the project sites

Types of business	Percentage (%)
Agro-business	12
Tourism	6
Handicrafts, shell money	6
General commerce and trade	6
Fisheries	15
Others sources	7
Subsistence farmers (no specific business)	48

The second majority of the respondents confirmed to depend on agro-business, while the rest of the respondents revealed that they depend on tourism, handicrafts, including shell money production and general trade. Table 17.9 show the percentage breakdown of the types of business villagers involved in at these sites.

The results for the individual site showed a similar trend as the overall observation for all the project sites (Table 17.10). It showed that Sairaghi and Oibola respondents depend more on fishing and marine ecology for their main sources of income, while Naro respondents depend more on agriculture (farming) for the same. Beside marine resources, Sairaghi respondents had depended on handicrafts, general commerce (trading) and other business ventures including tourism and paid beaches etc. The Oibola site on the other hand depended also on handy craft, shell money production and other sources for their alternate income to marine resources.

The analysis also showed that despite of the financial hardship the respondents experienced, there were some respondents who were able to save some money and invest for their future household uses at their respective project sites. The result has showed (Table 17.11) that, 33 % of the respondents have invested their money in consumable goods. The next 26 % of the respondents have invested in children's education while 16 % each invested in new family business ventures, 14 % in permanent houses while 11 % in financial institutions for various purposes.

Conclusion

From the analysis, we concluded that sources of income at these projects sites were similar and based mainly on marine resources. There was no huge difference between the levels of income and the sources at the three study sites. In regards of climate change impact, Oibola site has experienced adverse impacts on a relatively larger magnitude than the other sites, with consequences in the villagers' livelihood. For example, Oibola's respondents have indicated that they could not cultivate their traditional allocated farm plots because of the intrusion of salt water into those farm lands. This fact had increased the level of their dependency on marine resources and other trading activities.

(continued)

Table 17.10 Types of businesses owned by family members, three project sites

Types of business	Sairaghi (%)	Naro (%)	Oibola (%)
No business	26	10	13
Fishing and marine products	40	30	60
Handicraft, shell money etc.	36	13	13
Agro-business	8	42	7
Commerce and trading	12	17	10
Others	10	10	20

Table 17.11 Investment options at the three sites (combine)

Descriptions (Reasons)	Percentage (%)
Banking savings	11
Permanent house	14
Consumable goods	33
Education	26
New business ventures	16

We found that there was a general decline in agriculture sector across the three project sites. This fact was reflected from the decline of time spent in farms by respondents across the three project sites. As consequence, an increase in emigration of youth and the strong men to urban centers, searching for employment and better opportunities. However, this trend was observed to decrease in the last 2 years, as more villagers (especially men) have returned to participated in these projects.

We also noted respondents have engaged in commercial trading activities, besides fishing and farming, in order to support their families. This included setting up paid beaches and recreational areas at Sairaghi. Those who remained in agriculture and fishing activities constantly faced higher transportation costs for their produce, given the fact that they could eventually get better prices in Honiara. In spite this challenge, respondents continue to find alternate ways to sell their products at villages and surrounding urban centers.

In terms of development and management skills of these resources, the projects have contributed positively to the respective communities. Despite a slow economic impact on the community, it brought unity amongst the community such as the example of Naro and Oibola. The youth, church administration and the community village leaders at Naro have joint hand to ensure that the CTI is properly established and villagers build resilience to the impact of climate change. The introduction of MRP in Oibola area that is owned by the village was an example of successful self-governance where the villagers decided to take the leading role in deciding which type of project

(continued)

would be beneficial to the village and then they implemented it. Although the Sairaghi site has experienced some difficulties in the initial stage of the project implementation it was observed some progress as the project continued. Overall despite the implementation of these projects the respondents at these sites are yet to overwhelmingly experience the economic impact of these projects on the communities as expected and desired.

The Oibola successful experience on self-governance and autonomous decision on conservation actions at local level suggest that when villagers take conservation decisions, like the case of the MRP initiative, enabled villagers to own the project and ensure a higher sustainability level. The Naro project site has also shared the same perspective. The CTI has received a huge support from the community because of the involvement of youth and church leaders in managing and overseeing the establishment of the project at this site. In future projects, it would be important that villagers started to participate in the project identification, design, planning and implementation, to remove any distrust that may frustrate or delay the effective implementation of the project.

Acknowledgement We thank the PACE-SD for funding the research trips to the Solomon Islands and particularly to these project sites. We acknowledge the immense help received from the scholars whose articles are cited and included references of this paper. The authors are also grateful to authors / editors / publishers of all those articles, journals and books from where the literature for this article has been reviewed and discussed. A special thank you also extends to Professor Elizabeth Holland, Director of PACE-SD and Dr. Morgan Wairiu (external supervisor) for their editorial advice and guidance on writing the manuscript.

References

Albert S, Grinham A, Bythell J, Olds A, Schwarz A, Abernethy K, Aranani K, Sirikolo M, Watoto C, Duke N, Mckenzie J, Roelfsema C, Liggins L, Brokovich E, Pantos O, Oeta J, Gibbes B (2010a) Building social and ecological resilience to climate change in Roviana. The University of Queensland, Brisbane

Albert S, Tibbertts I, Uday J (2010b) Solomon Islands marine life: information on biology and management of marine resources. The University of Queensland, Brisbane

Albert JA, Trinidad AC, Cabral R, Boso D (2012) Economic value of coral reefs in Solomon Islands: case study findings from coral trade and non-coral trade communities. Report 1, Research project: Economic valuation of coral reefs and development of sustainable financing options in the Solomon Islands. World Fish, Honiara

Dudley N (2008) Guidelines for applying protected area management categories. IUCN, Gland

Dyoulgerov M, Bucher A, Zermoglio F (2011) Climate risk and adaptation country profile. Solomon Islands: vulnerability, risk reduction and adaptation to climate change. Global Facility for Disaster Reduction and Recovery, World Bank, Washington, DC

Gagahe N (2011) Report on 2009 population and housing census. Solomon Islands population and housing census. National Statistic Office, Honiara

Huhtala A, Ambrosi P (2010) Making the most of public finance for climate action. Group 1818 H Street, NW, Washington, DC

Kelman I, West JJ (2009) Climate change and small islands development states: a critical review. Ecol Environ Anthropol 5:1–16

Lal PN, Holland P (2010) Economics of resources and environmental project management in the Pacific. ICUN and SOPAC, Suva

Sanderson J, Islam SMN (2007) Climate change and economic development, SEA regional modelling and analysis. Palgrave Macmillian, Great Britain

Slade TN (2012) Pacific experience with modalities relevant for climate change financing. In: Pacific Islands forum leaders communiqué 2011, Suva

Sore R (2010) Solomon Islands national plan of action—coral triangle initiative on coral reefs, fisheries and food security. Ministry of Environment, Disaster Management and Meterology (MECDM), Honiara

van Berkering PJH, Scheri LM, Sultanian E, Lisher C (2007) The role of marine protected areas in contributing to poverty reduction: case study 2-Arnavon Community Marine Conservation Area, Honiara

Wickham F, Clarke J, Yee D, Pauku R (2012) National climate change policy: 2012–2017. Ministry of Environment, Disaster Management and Meteorology (MECDM), Honiara

Michael Otoara Ha'apio is currently a Ph.D. student in Climate Change (CC) at the Pacific Center for Environment and Sustainable Development ?(PaCE-SD),? University of the South Pacific, Suva Fiji. He was sponsored under the European Union Climate Change Pacific Leaders program at the ?PaCE-SD.? This paper was a chapter of the research he undertakes towards his M. Sc. CC qualification. Mr. Ha'apio also has a Degree and Post Graduate Diploma in Banking and Finance and a Master of Business Administration (MBA) from USP. He has more than 10 years' experience in the Banking and Finance sector. His area of interest for further study is to "explore the Economic Impact of Extreme weather events in the Pacific", such as recent flood in the Solomon Islands and Cyclones in Tonga and Vanuatu respectively early this year, 2014.

Ricardo Gonzalez is an Environmental and Resource Economist from Chile. He earned a Doctorate in Forest Science with major in Environmental and Resource Economics from University Austral of Chile in 2010, a Master of Business Economics in 2002, and a B.Sc. in Forestry in 1996, from University of Concepcion (Chile). He started his career as forest engineer in 1996 and as assistant professor in environmental and resource economics at University of La Frontera, Chile, in 2008. Dr. Gonzalez has been working as lecturer in the School of Economics at USP since February 2013. His research interests are in conservation policies under subsistence constraints; implementation of REDD+ policies and ?land-use? change modeling; valuation of environmental goods, and payments for environmental services. He also has experience in economics of ?land-use,? spatial panel data analysis and policy evaluation related to ?land-use? policies.

Chapter 18
The Significance of Contextual Vulnerability in Effective Adaptation to Climate Change on Tuvalu

Florent Baarsch and Lan Marie Nguyen Berg

Abstract There is wide recognition of the urgent need for climate change adaptation in developing countries. Tuvalu is a Least Developed Country, and a low-lying small island state and climate change adaptation has been at the center of its national interest for several years. This has resulted in a number of national plans, reports and funding schemes. Soil pollution, rapid urbanization and water table salinization caused by sea-level rise makes the population on Tuvalu solely reliant on rainwater catchment and storage for domestic water use. Moreover, current variability in rainfall patterns makes access to freshwater one of the most critical issues for Tuvalu, as highlighted by the historically long drought in 2011. According to most recent rainfall precipitation projections in the Pacific over the next decades, the current household water stress situation is expected to be further exacerbated in the face of climate change.

Data from a governmental household survey (2006) on rainwater catchment and storage facilities on Funafuti, in addition to a thorough analysis of funding schemes for water catchment between 2006 and 2010, is used to analyze the inventories of private and public rainwater catchment and storage equipment available in Funafuti in 2006 and in 2010. Moreover, a survey was conducted in one of the informal settlements to map freshwater access in some of the poorest and most vulnerable communities on the island. Despite a high volume of investment and a number of larger projects in water catchment and storage equipment on Tuvalu since the 1980s, the analysis of the surveys and funding schemes, combined with historical meteorological data, show that up to 70 % of inhabitants still lack access to adequate catchment and storage capacity to face historically long dry spells. This paper shows that: (1) On Funafuti, the lack of socioeconomic contextualization of projects has led to an inequitable allocation of equipment. Inequitable allocation is constraining effective climate change adaptation, especially for the poorest

F. Baarsch (✉)
Climate Analytics gGmbH, Friedrichstraße 231, Haus B 10969, Berlin, Germany
e-mail: florent.baarsch@climateanalytics.org

L.M.N. Berg
CICERO, Pb. 1129 Blindern, 0318 Oslo, Norway
e-mail: Lan.marie.berg@cicero.oslo.no

W. Leal Filho (ed.), *Climate Change in the Asia-Pacific Region*, Climate Change Management, DOI 10.1007/978-3-319-14938-7_18

populations, and (2) Pre-existing and contextual vulnerabilities should be taken into account in national and international planning as well as in the design and execution of projects to ensure that adaptation to climate change will be effective.

In a context of an increasing need for climate change adaptation, especially in Small Island Developing States and other developing countries, this study highlights the current barriers to effective adaptation for the 5,000 inhabitants of Funafuti. It thus contributes to the wider debate of how international organizations, national governments and development planners can ensure equitable climate change adaptation on a global scale.

Introduction

Climate change and variability threaten freshwater access in Small Island Developing States (SIDS) because of the islands' limited size and their constrained surface- and groundwater resources. The situation on the Pacific Island State Tuvalu provides a very concrete illustration of restricted access to freshwater resources and the need for adaptation measures on SIDS to meet freshwater needs in the context of climate change. Indeed, limestone islands and atolls have no fresh surface water and therefore rely fully on groundwater extraction and rainwater harvesting (Mimura et al. 2007). However, on Tuvalu the groundwater is unsuitable for human consumption and the inhabitants depend solely on rainwater harvesting (SOPAC 2007). Furthermore, the IPCC projects that increased frequency and intensity of droughts, modified rain patterns, and extreme weather events, combined with sea level rise, threaten the ability of many small island states to meet freshwater demand during periods of low rainfall (IPCC 2014b).

In this context of existing vulnerability and projected increasing exposure induced by climate change, adaptation in the water sector became a governmental priority (Ministry of Environment 2007). With the support of international donors, the Tuvaluan government planned and implemented several large-scale projects to increase freshwater access. Despite the significant amount of funding spent on these adaptation projects and the projects deployed, a lack of resilience in the freshwater sector became particularly apparent on Tuvalu's capital island Funafuti during the drought from October 2010 to September 2011. Due to severe freshwater shortage, the Tuvaluan Government declared a national state of emergency on 28 September 2011. Several countries including Australia, New Zealand, South Korea, and Japan provided emergency assistance.

The planning, funding and implementation of adaptation projects in the freshwater sector on Tuvalu can provide relevant lessons for other projects in the same sector and especially for projects carried out in low-lying areas exposed to similar risks and vulnerabilities. Although the Tuvaluan government and the international donor community were focusing on increasing freshwater access, the 2010–2011 drought unveiled the limits of the implemented measures. This paper contributes to the overall scientific debate on adaptation and vulnerability assessment in the view of improving future adaptation projects on Tuvalu and other SIDS, as well as low-lying areas in developing and developed countries.

Data from the Public Work Department (analyzed in this study) show that while some households had very high storage and catchment capacity prior to the implementation of these projects, the majority had low storage and catchment capacity. Furthermore, while the average monthly income was roughly AUD$ 800 (Government of Tuvalu 2010a), some of the poorest reported an average of only $200. As the monthly average cash spending on Funafuti is $1,218 per household, the poorest households struggle to meet their basic needs, including access to freshwater (Government of Tuvalu 2010a). Freshwater is mainly secured through private catchment and storage. Access to freshwater is thus dependent on the catchment and storage facilities of the respective households. Thus, people's relative wealth decides their ability buy equipment to catch and store water, and to purchase extra water from public catchment and desalination systems (PACC 2006). Despite pre-existing disparities in water catchment and storage, the adaptation projects supported by the European Union and Australia provided to each household on Funafuti one water tank of 10,000 L for storage and roof and gutter equipment to facilitate rainwater catchment. Based on the most recent IPCC definition of contextual vulnerability, these pre-existing disparities in capacity and income within the Tuvaluan population are interpreted as contextual vulnerability (IPCC 2014c) in this study.

This paper presents previous measures to increase private catchment and storage capacity and discusses the lack of contextual vulnerability assessment as a major constraint to ensure freshwater access on Funafuti. It argues that a social and contextual view on vulnerability is needed to ensure effective adaptation and illustrates the need to emphasize contextualized vulnerability assessments prior to implementing adaptation measures.

The first section of the study reviews the concept of contextual vulnerability and how it is currently being used in regards to adaptation to climate change. The second section analyses the present vulnerabilities of the Tuvaluan population through a contextual vulnerability framework. Finally, the third section examines how the lack of contextualization of the pre-existing vulnerability has led to ineffective adaptation in the water sector on Tuvalu.

Contextual Vulnerability to Address Adaptation Needs

Recently, the focus on effective adaptation has been growing in the scientific literature, although it still remains at its initial stage (see chapter 15.5.1.3 of the IPCC AR5). The factors determining "effective" still have to be fully inventoried and assessed in the light of the lessons learnt from implemented adaptation projects. The objective of this chapter is to demonstrate that the assessment and integration of contextual vulnerability in the planning and implementation of adaptation projects plays a strong role in ensuring effective adaptation. In a general review of the existing literature on adaptation to climate change and its barriers, Biesbroek et al. (2013) find that the majority of the barriers to effective adaptation are not only climate related. The barriers relate to other environmental problems or to the implementation of policies at different levels. Furthermore, frameworks and studies analysing barriers to adaptation as well as addressing the barriers are limited (Biesbroek et al. 2013). Moreover, studies on how to redesign adaptation related policy processes and institutions are still infrequent (Dovers and Hezri 2010). In this context a better understanding of how the assessment of contextual vulnerability enables more effective adaptation to climate change also contributes to deepen scientific understanding of effective adaptation barriers and enablers.

For many adaptation scholars the success of adaptation measures depends on effective understanding of human vulnerability (for example Gallopín 2006; Reed et al. 2013). The Fifth Assessment Report of the IPCC defines vulnerability as "the propensity or predisposition to be adversely affected. Vulnerability encompasses a variety of concepts and elements including sensitivity or susceptibility to harm and lack of capacity to cope and adapt." (IPCC 2014c). While contextual vulnerability is defined as "a present inability to cope with external pressures or changes, such as changing climate conditions. Contextual vulnerability is a characteristic of social and ecological systems generated by multiple factors and processes" (IPCC 2014c). Furthermore, people's capacity to adapt to climate change is shaped by factors such as wealth, access to technology, education, infrastructure, institutions, and governance (Engle 2011; Smit and Pilifosova 2007). As a consequence, climate change related vulnerability is not equally distributed between or within societies (Adger et al. 2007). Vulnerability can differ between individuals, between households, within a community and between communities (Kasperson et al. 2001).

People who are poor and lack access to education or proper infrastructure are vulnerable to climate change, as well as to other forms of stress (Kelly and Adger 2000). Kelly and Adger see vulnerability as a starting point and uses the analogy of the "wounded soldier" to explain that those who are already wounded, are susceptible to more harm. Consisting with the starting-point approach, O'Brien et al.'s (2007) framework of contextual vulnerability connects vulnerability to climate change with broader development challenges. Climate variability and change happens within a context of "political, institutional, economic, and social structures and changes" (O'Brien et al. 2007). The contextual conditions interact dynamically with exposure, which might be related to current climate variability and change or to other processes. Thus, contextual vulnerability conceptualizes vulnerability not only to climate change but also to various forms of change. Viewing vulnerability in a contextual framework enables us to understand the broader conditions in which climate change affects people differently across and within societies (O'Brien et al. 2007). As the drivers of vulnerability are created by the wider context, vulnerability to climate change cannot be isolated from vulnerability to other stressors. Therefore, changing the conditions for one contextual stressor can change people's overall capacity to adapt. Thus, potential measures to address vulnerability are closely related to issues such as poverty and development (O'Brien et al. 2009). Eriksen et al. (2011) found that in the implementation process of adaptation measures, project designers must take into account the specific vulnerability conditions, including climate change related vulnerability and poverty. Adding to this, Noble et al. saw poverty and persistent inequality as being the "most salient" conditions carving climate change-related vulnerability as vulnerability is influenced by the "capacity of groups and populations to reduce and manage the impacts of climate change".

Because vulnerable people and places are often excluded from decision-making, adaptive measures tend to benefit the relatively advantaged (Adger 2006). It is therefore important to analyze and recognize potential values and conflicts of interests within the individual societies and how these might influence the outcome of adaptation measures. To avoid maladaptation and to ensure that also the most vulnerable populations are targeted by adaptation measures, it is crucial to recognize the contextual conditions of various people and groups within societies. Jones (2010) proposes to mainstream social barriers within wider adaptation policy by ensuring that adaptation options recognise the impediments that represent social and economic barriers to the successful implementation of adaptation to climate change. Following the most recent literature, IPCC AR5 integrates an inclusive definition of the concept of risk, which involves both climate change and non-climate change related aspects such as socioeconomic risks (e.g. fluctuations in commodity prices), and environmental risks that are not climate change related (e.g. local pollution).

In the context of recent literature on contextual vulnerability and effective adaptation, the current chapter investigates how the limited assessment of pre-existing contextual vulnerability of the water sector adaptation projects in Tuvalu lead to less effective adaptation and further increased vulnerability gaps between different socioeconomic classes. The following section analyses the climate and non-climate related risks to which Funafuti households' water access are exposed and investigates the sensitivity and adaptive capacity on the household level.

Vulnerability and Risk for Funafuti Households

Household water access on Funafuti is exposed to a double risk. Firstly, human activities cause groundwater contamination and leave the population solely dependent on rainwater catchment and storage. Secondly, climate change contributes to modification in precipitation patterns and therefore reduces reliability on rainwater catchment and storage as a source of safe potable water. Household fresh water access is met through private individual catchment and storage facilities. Income level and catchment and storage capacity of the individual household decide the level of vulnerability. The lack of measures to meet the needs of the poorest and most vulnerable households has limited the effectiveness of the wide range of adaptation projects in the water sector on Funafuti.

Contaminated Water Resources

The Government of Tuvalu recommends that groundwater on Funafuti be only used for secondary purposes, as it is contaminated by leaking septic tanks, animal waste, heavy metals, and saltwater intrusion, and thus unsafe for human consumption.[1] Leakage from septic tanks and soak pits easily enters the shallow groundwater resources situated only between 1 to 1.3 m below the ground. As the islands are partly built by cobble and boulder size sediments, pollutants travel easily between the groundwater system, the lagoon and the reef (Crennan 2004).

Most households in Funafuti use septic tanks as part of their basic sanitation systems (Crennan 2004). However, to minimize expenditures the construction was simplified and in 2001, 96 % of the tanks were found to be inadequately constructed and not suited for Funafuti's geology (Government of Tuvalu 2007). Crennan (2004) found that the tanks are leaking directly into the groundwater. Moreover, feces from pigpens also contaminate the groundwater and lagoons (Government of Tuvalu 2007) and the lack of proper waste management causes high concentrations of heavy metals in the groundwater resources (PACC 2006). The copper, lead, arsenic and chromium content are well above the recommended World Health Organization (WHO) standard of noxious heavy metals in drinking water (Ministry of Environment 2005).

Fresh groundwater in low atolls, like Funafuti, is balanced between episodic rainfall, replenishment, and continuous depletion by evaporation, extraction and outflow to and mixing with seawater (White et al. 2007). Seawater intrusion is a significant threat to fresh groundwater supplies of small island nations due to the transmissive qualities of the geology and the limited sizes of the water lenses (White et al. 2007). The IPCC lists saline intrusion into freshwater lenses as one of the key climate change related issues on Small Island Developing States (Mimura et al. 2007). This is already an issue on Funafuti. Webb found that the groundwater on Fongafale has a conductivity of $4{,}252 \pm 1{,}741\ \mu S.cm^{-1}$, which is not safe for human consumption.[2] The population on Funafuti is thus totally dependent on rainwater catchment and storage for freshwater consumption. However, changes in precipitation patterns make rainwater access unreliable.

Changing Precipitation Patterns

Despite a high average annual rainfall of 3,500 mm/y on Funafuti, droughts and water shortage occur regularly (SOPAC 2007). Lal et al. (2006) states that Tuvalu experiences dry weather up to 3–4 months a year, as well as prolonged droughts

[1] Interview with Dr. Nese Ituaso-Conway, Chief Public Health, Princess Margaret Hospital, Ministry of Health, Funafuti, Tuvalu. 23.3.2011.

[2] According to Falklands guidelines for coral atoll groundwater lenses (Webb).

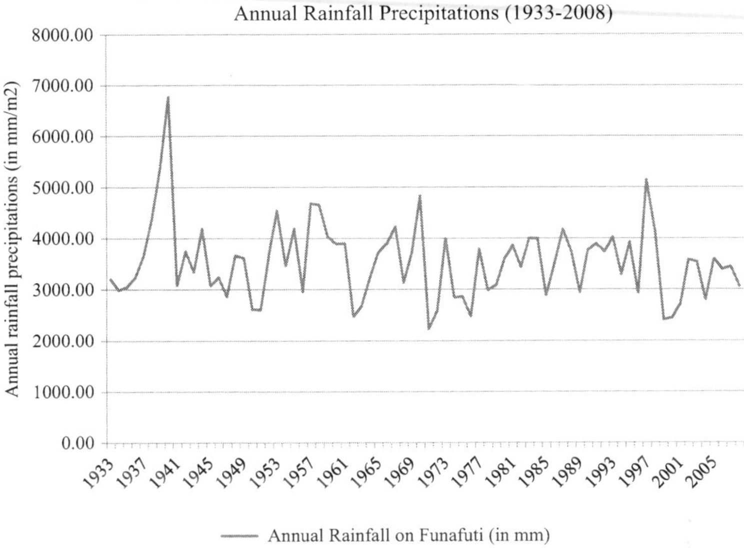

Fig. 18.1 Annual rainfall precipitations on Funafuti from 1933 to 2008. *Source*: Meteorological Office of Tuvalu 2008

(Lal et al. 2006). Annual rainfall precipitation measured by the Meteorological Office of Tuvalu between 1933 and 2008 (Meteorological Office of Tuvalu 2008) show a slight decrease, which could be regarded as consistent with most recent IPCC AR5 projections (IPCC 2014b) (Fig. 18.1).

According to the recent IPCC Working Group II report (IPCC 2014b), precipitation in the southern tropical Pacific are projected to increase up to 4 % compared to 1981–2005 level by the end of the twenty-first century.[3] However, the current trend and projections show that extreme rainfall events are increasing, leading to an increased probability for longer droughts on Tuvalu (IPCC 2014b; Mimura et al. 2007). As a period of 2–3 weeks can reduce water levels in many tanks and reservoirs by up to 50 %, even a slightly higher occurrence of drought can cause serious shortage of water on Funafuti on a regular basis (PACC 2006).

[3] Projections were made in the RCP4.5 scenario, which corresponds to a 2.5 °C warming above preindustrial level, and a 1.97°C warming above 1986–2005 base period. Precipitation projections are based on the average of an ensemble of 42 CMIP5 global models.

Pre-Existing Inequalities Reflect in Water Access

Defining a Vulnerability Threshold

Sufficient catchment surface and storage capacity are the two main requirements for ensuring water access on Funafuti, and therefore determine households' adaptive capacity.

In 1986 the Tuvaluan Government launched a national objective to ensure a minimum of 50 L of freshwater per person per day. To meet this objective through private rainwater harvesting, every house on Funafuti has to be equipped with sufficient roof surface and water storage capacity per person.

In annual consumption, 50 L per person per day amounts to about 18,000 L per person. With an average precipitation of 3,500 L a minimum roof surface of about 5.2 m^2 is required per capita, assuming that consumption does not exceed 50 L per day and no water is wasted. Taking these margins into account, this paper assumes 6 m^2 (approximately 15 % more) as a reference for the adequate level of equipment required to meet the 50-L governmental target. However, as it is projected that climate change will increase rainfall precipitation seasonality, it is more reasonable to take the lowest recorded consecutive and annual precipitations as potential future patterns for extreme droughts. Based on rainfall precipitation of 1950 for the most consecutive no-rain day on Funafuti and 1971 for the lowest annual precipitation in record on Funafuti, we calculated that 9 m^2 of catchment surface per capita would have the capacity to supply 50 L of water per capita per day during most extreme experienced droughts on Tuvalu.

Droughts are often referred to as a creeping phenomenon. As the occurrence of a drought is highly contextual, it is challenging to define what a drought is, and when it starts and ends. The National Adaptation Programme of Action of Tuvalu (2007) states that: "*Droughts of up to three months and longer had occurred in Tuvalu.*" (Ministry of Environment 2007) Under the assumption that 3 months represents a significant drought period on Tuvalu, we use this time frame to calculate the minimum storage capacity per person on Funafuti: 50 L * 90 days = 4,500 L per person. Taking a runoff coefficient into account, this paper takes 5,000 L per person (11 % higher) as reference for minimum storage capacity.

As the above shows, ensuring access to 50 L per capita per day require each household to have a minimum of 6 m^2 of functioning roof surface and 5,000 L of storage capacity per capita. It is important to note that the current figures are calculated per capita but the adaptation projects implemented in Tuvalu address the needs of entire households. The IPCC defines sensitivity as "the degree to which a system is affected by climate variability". At the household level, the degree of affectedness is dependent on the number of people living in the household (the system). Furthermore, adaptive capacity also depends on household education level, income and other socioeconomic aspects (e.g. Hughes et al. 2012). Thus, socioeconomic aspects, as well as the number of people living in the households are measures of household sensitivity.

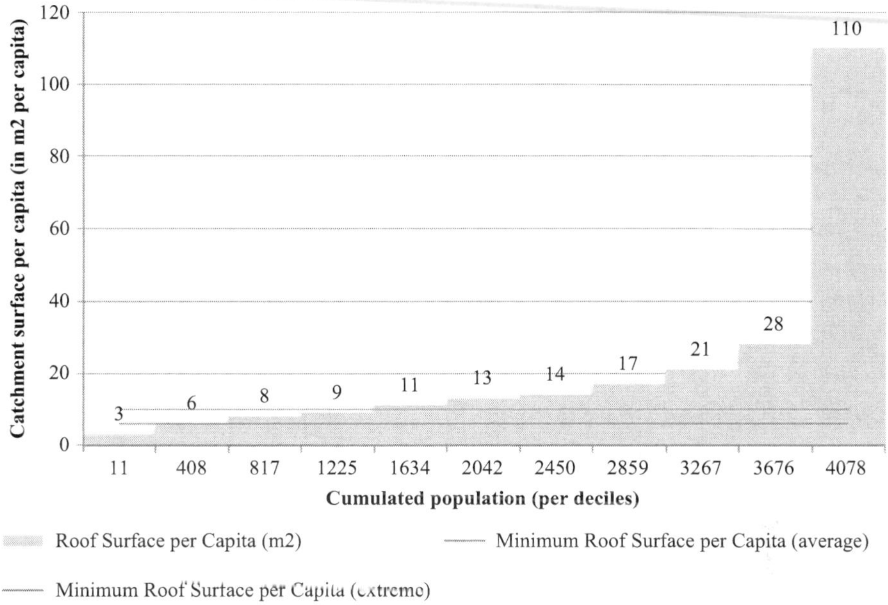

Fig. 18.2 Catchment surface per capita in 2006 on Funafuti. *Source*: Public Works Department 2006

Adaptive Capacity for Funafuti Households

In 2006 the Public Works Department (PWD) conducted a survey inventorying roof surface and quality as well as storage capacity in Funafuti. A total of 526[4] residential households were interviewed, representing 4,078 persons. The results of the catchment surface per capita as of 2006 are displayed by population deciles in Fig. 18.2 below.

As Fig. 18.2 shows, there are significant disparities in roof surface per capita within the Funafuti population. In 2006, the least equipped 10 % of the population on Funafuti share approximately 2,030 m^2 of catchment surface. In contrast, the most equipped 10 % of the population share 18,000 m^2, on average 43 m^2 of catchment surface. According to the survey, 337 persons had roof surface of less than 6 m^2, and thus, did not have enough catchment surface area to meet their water needs. These are the most exposed to low rainfall precipitation events and the most vulnerable in terms of water access on Funafuti. The 1,421 persons with less than 10 m^2 of catchment surface per capita, can be seen as vulnerable to extremely low rainfall precipitations events and future climate change impacts. Depending on

[4] This number is an approximation, as the results received from the Public Works Department contain incomplete information.

what baseline we use, between 8 and 35 % of the total Funafuti population have a low adaptive capacity to the effects of drought.

Thus, already in 2006, the majority of households on Funafuti already had access to sufficient water catchment surface while others had less than they needed. The disparity can originate from a range of factors such as poverty or limited social networks, which increase household vulnerability to climate related stress. Furthermore, disparities in water storage in particular are more significant and are studied in the following section. However, it appears in the current assessment that this disparity in adaptive capacity was not taken into account when the rainwater catchment and storage projects outlined below were designed.

Lack of Vulnerability Assessment: Ineffective Adaptation

The Government of Tuvalu, supported by different international and regional, public and private organisations, has implemented several adaptation projects in the water sector. Despite significant funding, the majority of inhabitants of Funafuti do not have access to sufficient water catchment and storage capacity. Moreover, in-country disparities have increased, leaving the poorest people of Funafuti highly vulnerable to climate related risks.

Significant Adaptation Spending in Fresh Water Access

The Government of Tuvalu has focused on meeting freshwater needs for several decades. In 1986 it announced a national goal of providing a minimum of 50 L per person per day, in line with the recommendation by the World Health Organization (Howard and Bartram 2003). National commitment is expressed in Tuvalu's National Adaptation Program of Action and in the regional UNDP project Pacific Adaptation to Climate Change (PACC).

The past 40 years saw a vast number of projects carried out, aiming predominantly at securing public and private catchment and storage facilities. Figure 18.3 provides a timeline of water projects implemented between 1980 and 2013.[5]

[5] Comprehensive information about the water projects implemented between 1980 and present is difficult to obtain. Therefore, the timeline shown in Fig. 18.3 might not be completely inclusive.

Year	Program Funders	Description and costs (when available)
1981-1984	European Economic Community (EEC) - Save the Children Federation	Provided most households on Tuvalu with a 3,760 liter tank each (Government of Tuvalu, 1988)
1984-1992	United Nations Development Program (UNDP)	Construction of six hundred 6,000 liter tanks on Funafuti (Government of Tuvalu, 1992)
1990s	USAid and Norway	Project providing cisterns and water tanks for each community's *maneapa* (meeting hall) and additional storage for the government buildings on Funafuti (Government of Tuvalu, 1992)
2002-2004	Taiwan	A 1,362,748 liter underground cistern was constructed under the new government building in order to enlarge the publicly accessible catchment area
2006	Global Environmental Facility (GEF)	Construction of a minimum of forty composting toilets within the homes of community members to help reduce wastewater pollution and increase water conservation within Funafuti (Seleganiu & Moulogo, 2010)
2007-2008	AusAid	Three hundred 10,000 liter tanks for households in Funafuti are provided
2008	European Development Fund (European Union)	One hundred 6,000 liter tanks for households on Nui, 310 10,000 liter tanks for households on Funafuti and a 4,000 liter capacity water truck supplied to the Ministry of Works. In addition it financed improvements of gutters and rooftops and community awareness programs (European Development Fund, 2010) Cost: EUR 700,000
2009	AusAid	300 10,000 liter tanks for Funafuti households. AusAid also funded improving rainwater harvesting systems (roof and gutters).
2009	European Development Fund (European Union)	One water tank to each household in the outer islands in addition to new public water storage capacity both in the outer islands and Funafuti (Government of Tuvalu, 2010b) Cost: EUR 4.4 million
2009-2013	Pacific Adaptation to Climate Change (PACC)	Long-term adaptation measures in order to increase resilience by focusing on three key development sectors: Water resource management, food production and security and coastal zone and associated infrastructure (PACC, 2006) Cost: USD 500,000

Fig. 18.3 Projects related to freshwater access on Tuvalu from 1981 to 2013

Adaptation Programs Increased Inequality

Since 2006 water storage capacity on Funafuti increased significantly due to EDF and AusAid projects. Between 2007 and 2010 these projects supplied 989 tanks to Funafuti households, allocated on a one-per-household-system (Salesa 2011). As

Fig. 18.3 shows, between 2008 and 2013 the total funding for water projects on Tuvalu was more than USD 7 million. This is approximately USD 650 per capita. However, these projects contributed to increase the water access disparities between households on Funafuti.

For this study it was not possible to obtain the results of the roof quality and storage capacity survey conducted by the Public Work Department in 2010. As an alternate we integrated an additional amount of 10,000 L storage capacity to all the households covered in the 2006 survey. This method represents three major weaknesses; Firstly, it is assumed that all the households received a tank; Secondly, it does not allow to take into account the additional storage capacity purchased by the richer households in this time period; Thirdly, it does not account for demographic growth within the households, households' population is assumed constant at 2006 level. However, adding the equal repartition of the tanks from EDF and AusAid, gives a better overview of the current situation and evolution. Figure 18.4 shows the growth of water storage capacity per capita in 2006 and in 2010.

Figure 18.4 illustrates significant disparities in water storage per capita between Funafuti households, assuming the equal distribution of water tanks as of 2010. While the average storage capacity on Funafuti is about 5,500 L per capita, 72 % of the population is currently not able to meet the 5,000 L requirement. The most equipped 10 % of the population share 30 % of the total water storage capacity with an average of 13,971 L per capita, while the least equipped 10 % share only 3 %, or on average 1,487 L per capita.

The EDF and AusAid projects allocated one tank to each household on Funafuti, regardless of existing equipment or number of household members. However, according to the 2006 survey, one hundred of the households already had access to 5,000 L or more per capita. The projects increased the ability of the population to meet the 5,000-L requirement from 17 (2006) to 28 % (2010). However, assuming the same number of water tanks provided by donors, we virtually allocated tanks per households by prioritizing households with a storage capacity of less than 5,000 L per capita in 2006, the adaptation projects would have increased the ability 31 % to meet the 5,000 L per capita requirement, 3 % more than with the equal distribution of the EDF and AusAid projects (see Fig. 18.4). The projects would clearly have been more efficient if the allocation of the tanks had been based on the number of liters per capita, reflecting the current vulnerabilities of the individual households, instead of a one-per-household system. The increase in the percentage of population having access to 5,000 L storage capacity is rather limited but the vulnerability based allocation is particularly relevant for the least equipped households (the first 10 percentile in the calculations), whose average storage capacity increases by about 300,000 L of capacity compared to the one-per-household allocation rule.

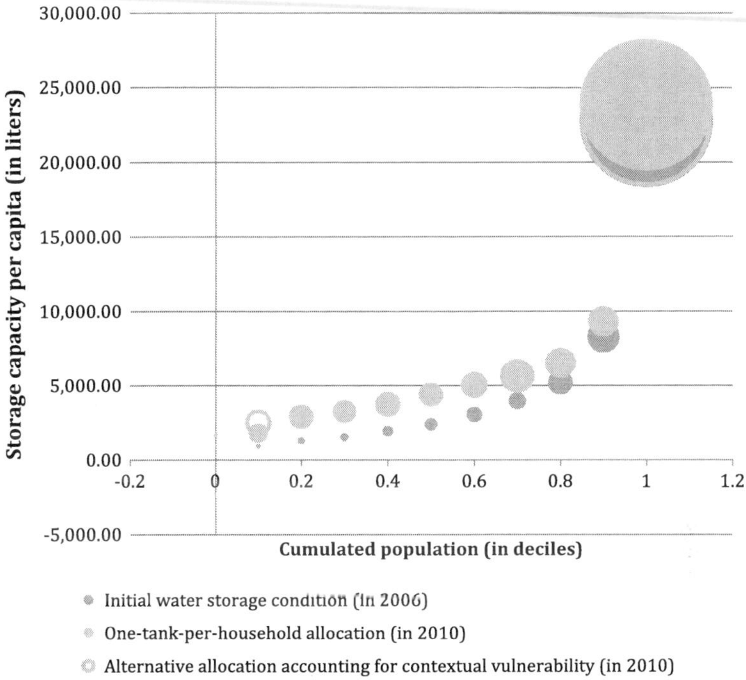

Fig. 18.4 Water storage capacity per capita in 2006 and in 2010, assuming the equal distribution of water tanks as of 2010. The size of the *circles* represents the total storage capacity per decile. *Source*: authors based on public works department 2006

The Poor Remain Highly Vulnerable

To estimate how the most vulnerable populations of Funafuti had benefited from the implementation of the adaptation projects, a survey was conducted in one of the poorest areas of the city. The results of the survey are then compared to the level of equipment of the rest of population.

In May 2011 a survey of water systems and use in the settlement along one of the borrow pits on Funafuti (Fogafale) was realized for this study.[6] The inhabitants living along the borrow pits of Fogafale are among the poorest on Funafuti. Most of the poorest populations on Tuvalu live on the outer islands. However, the urban

[6] "Borrow pit" is the name given to the large holes dug out by the Americans in the construction of the airstrip and other infrastructure during World War II. The borrow-pits on Fogafale are now used as informal waste dumps, building and cleaning of pigpens, construction of houses, swimming and bathing (children especially). Both water quality assessments carried out in 2005, revealed that all borrow-pits on Fogafale are highly contaminated by faecal bacteria and has low dissolved oxygen.

SOPAC (2007). *Sustainable Integrated Water Resources and Wastewater Management for Tuvalu. Diagnostic Report*. August 2007. Page 20.

poor living on Funafuti, are seen as the worst off. In contrast to the low-income households on the outer islands, those on Funafuti often lack access to arable land on which to carry out subsistence agriculture activities. Moreover, the urban poor typically live in substandard housing with insufficient access to water, electricity and waste management services as well as more than likely, tenure insecurity (SOPAC 2007). The inhabitants of the particular borrow pit between the Nanumea maneapa (meeting house) and the Vaitupu maneapa are among the poorest on Funafuti. Many live in shack-like conditions and have few sources of monetary income even though kinship support plays a significant role in supporting these households.

For the purposes of this study a total of 26 households, or 158 persons, were surveyed. Two households did not participate in the interview, as the owners were absent at the time the survey was conducted. However, their catchment and storage facilities were measured and studied alongside the other households.

According to survey results, the poorest population remains highly vulnerable to climate change related impacts despite adaptation projects. The findings show that average daily water consumption per capita in the borrow pit was 40.7 L. Around 80 % of the households report that they often suffer water shortages.[7] In the borrow pit, 60 % of the 26 households surveyed presented a significant lack of roof, gutter and pipe maintenance. Figure 18.5 compares the roof quality of the households of the borrow pit to all households of Funafuti.[8]

The lack of maintenance and repair primarily affects the household's capacity to harvest or store enough water to meet its needs. To enhance efficient water harvesting and use, a focus needs to be put on maintenance of the water tanks, roof and pipe systems. Similarly, average water storage capacity per capita is lower among this population than among the rest of Funafuti population. On average it was measured that people living in the borrow pit had access to approximately 1,795 L, which correspond to the 10 first percentiles of the Funafuti population (1,807, following the one-per-household tank allocation rule).

The results of the survey conducted among the most vulnerable population of Funafuti highlight the lack of contextualization of the adaptation projects implemented on Funafuti. The households of the known poorest and least equipped population of the capital city did not receive special consideration in order to overcome their preexisting vulnerability. Of the 26 households surveyed, 13 had only one 10,000 L tank. Furthermore, only two had sufficient water access to cope with the consequences of a drought longer than 3 months. For the rest of the borrow

[7] When the survey was realised on the 4 and 5 of May 2011, 19 of the 24 households reported being in a situation a shortage between every day and twice a year.

[8] The survey focused on roof surface, roof quality and storage capacity was conducted by the Public Works Department (PWD) in 2006. 526 (This number is an approximation, considering that the results we received from the PWD contained incomplete information) residential households of Funafuti were interviewed representing a population of 4,078 persons. This first survey was realised thanks to the research assistance from the Red Cross Tuvalu and Tuvalu Familial Health Association.

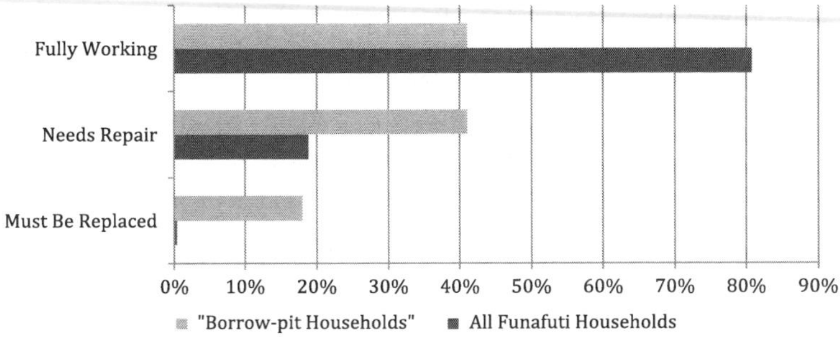

Fig. 18.5 Roof quality of the households of the borrow pit compared to all the households of Funafuti for which data were available. *Source*: authors (Borrow pit households data) and public work department 2006 (all Funafuti households)

pit population, the average number of days of water storage capacity is only 40 days, assuming a per capita consumption of only 50 L per day.

Conclusion

Tuvalu and other SIDS are highly exposed to the impacts of climate change. Therefore, they are often referred, rightly or not, as the canaries in the coal mine (Gemenne 2010). Not only are these communities experiencing the early effects of climate change, many of them are also at the forefront in implementing adaptation measures. Thus, these countries and communities are not just canaries, warning humankind of the dangers to come, but they enrich common understanding and knowledge on how to effectively adapt to the impacts of climate change.

The study shows that on Funafuti, the lack of socioeconomic contextualization of projects has led to an increase in disparities in the allocation of water catchment and storage equipment, limiting effective climate change adaptation for the poorest populations. This study highlights that pre-existing and contextual vulnerabilities should be taken into account in national and local planning as well as in the design and execution of projects to ensure that adaptation to climate change is effective. The poorest populations of Funafuti, specifically the ones living by the borrow pits are still extremely vulnerable to the effects of droughts. Their access to equipment to catch and store rainwater is largely insufficient to meet the freshwater needs of the household members. Despite the implementation of several adaptation projects, some of the persons interviewed in the borrow pit area reported to be in a chronic state of water shortage; 19 of the 24 households reported experiencing water shortage between every day and twice a year. Access to safe freshwater is a prerequisite to human, social and economic development, and therefore

(continued)

constraints on access by certain sections of the population contribute to further increase current and future vulnerabilities.

In a context of an increasing need for climate change adaptation as the temperature rises and the impacts are felt, especially in SIDS and other developing countries, this study highlights that the limited account of contextual vulnerabilities constitutes a barrier to effective adaptation for the 5,000 inhabitants of Funafuti. This study thus contributes to the wider debate of how international organizations, national governments and development planners can ensure equitable climate change adaptation on a global scale.

References

Adger WN (2006) Vulnerability. Glob Environ Chang 16(3):268–281. doi:10.1016/j.gloenvcha.2006.02.006

Adger WN, Agrawala S, Mirza MMQ, Conde C, O'Brien K, Pulhin J, Takahashi K et al (2007) Chapter 17. Assessment of adaptation practices, options, constraints and capacity. In: Parry ML, Canziani OF, Palutikof JP, van der Linden PJ, Hanson CE (eds) Climate change 2007: impacts, adaptation and vulnerability. contribution of working group II to the fourth assessment report of the intergovernmental panel on climate change. Cambridge University Press, Cambridge, pp 717–743

Biesbroek GR, Klostermann JEM, Termeer CJAM, Kabat P (2013) On the nature of barriers to climate change adaptation. Reg Environ Chang 13(5):1119–1129. doi:10.1007/s10113-013-0421-y

Crennan L (2004) Solution report: waste management, pollution prevention and improved sanitation in Tuvalu. Funafuti, Tuvalu

Dovers SR, Hezri AA (2010) Institutions and policy processes: the means to the ends of adaptation. Wiley Interdiscip Rev: Clim Chang 1(2):212–231, Retrieved from http://dx.doi.org/10.1002/wcc.29

Engle NL (2011) Adaptive capacity and its assessment. Glob Environ Chang 21(2):647–656. doi:10.1016/j.gloenvcha.2011.01.019

Eriksen S, Aldunce P, Bahinipati CS, Martins RD, Molefe JI, Nhemachena C, Ulsrud K et al (2011) When not every response to climate change is a good one: identifying principles for sustainable adaptation. Cli Dev 3(1):7–20. doi:10.3763/cdev.2010.0060

Gallopín GC (2006) Linkages between vulnerability, resilience, and adaptive capacity. Glob Environ Chang 16(3):293–303. doi:10.1016/j.gloenvcha.2006.02.004

Gemenne F (2010) Tuvalu, Un Laboratoire du Changement Climatique. Revue Tiers Monde, 1–22

Government of Tuvalu (2007) Integrated water resource management. Funafuti, Tuvalu

Government of Tuvalu (2010a) Household Income Expenditure Survey (HIES) 2010. Funafuti, Tuvalu

Howard G, Bartram J (2003) Domestic water quantity, service level and health. WHO, Geneva, Switzerland

Hughes S, Yau A, Max L, Petrovic N, Davenport F, Marshall M, Cinner JE et al (2012) A framework to assess national level vulnerability from the perspective of food security: the case of coral reef fisheries. Environ Sci Pol 23:95–108. doi:10.1016/j.envsci.2012.07.012

IPCC (2014b) Chapter 29 small islands. In: Nurse L, McLean R, Agard J, Briguglio LP, Duvat V, Pelsikoti N, Webb A et al (eds) Climate change 2014: impacts, adaptation, and vulnerability. Part A: global and sectoral aspects. Contribution of working group II to the fifth assessment report of the intergovernmental panel on climate change. Cambridge University Press, Cambridge, pp 1–60

IPCC (2014c) WGII AR5 Glossary. In: Agard J, Schipper L (eds) Climate change 2014: impacts, adaptation, and vulnerability. Part A: global and sectoral aspects. Contribution of working group II to the fifth assessment report of the intergovernmental panel on climate change. Cambridge University Press, Cambridge, pp 1–30

Jones L (2010) Overcoming social barriers to climate change adaptation. London, pp 1–7

Kasperson RE, Kasperson JX, Dow K (2001) Vulnerability, equity and global environmental change. In: Kasperson RE, Kasperson JX (eds) Global environmental risk. United Nations University Press, Tokyo, pp 247–272

Kelly PM, Adger WN (2000) Theory and practice in assessing vulnerability to climate change and facilitating adaptation. Clim Change 47:325–352

Lal P, Saloa K, Uili F (2006) IWP-Pacific technical report (International waters project) no. 36: economics of liquid waste management in Funafuti, Tuvalu. Apia, Samoa

Meteorological Office of Tuvalu (2008) Monthly rainfall precipitation 1933–2008. Government of Tuvalu, Funafuti, Tuvalu

Mimura N, Nurse L, McLean RF, Agard J, Briguglio L, Lefale P, Payet R, Sem G (2007) Small Islands. In: Parry ML, Canziani OF, Palutikof JP, van der Linden PJ (eds) Climate change 2007: impacts, adaptation and vulnerability. Contribution of working group II to the fourth assessment report of the intergovernmental panel on climate change. Cambridge University Press, Cambridge

Ministry of Environment (2005) Water quality report. Funafuti, Tuvalu

Ministry of Environment (2007) National adaptation programme of actions—Tuvalu. Bonn, Germany, pp 1–55. Retrieved from http://unfccc.int/resource/docs/napa/tuv01.pdf

O'Brien K, Eriksen S, Nygaard LP, Schjolden A (2007) Why different interpretations of vulnerability matter in climate change discourses. Clim Pol 7(1):73–88

O'Brien K, Quinlan T, Ziervogel G (2009) Vulnerability interventions in the context of multiple stressors: lessons from the Southern Africa Vulnerability Initiative (SAVI). Environ Sci Pol 12 (1):23–32. doi:10.1016/j.envsci.2008.10.008

PACC (2006) Report of in-country consultations. Funafuti, Tuvalu

Reed MS, Podesta G, Fazey I, Geeson N, Hessel R, Hubacek K, Thomas AD et al (2013) Combining analytical frameworks to assess livelihood vulnerability to climate change and analyse adaptation options. Ecol Econ 94:66–77. doi:10.1016/j.ecolecon.2013.07.007

Salesa UM (2011) Water tank (2000 gal) report. Funafuti, Tuvalu

Smit B, Pilifosova O (2007) Chapter 18. Adaptation to climate change in the context of sustainable development and equity. Geneva, Switzerland, pp 877–912

SOPAC (2007) Sustainable integrated water resources and wastewater management for Tuvalu. Diagnostic report

White I, Falkland T, Perez P, Dray A, Metutera T, Metai E, Overmars M (2007) Challenges in freshwater management in low coral atolls. J Clean Prod 15(16):1522–1528. doi:10.1016/j.jclepro.2006.07.051

Chapter 19
Pathways for Climate Resilient Livelihoods: The Case of a Large Cardamom Farming in the Dzongu Valley of the Tista River Basin, Sikkim Himalaya

Vimal Khawas

Abstract Sikkim is primarily a rural and agricultural economy, where over 60 % of its population is directly engaged in agriculture. Principal food crops cultured in the region include maize, paddy, barley, millet, wheat, buck wheat, beans etc. Important cash crops are cardamom, ginger, potato, soybean, fruit crops and vegetables etc. Large cardamom is raised on commercial basis and is exported both within and outside the country. In fact, Sikkim has the largest area and the highest production of large cardamom in India. It is a foreign exchange earner crop of Sikkim. The state has limited industrial potential due to its geologic and geomorphic constraints. Large cardamom farming has been suffering from decline in production and gradual drying and subsequent death of the plant in the last 1–1.5 decades.

There has been a steady decline in the yield of large cardamom over the years. One of the important factors of declining yield is ascribed to old age of cardamom bushes. A section of the policy makers and farmers also believe that monoculture of cardamom plants in the same fields for generations, is perhaps, an added cause of declining productivity of the cardamom. However, a further more important and serious factor in this regard has been the destruction of cardamom orchards by viral diseases in recent years. There are increasing warnings from the scientists, policy makers and people on field that change in temperature and rainfall pattern in the region over last many years could be the potential factors of the disease spreading viruses and subsequent destruction and declining productivity of large cardamom plantations across Sikkim.

V. Khawas (✉)
Department of Geography, Sikkim University, Samdur, Tadong, Gangtok 737102, Sikkim, India
e-mail: vimalkhawas@gmail.com

W. Leal Filho (ed.), *Climate Change in the Asia-Pacific Region*, Climate Change Management, DOI 10.1007/978-3-319-14938-7_19

Introduction

There has been a steady decline in the yield of large cardamom in Sikkim over the years. The yield used to be 250 kg per hectare in 1975–1976 which came down to as low as 153 kg per hectare in 1996–1997. One of the important factors of declining yield of the crop is attributed to old age of cardamom bushes. However, even more important factor in this regard has been the destruction of cardamom orchards by diseases in recent years. There are increasing indications from the scientists, policy makers and people on the field that change in temperature and rainfall pattern in the region over last many years could be the potential factors of destruction and declining productivity of large cardamom plantations in Sikkim.

The question that rise in temperature will have a net positive, negative, or negligible impact on crop yields takes on a larger significance because additional hundreds of millions of people could be at risk of hunger or the window of opportunity (Cassman 2007). To answer this question, Lobell and Field (2007) used an innovative empirical/geostatistical approach to estimate the impact of increased temperature since 1980 on crop yields—a period when global mean temperature increased by 0.4 °C. For three major crops—maize, wheat, and barley—there was a significant negative response to increased temperature. For all six crops evaluated (also including rice, soybean, and sorghum), the net impact of climate trends on yield since 1980 was negative (see Cassman 2007; Lobell and Field 2007). While the approach used by Lobell and Field can be questioned on several points, the body of their work represents an ambitious global assessment of recent climate impact on crop yields (Cassman 2007). Such scientific findings may be very relevant and informative for a largely agrarian economy like India. India is a second largest producer of wheat and the fifth largest producer of maize in the world. While the impact on yields may go unnoticed at a micro level the outcomes at global production may be substantial.

The third assessment report of the IPCC specifically considers agriculture as highly vulnerable to impacts of climate change in the South Asian region. In South Asia food security therefore becomes a primary concern. Crop production and aquaculture would be threatened by thermal and water stresses, sea-level rise, increased flooding, and strong winds associated with intense tropical cyclones. Acute water shortages combined with thermal stress should adversely affect wheat, and more severely, rice productivity in India given even a positive 'fertilisation' effect through an elevated atmospheric CO_2 level in the future.

The purpose of this paper is, therefore, to examine and understand the relationship of climate change with the overall health of large cardamom plantation and rural economy in the Sikkim Himalaya. The study becomes relevant and timely, as Sikkim is primarily a rural and agricultural economy where over 60 % of its population is directly engaged in agriculture. Principal food crops cultured in the region include maize, paddy, barley, millet, wheat, buck wheat, beans etc. Important cash crops are cardamom, ginger, potato, soybean, fruit crops and vegetables etc. Large cardamom is raised on commercial basis and is exported both within and outside the country. In fact, Sikkim has the largest area and the highest production of large cardamom in India. It is a foreign exchange earner crop of Sikkim. The state has limited industrial potential due to its geologic and geomorphic constraints.

Notably, since the mid-1970s the average air temperature measured at 49 stations of the Himalayan region rose by 1 °C with high elevation sites warming the most (Hasnain 2000; WWF 2005). This figure is twice as fast as the 0.6 °C average warming for the mid-latitudinal northern hemisphere over the same time period (IPPC 2001; WWF 2005). Studies in Nepal (Shrestha et al. 1999) and Tibetan Plateau (Liu et al. 2002) indicate the rising temperature in recent times, with the warming being consistent and continuous after the mid-1970s in Nepal. The average warming in Nepal in its annual temperature between 1977 and 1994 was found to be 0.06 °C per year. Incidentally, in both the countries the rate of warming is found to be more pronounced in the high altitude regions than the lower ones. Similarly, the lowland areas of India do not show significant warming trend (WWF 2005) indicating that the Himalaya is more sensitive and profoundly affected by climate change.

This work is mainly based on published/unpublished reports of the Government of India and Sikkim, local academic publications, NGO reports, reports in the regional news dailies, interaction with the policy makers/policy planners and grassroots workers, field observations and focused group discussion with the farmers. Dzongu valley, one of the major cardamom producing areas of Sikkim, located in the North district of Sikkim was selected for focus group discussion. The study, therefore, is largely descriptive and suggestive in nature.

Climate Change in the Sikkim Himalaya

With the impacts of global warming becoming more apparent, Sikkim Himalaya is perhaps, most in tune to the signs of change brought about by climate warming. The people across the towns and villages of Sikkim narrate revealing insights on how global warming is affecting their lives and livelihood. Metrological Department records reveal that between 1958 and 2005, there had been a slight change in the climate of Gangtok. According to K. Seetharam, Director of the Meteorology Centre, Gangtok[1] maximum temperature has been rising by 0.2 °C per decade and minimum temperature has been falling by 0.3 °C per decade. The annual rainfall has been increasing by 49.6 mm per decade.

Further, officials at the Gangtok Meteorological office inform that for the past 3 years, the temperature in the state capital—located at an altitude of 5,480 ft—have always been on a higher side with the nights becoming warmer with each passing winter. The impact of global warming on the hill town is, further, evident from the temperature recordings of 30 years—from 1957 to 1997. The average

[1] Reported by Dionne Bunsha, *Frontline*, July 5—18, 2008.

minimum temperature for January was 4 °C for this period while the maximum temperature was 12 °C. From 1997 onwards, the maximum temperature for January has been up by 3°C and the minimum by 4.[2]

The state has also been witnessing prolonged periods of dry spell during winters often triggering rampant bush fires. In 2006–2007, forest fires were witnessed all over the state from January to March, resulting in an extensive loss of flora and fauna.[3] Even the higher altitudes of Lachung and Lachen (both above 8,000 ft) in North Sikkim were not spared by the bush fires and the army had to be called in. In short, less moisture is entering Sikkim during winter but there has been a lot of evaporation. Sikkim experienced its longest ever recorded, 7 month long, dry and warm winter during 2008–2009. The physical effects of it included record breaking forest fires incidents, failure of winter crops and forced changes in the timing of subsequent crops.[4]

Higher temperatures have brought new crops to the mountains. Most of the seasonal flowers which were supposed to bloom between March end and early April are already blooming before February end. There are marked evidences of various birds, animals and insects changing their habitation. Amazingly, they are adapting to their new habitations and becoming habitual in the high altitude regions. Moreover, not only are the animals, birds and insects are changing their habits and habitats, the tree line is also displaying signs of global warming by shifting upward.

There are copious evidences of the low altitude tree species that are gradually shifting towards high altitude zone as the climate there are becoming salubrious enough for them to survive. *Uttis*, a hot climate species, is now abundantly found in places above Gangtok. If the low altitude species are replacing the high altitude species, naturally the latter will reach the state of extinction. Further, the low altitude trees are being replaced by shrubs like *banmara* and others kinds of weeds known to destroy the forests.

It has also been found that Glaciers in Sikkim Himalaya are not behaving normally in recent years. For instance, the Jemu Glacier [located in North Sikkim] retreated by around 20 m per year during 1975–1990 (Bahadur 2004: 53). Small streams that feed the large rivers are drying up more recently in and around the major watersheds of the region. This has not only affected the volume of the major rivers but also impacted the delicate relationship of flora & fauna and human habitation, particularly the livelihood of the poor rural hill folks.

Keeping in mind the world wide debate on global warming and its likely impact on the glaciology and other natural resource base of the state, the Government of Sikkim recently appointed a high level national expert Group/ Commission headed by glaciologist Prof Syed Iqbal Hasnain. The Commission is expected to examine

[2] Reported by *The Telegraph* Wednesday, January 9, 2008.

[3] Reported by *The Telegraph* Wednesday, January 9, 2008.

[4] Reported by Rai Subash and Parvinder Kaur, 'Get Rational about Climate Change not just Fashionable', *Now*, April 12, 2009.

and report all the significant issues related to glaciers in and around Sikkim Himalaya.

> Sikkim is a mini theatre which in a way displays how climate change triggered by non-natural forces at the global level could bring disastrous natural calamities. We are worried in Sikkim as we have seen warm winters, increased flooding, landslides and rock avalanches from destabilised slopes. The torrential rains and unusually prolonged monsoon in 2007 caused extensive damage due to landslides. This was definitely not the case when I was young,
>
> **Chief Minister of Sikkim, Dr Pawan Kumar Chamling, while inaugurating the Commission's first meeting in January 2008.**

Agriculture in the Sikkim Himalaya

Sikkim is primarily a rural and agricultural economy where over 60 % of its population is directly engaged in agriculture. Principal food crops cultured in the region include maize, paddy, barley, millet, wheat, buck wheat, beans etc. Important cash crops are cardamom, ginger, potato, soybean, fruit crops and vegetables etc.

Horticulture is one of the major economic activities of the people of Sikkim. Large cardamom, ginger and turmeric are the principal crops while Mandarin orange, guava, mango, banana etc are the principal fruits grown in the state. Sikkim is also a paradise for flowers. Gladioli, anthuriums, lilliums, primulas, rhododendrons, orchids as well as many other floral species thrive here. The state is home to an amazing 450 species of exotic orchids. There is immense potential for developing floriculture on a commercial basis here. The Government of Sikkim is making concerted efforts to turn this sector into an export-oriented industry.

Large Cardamom: A Major Cash Crop

Large cardamom (*Amomum sublatum Roxb*) locally known as "Allainchi" is the most important cash crop of Sikkim. It has been the main cash crop of Sikkim for generations and is an important contributor to the local economy. Large cardamom is planted between June and July when there is enough moisture in the soil. It is normally harvested in August. The crop grows well under the shade of forest trees at altitudes ranging from 1,000 to 2,000 m with a rainfall of 3,000–3,500 mm per annum. Deep and well-drained soils with a loamy texture are best suited for cardamom (Pathak 2008). Compared to traditional crops like maize and millet, the income from cardamom is much higher and therefore facilitates to alleviate many smallholder households from poverty.

Over 16,949 farmers cultivate large cardamom in the Sikkim Himalaya of which 15, 209 are small and marginal farmers, while around 30 % are totally dependent on this crop for their survival. Large cardamom plantation covers about 26 thousand

Region	Percentage Share	Production (MT)
India	53.3	4000
Nepal	33.3	2500
Bhutan	13.3	1000

Fig. 19.1 Production of large cardamom in the world. *Source*: For both tables Berrig et al. (1993) and Mande et al.

Region	Percentage Share	Production (MT)
Sikkim	87.5	3500
Darjeeling	12.5	500

Fig. 19.2 Production of large cardamom within India

hectares of land under cultivation. With about 54 % of the global market for large cardamom, India is the largest producer in the world (Fig. 19.1), followed by Nepal with 33 % of the market and Bhutan with 13 %.[5] Of the Indian output, almost 88 % used to come from Sikkim till about a decade ago (Fig. 19.2), making it the largest producer of large cardamom in the world. However, it is believed, Sikkim today produces only about 30–40 %. Besides Sikkim, Darjeeling district of West Bengal and other North-Eastern states of India like Mizoram, Manippur, Meghalaya, Nagaland and Arunachal also culture large cardamom.

Depletion and Destruction of Large Cardamom in Sikkim

There has been a steady decline in the yield of large cardamom in Sikkim over the years. The yield used to be 250 kg per hectare in 1975–1976 which came down to as low as 153 kg per hectare in 1996–1997 (Figs. 19.3 and 19.4). One of the important factors of declining yield of the crop is attributed to old age of cardamom bushes (Lama 2001, 2007). According to the officials of the Spice Board the crop has been ailing since the 1980s. The decline began with two successive and prolonged bouts of viral infections (chirkey and foorkey[6] virus) that hit large cardamom growing

[5] It is used as a spice in several Ayurvedic preparations. It contains 2–3 % essential oil and possesses medicinal properties. Large cardamom has a pleasant aromatic odour, due to which it is extensively used for flavouring vegetables and many food preparations in India. It is also used as an essential ingredient in mixed spices preparation. Apart from aroma, large cardamom also has high medicinal value. The decoction of seeds is used as a gargle in infection of teeth and gums. Large cardamom seeds are considered as an antidote to either snake venom or scorpion venom. It is also reported that large cardamom seeds are used as preventive as well as curative measure for throat troubles, congestion of lungs, inflammation of eyelids, digestive disorders and in the treatment of pulmonary tuberculosis.

[6] Also known as Cardamom Bushy Dwarf Virus.

District	2004-05			2005-06		
	Area ['000 Hectare]	Production ['000 Tonnes]	Average Yield [per Hectare]	Area ['000 Hectare]	Production ['000 Tonnes]	Average Yield [per Hectare]
North	**8.99**	**1.38**	**154**	**6.73**	**0.98**	**145**
East	6.77	1.00	148	5.80	1.03	177
South	4.52	0.66	146	3.84	0.61	160
West	4.53	0.70	155	3.60	0.54	150
State	24.80	3.74	151	19.97	3.16	158

Fig. 19.3 District-wise area, production and yield of large cardamom. *Source*: Sikkim: a statistical profile 2006–2007, p. 110

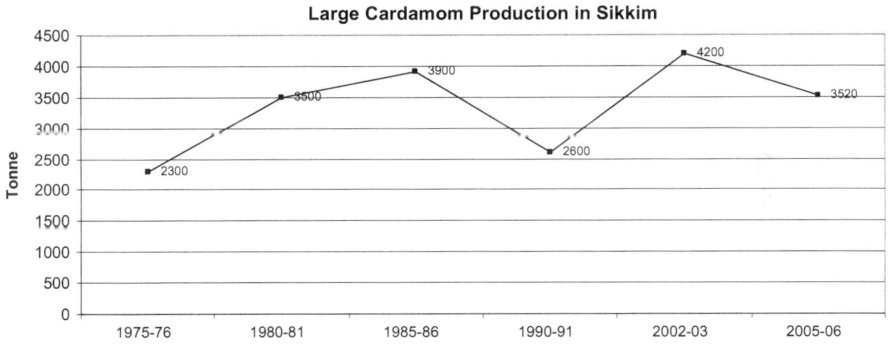

Fig. 19.4 Large cardamom production in Sikkim. *Source*: Computed from Lama (2007)

tracts in Sikkim, North Bengal, Nepal and Bhutan. In surveys across the Darjeeling-Sikkim Hill area during 2002 and 2003, as much as 27 % of plants were found to be affected (Mandal et al 2004).

The already weakened crop was also affected by a disease caused by a soil-borne fungus, *Fusarium oxysporum.* Though this fungus has always been present in the soil here, botanists declare that a mix of environmental conditions, like deforestation and acidification of soil with old plantations and poor management, allowed the fungus to spread, affecting more than 60 % of the crop. The diseases have ruined the farmers' livelihood and local as well as regional economy (Fig. 19.5).

Pathak (2008) observes:

Fungal or bacterial diseases are seldom reported in large cardamom. Only minor diseases like leaf streak or rot diseases are found in isolated areas. The major threat to large cardamom is the widespread occurrence of viral diseases, viz., chirkey and foorkey. These diseases are seen throughout the large cardamom growing tracts of Sikkim and Darjeeling and cause considerable crop loss. *These diseases have spread due to drastic change in the ecosystem, inadequate rain in dry months* **[emphasis mine]** and absence of good agricultural practices by the farmers. Many cardamom farmers failed to plant varieties suitable to their altitude.

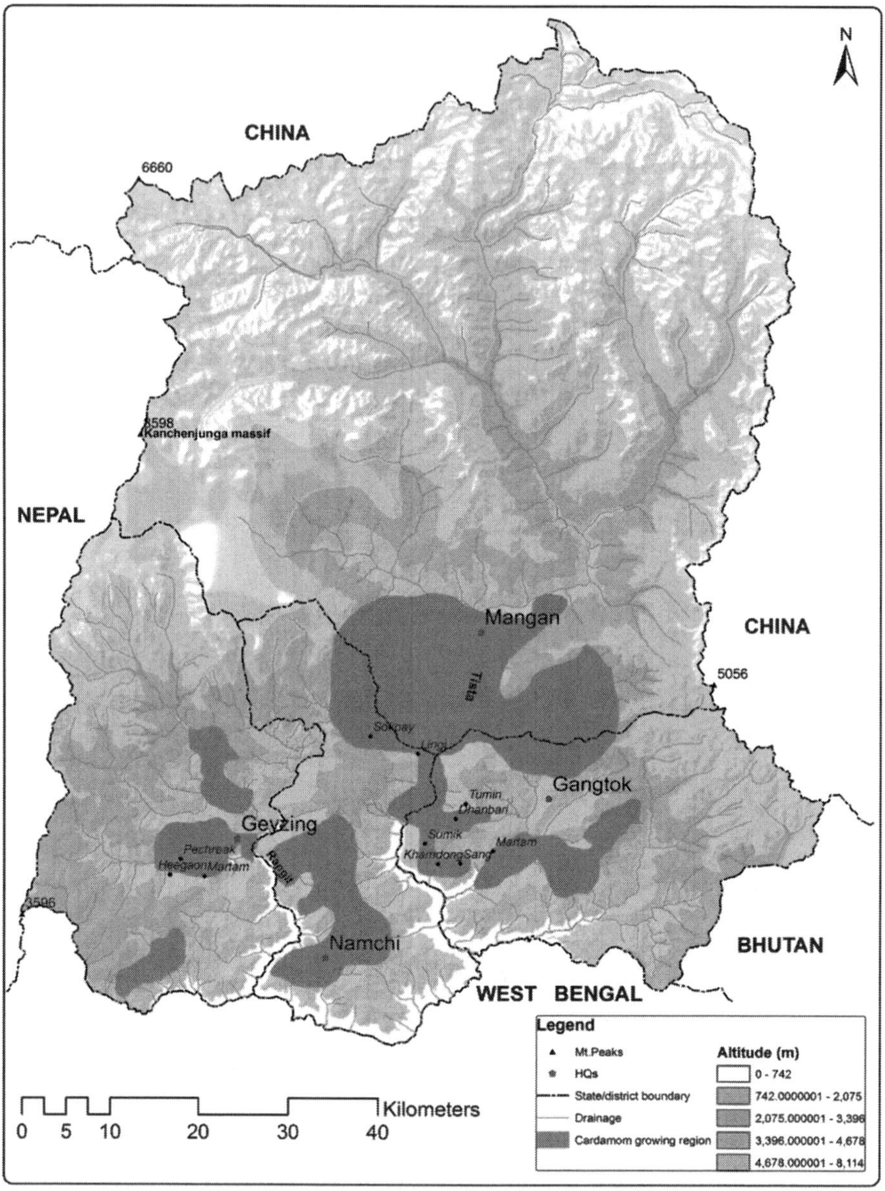

Fig. 19.5 Cardamom growing areas in the Sikkim Himalaya. *Note*: Map prepared by Dilli Ram Dahal, Senior Research Scholar, Viswavarati University, Bolpur, India

Chirkey The symptoms are characterized by mosaic appearance on the tender leaves with pale streaks, which slowly turns into brown, resulting in withering and drying of the plants. Growth and

	yield of the affected plants gradually declines and ultimately they perish. The disease is transmitted by aphids. It also spreads by planting infected suckers. Transporting of infected suckers from one area to another leads to the spread of this disease. The disease is also transmitted mechanically through the knife used for harvesting.
Foorkey	Numerous small tillers appear at the base of the affected plants which become stunted and fail to give any yield. Even the inflorescence is noticed to produce unproductive spikes.
Leaf eating Caterpillar	Initially the caterpillar of the moth Artona Chorista feeds on the leaf lamina from under the surface of the leaf and finally defoliates the leaf completely leaving only the midribs. Their incidence is noticed in May-July and October-March. At present these insects are kept under control by their natural enemies. If insecticides are used to control them, then their natural enemies will also disappear which may lead to an outbreak of these pests in epidemic form. The best method of control is to inspect the plantations during May-July and October-March, to handpick the infected leaves along with the caterpillars and destroy them by burning.

Pathak (2008) suggested following measures to check the spread of viral diseases in cardamom plants:

i) Keep a constant vigil to detect disease affected parts.
ii) Uproot and destroy affected plants as soon as symptoms appear. Repeat detection and uprooting at regular intervals.
iii) Use seedlings produced in certified nurseries.
iv) Propagation through suckers is recommended only through certified multiplication nurseries.

A Case of Dzongu Valley in North Sikkim

Dzongu Valley is a place located in the northern part of Sikkim and is believed to be oldest inhabited region of the Lepchas. Not very far away from Mangan, the headquarters of North Sikkim District, Dzongu is still an untainted and least populated area of the Himalayan Sikkim. Earlier, Dzongu was restricted to the Lepchas, but in the recent years, it has been made accessible to the tourists. Lying at the close proximity to the Kangchendzonga Biosphere Reserve at an elevation ranging between 3,000 and 20,000 ft above sea level, Dzongu is an unpolluted and least trodden territory in Sikkim Himalaya. Still uninfluenced by modern civilisation, the traditional Lepcha Communities in Dzongu live in relative isolation from the outside world with their own traditional practice and believe. Most part of Dzongu is forested region and fed by two important rivers, the Teesta and Tolung.

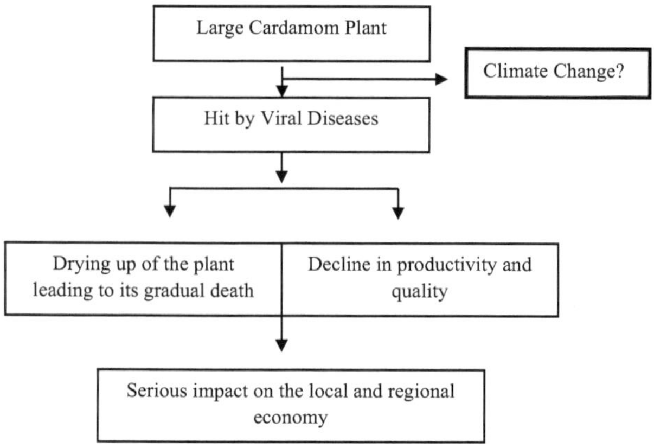

Fig. 19.6 Demise of large cardamom in Dzongu

The region is still unexplored because of its inaccessibility and the forests in this area have a rich variation of flora and fauna.

The region used to be traditionally called the home of large cardamom. Although many different varieties of cardamom are grown across Sikkim and other adjacent neighboring states, large cardamom grown in Dzongu occupy unique in and around the region and also commanded higher prices from the market. The affluence and economic status of the family in Dzongu used to largely depend on the production of large cardamom. All families traditionally inherited large cardamom farms from their fore fathers.

A typical household in Dzongu used to produce about 80–90 sacks of large cardamom and sell around 1,000 kgs annually. Cardamom plants have been seriously attacked by diseases called *chirkey* and *foorkey* in the last 8–10 years.[7] The plants gradually dry up, gives drastically reduced productivity and quality of cardamom seeds and ultimately die (Fig. 19.6). Leaf eating caterpillar and stem borer have also been reported to be present in cardamom plants in recent times. The production of large cardamom has gone down by more then 80 % in and around Dzongu region during the period. The quality of the cardamom seed has also drastically gone down. The cardamom seeds do not get fully ripped and matured during the harvesting period leading to low market price. People in the area had never anticipated such sorry state of the crop in the wildest of their dreams (See Figs. 19.7 and 19.8).

[7] As reported by the farmers in the year 2010.

Fig. 19.7 Relatively healthy cardamom plantation: Dzongu, North Sikkim. [Photo: Vimal Khawas, May 2004]

Fig. 19.8 Disease laden cardamom plantation: Dzongu, North Sikkim. [Photo: Vimal Khawas, June 2009]

Major Causes Attributed to the Devastation of Large Cardamom

Several reasons have been attributed both by the farmers and the policy makers regarding the increasing ill health and declining productivity of large cardamom orchards in Sikkim. Besides other causes cited by the farmers like monoculture in the same field for too long a period, the farmers, policy makers and the researchers largely support following two factors.

Old Age of Cardamom Orchards

There has been no replantation in the age-old cardamom growing areas of Sikkim. The usual economic bearing of cardamom orchard is for about 12 years. After 12–15 years orchards need re-plantation. The peak age of the large cardamom is from 4 to 8 years (Lama 2001: 51). The decline in the productivity of large cardamom is mainly attributed to very old age of cardamom bushes as the production life span of the orchards is gone. The fertility of the soil has also reduced as the mother plants have lived for more than 50 years now. Over 80 % of the bushes require immediate replantation. The absence of gap filling plants destroyed by chirkey, foorkey and other diseases and pests annually is equally detrimental.

Climate Change

There are no tangible scientific research findings on the ground as yet that validates the linkage of devastating blight epidemic on the cardamom crop to rising temperatures and fluctuating rainfall. However, the situation fits in, by and large, with the projections/predictions, on the plight of the agriculture due to climate change, made by the International Panel of Climate Change (IPCC) reports.

> In 1998-9, there was a very long spell of dryness in Sikkim. According to the estimates by the Agriculture Department, Government of Sikkim, the total rainfall was 93.34 percent lower that that recorded over the last 25 years. This had visible adverse effect on the economy, especially on drinking water, rabi crops and cash crops (like cardamom, ginger and orange). Because of this prolonged drought farmers had to postpone the sowing of seeds and in many cases had to resort to not sowing the same. Farmers from the North district were the most affected. It is in fact estimated that 60 percent of the State's large cardamom has been lost in this unprecedented dry spell. (Lama 2001: 52)

The blight outbreak was, however, reported after 2002. According to Mr. Sudarshan, deputy director of the Spice Board:

> There was a dry period from 2000–2002, which could have made the plants weak and could have resulted in the spread of the disease. It could be partly due to climate change. We don't have reliable data in all the cardamom growing areas—from low to high altitudes—to be able to make the connection. The spores of the fungus are carried through wet wind. The epidemic has become unmanageable in some places. The yield has fallen between 30 to 50 percent.

Scientists are, nonetheless, still investing the concrete reasons for the outbreak of diseases across the large cardamom plantations in Sikkim. They are struggling to scientifically prove the linkage of climate warming with increasingly dwindling health of the large cardamom in the state. However, large section of the farmers and policy makers believe that rise in temperature and variation in the monsoon and winter rainfalls are the major causes of cardamom's ill health. According to the farmers, cardamom orchards in the lower hills (*awul*) are severely affected by the diseases while those in the higher hills (*lekh*) are comparatively healthy even today.

Measures Taken by the Farmers

Adaptive Measures

Today the farmers have lost their hopes in large cardamom farming and have gradually resorting to diversify their economic activity. They are shifting to alternative farming practices such as cultivation of ginger, orange and other horticulture crops although the economic returns of these crops are relatively lower than that of large cardamom. In fact, these crops were traditional cultivated by the farmers at the subsistence level while large cardamom used to be the major income generator. With the gradual demise of the cardamom orchards, the farmers are now trying out crops like ginger and orange at commercial level to supplement the income that used to be generated by large cardamom. Horticulture crops like organic vegetables are also being tried out more recently. The farmers have also started increasing their livestock population.

Mitigative Measures

An important mitigative measure tried out by the farmers to save the dying cardamom plant was spraying of medicine on the cardamom orchards both at their own initiative and with the support of the government. However, spraying of medicine did not prove to be fruitful.

On the advice of the spice board and other government agencies, the farmers are also cleaning and burning the fields where cardamom orchards are totally destroyed. The government has advised the farmers to keep such fields fallow for some[8] years and to properly manure the cardamom field.

Further, replantation of the cardamom plants both on the weak/old orchard areas and virgin areas are being carried out by the farmers often with the help of the Spice Board and Agriculture/Horticulture Department of the state government.

Support from the Government

The Spices Board[9] has recently taken up various programmes to rejuvenate large cardamom plantations in the state through the following schemes:

---

[8] At least 2 years.

[9] Spices Board was constituted on 26th February 1986 under the Spices Board Act 1986 (No. 10 of 1986) with the merger of the erstwhile Cardamom Board (1968) and Spices Export Promotion Council (1960). Spices Board is one of the five Commodity Boards functioning under the Ministry

1. Certified nursery scheme: The Board supports the raising of nurseries in farmers' fields by offering a grant-in-aid of Rs. 10,000 per nursery producing 10,000 suckers.
2. Replanting scheme: This scheme is intended to encourage growers to take up replantation of senile and uneconomic gardens. A subsidy of Rs. 6,000 per hectare is offered to the growers.
3. Supply of sprinkler irrigation units: To tide over the drought situations during summer, the Board assists growers in procuring and installing sprinkler units and accessories by providing 50 % of the cost as subsidy with a ceiling of Rs. 2,500 per set.
4. Low cost driers: In order to improve the quality of the cured large cardamom the Board helps the growers to replace their traditional *bhatti* curing system with improved driers by providing a subsidy of Rs. 10,000 per drier.
5. Processing units: The conventionally cured cardamom does not fetch the right price because of unscientific processing and packaging. In order to overcome this situation, the Board proposes to assist the growers in setting up their own processing/powdering/packaging units by providing assistance to the tune of 50 % of the cost, subject to a maximum ceiling of Rs. 50,000 per unit.

Production of large cardamom has been, however, reducing in Sikkim despite the efforts of the Spices Board. About Rs. 23 crore has been provided in the 11th Five Year Plan for the production of large cardamom in the state of Sikkim and in Darjeeling district of West Bengal.[10] For the past few years, the Spices Board is concentrating on disease free planting material and replanting of cardamom bushes. About 1,600 nurseries have been proposed for developing healthy seedlings to replace the old bushes in Sikkim. A subsidy of Rs 1.2 crore is expected to be disbursed among cardamom growers. The State Horticulture and Cash Crops Development Department is planning to replace the ailing cardamom plantations with organic vegetables or bamboo plantations.

Conclusion

Sikkim Himalaya as an integral part of the Eastern Himalayan region has witnessed notable variation in its temperature and rainfall pattern over the years. The paper highlighted the physical effects of such meteorological deviations in the region including glacier retreat, glacial lake outburst floods, record breaking forest fire incidents, failure of agricultural crops and forced changes in the timing of crops and changing habits and habitats of the animals, birds and insects.

(continued)

of Commerce & Industry. It is an autonomous body responsible for the export promotion of the scheduled spices and production development of some of them such as Cardamom.

[10] Reported by ***The Statesman*, February 17, 2009**.

With the help of state (province) level data analysis and case study of the Dzongu Valley, the paper reported the decline in production and gradual drying and subsequent death of the large cardamom plant in the last couple of decades. There has been a steady decline in the yield of large cardamom over the years as well. One of the important factors of declining yield is ascribed to old age of cardamom bushes. A section of the policy makers and farmers also believe that monoculture of cardamom plants in the same fields for generations, is perhaps, an added cause of declining productivity of the cardamom. However, a further more important and serious factor in this regard has been the destruction of cardamom orchards by viral diseases in recent years. There are increasing warnings from the scientists, policy makers and people on field that change in temperature and rainfall pattern in the region over last many years could be the potential factors of the disease spreading viruses and subsequent destruction and declining productivity of large cardamom plantations across Sikkim.

Both adaptive and mitigative measures are being sought by the farmers to sustain their livelihood. These measures have been taken both at the household/community initiative as well as with the support of the government. In the first case, the farmers are resorting to alternative livelihood opportunities like experimenting other cash crops like ginger, orange, vegetables at commercial level. With regard to the mitigative measures, the farmers are spraying pesticides, cleaning the fields, keeping the land fallow for some years and replanting the fields with fresh cardamom saplings. The government, especially Spice Board of India, has been helping the farmers since last many years to salvage their livelihood. The results of the mitigative measures have not been very fruitful, nonetheless.

One of the cardinal points highlighted by the paper is that adaptation measures like alternative livelihood strategies by diversifying agricultural practices adopted by the farmers themselves have been far successful than majority of the mitigation measures and programmes pushed by the government agencies.

However, since a large chunk people in the rural areas still dependent on cardamom as the main cash crop providing them with a significant part of their livelihood, the impact of steadily falling health of cardamom orchards and declining yield is likely to be disastrous. Such a scenario calls for a need to workout and evolve a comprehensive strategy to revitalise the health of cardamom plants and sustain the yields. Further more, plants affected by the viral diseases cannot be cured but the losses may be minimised by adopting appropriate management practices.

References

Bahadur J (2004) Himalayan snow and glaciers: associated environmental problems, progress and prospects. Concept Publishing Company, New Delhi

Berrig CH, Koechilin R, Tarnutzer A (1993) Sikkim: a preliminary profile with particular emphasis on horticulture and animal husbandry. Study Group on Institutions, Human Actions and Resource Management, Institute of Geography, University of Zurich, pp 47–51

Cassman KG (2007) Climate change, biofuels, and global food security. Environ Res Lett 2 (011002):1–3

Hasnain SI (2000) Status of Glacier Research in the HKH region 2000. ICIMOD, Kathmandu

Lama MP (2001) Human development report: Sikkim. Government of Sikkim, Gangtok

Intergovernmental Panel on Climate Change (2001) Climate change 2001: the scientific basis. Contribution of group I to the third assessment report of the intergovernmental panel for climate change

Lama MP (2007) Sikkim economic survey 2006–2007. Government of Sikkim, Gangtok

Liu S et al (2002) Glacier fluctuation and the inferred climatic changes in the A'nyemaqen Mountains in the source area of the Yellow River, China. J Glaciol Geocryol 24(6):701–707

Lobell DB, Field CB (2007) Global scale climate-crop yield relationships and the impacts of recent warming. Environ Res Lett 2(014002):1–7

Mandal B, Mandal S, Pun KB, Varma ACPV (2004) First report of the association of a nanovirus with foorkey disease of large cardamom in India. Plant Disease 88(428), [published online]

Pathak A (2008) Cultivation of large cardamom in Sikkim. Ishani: Tribal World of the North East India 2(6), November

Shrestha AB et al (1999) Maximum temperature trends in the Himalaya and its vicinity: an analysis based on temperature records from Nepal for the period of 1971–1997. J Clim 12:2775–2787

WWF (2005) An overview of glaciers, glacier retreat and subsequent impacts in Nepal, India and China. World Wide Fund for Nature, Kathmandu

Chapter 20
Progress Towards Low Carbon Fuels in the Pacific: Prospects and Challenges

Pritika Bijay and Anirudh Singh

Abstract The use of first generation biofuels in the Pacific has been receiving increasing attention recently. Apart from their contribution to the global emissions reduction effort, they provide the region with one of the most available means of reducing dependence on costly fossil fuel imports. If implemented properly, an indigenous biofuels production programme can contribute to social and economic development, improve energy access as well as provide a valuable means for satisfying the region's emissions reduction obligations. The challenges with first generation biofuels such as ethanol and vegetable oil-derived fuels are well known. The two main issues commonly aired, are the food versus energy debate and the costs of production. The region however offers important advantages not available elsewhere that allows one to reconsider these alternative fuel candidates. These include favourable climate and the apparent availability of marginal land where feedstock such as jatropha and pongamia can thrive. Recent technological advantages in fuel conversion is now stimulating new interest in the potential of these biofuels for the region. However, the successful realization of a biofuels industry for the region must seek answers for some of the barriers to its development that have become increasingly obvious recently.

This chapter discusses the present status and future prospects of ethanol and biodiesel as possible transportation and power generation fuels for the region. The range of indigenously available feedstock, including sugar juice, molasses, cassava, for ethanol and coconut oil, pongamia and jatropha for biodiesel are considered, and the possibilities of the commercial realization of hydrotreated vegetable oils (HVO) explored. Barriers to the successful development of a biofuels industry are considered and possible solutions evaluated.

P. Bijay (✉)
Development, Marketing and Communications, The University of the South Pacific, Suva, Fiji
e-mail: bijay_p@usp.ac.fj

A. Singh
School of Engineering and Physics, The University of the South Pacific, Suva, Fiji
e-mail: singh_ag@usp.ac.fj

W. Leal Filho (ed.), *Climate Change in the Asia-Pacific Region*, Climate Change Management, DOI 10.1007/978-3-319-14938-7_20

Introduction

According to the WGII contribution to the IPCC fifth assessment report, the current and future risks to small islands due to climate change include sea level rise, tropical cyclones, increase in sea surface temperature and changing rainfall patterns (IPCC 2014). Sea level rise presents special threat to the low-lying coastal areas on islands and atolls, and contribute to severe erosion and sea-flood risks to these islands, with wave over-wash of sea water degrading fresh ground water resources. Rise in sea temperature brings about increased coral bleaching and reef degradation.

The Pacific Island Countries (PICs) consist of a mixture of coral atolls and volcanic islands, most of which are already feeling the adverse effects of climate change predicted above. Coastal inundation is leading to village relocations in Fiji (Wilson 2014). Rising sea levels and sea water over-wash during king tides and tropical cyclones are giving rise to fresh ground water degradation in Kiribati, Marshall Islands and Tuvalu (Weir and Virani 2010). While the contribution to the global carbon inventory by the PICs is insignificant, many of these countries are nevertheless committed to the UNFCCC and are determined to meet their obligations. Fiji signed the Convention in 1992 at Rio and ratified it in 1993. It submitted its Initial National Communication on 18 May 2006, and is currently in the process of completing its Second National Communication. Other Pacific Island Countries that have also submitted their Initial National Communications are Kiribati (October 1999), Marshall Islands (November 2000), the Federated States of Micronesia (December 1999), Nauru (October 1999), Palau (June 2003), Papua New Guinea (February 2002), Samoa (October 1999), Tonga (July 2005), Tuvalu (October 1999), and Vanuatu (October 1999). Of these, Kiribati, Samoa and Tonga have also submitted their Second National Communication to date (UNFCCC 2014).

The Pacific Island Countries (PICs) are committed to the global carbon reduction efforts but they are also faced with several challenges relating to the supply of fuel for transportation and power generation. Not the least amongst these are the lack of indigenous fossil fuel sources for their transportation and power generation needs, and the geographical remoteness of these nations from the major global fuel supply lines (Johnston 2012).

A possible way in which both these requirements can be met lies in the production of first generation biofuels. Chief amongst these are vegetable oils and their chemical derivatives such as biodiesel, as well as bioethanol produced from the fermentation of feedstock containing sugar and starch. Issues commonly raised about the suitability of these fuels as alternatives for fossil fuels, include the food versus fuel debate and their carbon-neutrality (Crutzen et al. 2007). However, such arguments can be easily addressed with full life cycle analyses of the environmental impacts as well as the possible availability of land resources that have previously not been used in any productive manner.

The larger volcanic Pacific island nations such as Papua New Guinea (PNG), Vanuatu, Solomon Islands and Fiji should possess the required resources such as under-utilized marginal land area as well as the appropriate climatic conditions that provide suitable pre-requisites for the establishment of a biofuels industry. In the case of Fiji, the required feedstock for such a biofuel industry has already been available in the country for some time, or is planned for cultivation in the near future. They include sugarcane and cassava for the production of bioethanol, and coconut oil (CNO) and non-edible oils such as those produced by the FJ $174 million pongamia project (Biofuels International n.d.) and possible jatropha projects (Biofuels International n.d.) for biodiesel production.

Ethanol is a biofuel that can be used in either blended or neat (unblended) form for spark-ignition engines used in transportation and power generation. Current vehicle engine technology can use the E10 blend (10 % ethanol, 90 % petrol blend) without the need for engine modification. The bioethanol may be derived from sugarcane juice, or molasses, a by-product of the sugar industry. Another possibility is the starch-content of tropical root crops such as cassava. The technologies for bioethanol production from these two feedstock, though different in detail, are well-known and easy to establish on a commercial basis in the PICs.

Coconut oil (CNO) provides a readily-available feedstock for biodiesel production in much of the PICs, including Fiji. It can be used to produce coconut methyl ester (CME) via the process of trans-esterification which requires methanol and an alkali such as sodium hydroxide as the basic ingredient in addition to the CNO. CME can be used to form blends such as B5 (5 % CME blended with 95 % petroleum diesel) for current generation compression-ignition engines (i.e., diesel engines).

Feedstock production for a biofuel industry in Fiji has not been without its problems. Up to a few years back, the sugar industry had been in a steady decline in Fiji. The challenges of CNO production over the last few decades have been well-documented (Singh 2012a). The latest figure released by the Fiji Bureau of Statistics indicates that in 2012, 4,977 tonnes of coconut oil was produced. The copra industry as a food industry has also been in decline for some time (Singh 2012a).

Other barriers to biofuels production relate to the institutional and legislative mechanisms and tools necessary to support a viable industry. Nonetheless the problem is not unassailable. New interest by government and business entrepreneurs are now providing a fresh impetus to the prospects of establishing a viable biofuels industry based on CNO in Fiji.

In this chapter, after enumerating the general requirements for a biofuels industry, we review in detail the current situation relating to feedstock production for both types of biofuels. The full range of indigenously available feedstock for ethanol and biodiesel production are considered. Issues pertaining to the commercial utilization of the fuel are discussed, and the technical and human resource capacity as well as other requirements to support a biofuels industry are evaluated. We then consider the future prospects of ethanol and biodiesel as possible transportation and power generation fuels for the region.

Requirements for a Biofuels Industry

To establish a sustainable biofuels industry in Fiji and the PICs, there are certain essential requirements that cannot be overlooked. Not least amongst these are:

- the availability of resources;
- the technology infrastructure;
- human capacity;
- the institutional and legislative framework; and
- the required standards and regulations.

For a viable biofuel industry to thrive, there must first be sufficient resources, both in the form of crops to provide the feedstock for the industry, as well as the land resources to support such plantations. With its tropical climate, Fiji and the PICs are able to support a range of edible and non-edible crops that can serve this purpose. These include sugarcane, cassava, and non-edible crops such as pongamia and jatropha. It has yet to be fully ascertained whether there are sufficient land resources available to grow the required volumes of feedstock, especially since this latter resource will be shared with food crops.

The establishment of a biofuels industry will also require a minimum level of extraction, milling and production technology. These in turn will depend on a basic technological and supply support structure to become viable. While the former technologies may be set up expeditiously, the existence of the latter will in general be decided by the broader economic parameters of the country.

Human resources are required at all levels and stages. The biggest challenge for developing countries is the ability to attract not only technical people of the required expertise, but also to entice entrepreneurs with sufficient experience and financial support to initiate the action by setting up start-up companies.

To formalize the establishment of a new industry will require the enactment of the appropriate legal and institutional framework, beginning at the policy level. This is an integrated development process that requires the appropriate political will, sufficient entrepreneurial enterprise and financial support. The emergence of a new fuel on the market will necessarily have to undergo the test of quality assurance measures. A pre-requisite to this is the establishment of fuel standards that incorporate the fuel characteristics of the new fuels, and regulations to enforce these standards.

It is thus of interest to examine the status of Fiji and the region in relation to the basic requirements outlined above, and to see how they are met. This task is briefly addressed in the sections that follow. But the factors that influence and ultimately determine the development of the industry may extend well beyond the obvious requirements outlined above. Issues such as start-up financing, the reluctance of the car industry to embrace new fuel standards and to commence using new biofuels, the novelty of the whole scheme and the general state of development of the country can all come together and campaign against the development of any new biofuels industry, as will be illustrated more fully in a later section of this chapter.

The following section considers the production science and technology, available resources and the production potential of bioethanol, one of the two main contenders for a new biofuel for Fiji and the Pacific region.

Bioethanol

Bioethanol or ethanol (ethyl alcohol) is considered renewable when produced from sustainable agricultural sources. Bioethanol's greatest benefit lies in its potential to reduce greenhouse gas emissions by partial replacement of oil as a transport fuel (IEA 2004). Reduction in the greenhouse gas emissions is significantly dependent on the feedstock and the technology used in the production process of ethanol as well as the distribution and blending procedures. This biofuel can also reduce the burden of foreign currency expenditure for poor countries that import petroleum products but have the potential to produce and use bioethanol (WWI 2007).

Brazil's success story in using sugarcane based ethanol is quite well known and the PICs can learn from this. Many advocates of biofuel subsidies and mandates frequently cite Brazil's experience. Brazil is the world's number two ethanol producer and the leading ethanol exporter, using sugarcane as its feedstock (Hofstrand 2009). In 2011, Brazil contributed to 24 % of global ethanol production, when compared to 30 % in 2010 (REN21 2012). After being number one exporter for many years, the decline in ethanol production is attributed to the decline in investment in new sugarcane assets and plantations since the 2008 financial crisis, poor sugarcane harvests due to unfavorable weather and high world sugar prices (OECD 2011; Colitt and Nielsen 2012). Authorities in Brazil had made it mandatory to have 20–25 % ethanol blended in petrol. However, since the decline in ethanol production Brazil announced new policies to stimulate sugar production and to reduce the amount of ethanol required in gasoline to 20 % (Biofuels Digest 2012). Bioethanol is an excellent substitute for gasoline, the main car fuel used by spark ignition (SI) engines around the globe. Bioethanol can be used in SI engines, either in its pure form or blended with conventional petroleum-derived fuels.

Production Science and Technology

Bioethanol is produced by fermenting sugars or substances that can be converted to sugars, such as starch and cellulose. Sugar and starch based feedstock are currently predominant at the industrial level and they are so far economically favourable (Krishna et al. 2001; Shen et al. 2008; Sukumaran et al. 2009). The use of lignocellulosic materials for second generation bioethanol production is currently not considered commercially viable as lignocellulosic materials contain more complex sugar polymers, such as cellulose and hemicellulose, which are more difficult to break down chemically, together with lignin (E4tech 2008).

Bioethanol is produced from a biological process known as fermentation using yeast such as *Saccharomyces cerevisiae* in which sugars such as glucose, fructose, and sucrose are converted into cellular energy and thereby produce ethanol and carbon dioxide as metabolic waste products. Fermentation by yeast using sugar cane and molasses as the main sources of sugar for bioethanol has a comparative advantage when compared to starch and lignocellulose as they can be converted into ethanol directly without any additional process. Starches from corn or cassava feedstocks need to be hydrolysed with amylase enzymes to produce sugar that can be fermented. A disadvantage is that many of these raw materials are considered to be a human food resource and will therefore be too expensive to use for fuel ethanol production (Badger 2002).

Starch based materials that commonly are used for ethanol production include corn, potatoes, cassava, and various cereal grains. Prior to fermentation, the starch must be converted to sugars. Starches from corn or cassava feedstocks need to be hydrolysed with amylase enzymes to produce sugar that can be fermented. Starch substance comprises the major part of human diet and is synthesized naturally in plants. Some of these plants are corn, potato, rice, sorghum, wheat and cassava. Starch molecules are glucose polymers linked together by α-1,4 and α-1,6 glucosidic bonds (Kearsley and Dziedzic 1995). Starch must first be hydrolysed (i.e., broken up into smaller units) into glucose units prior to alcohol fermentation by ethanologenic micro-organisms such as yeast (Lee et al. 1992).

Starch hydrolysis can be achieved by using two enzymes: α-amylase and amyloglucosidase or glucoamylase. Starch hydrolysis is achieved in two stages—liquefaction and saccharification. During the liquefaction process, the α-amylase enzymes work on the gelatinized starch slurry to partially hydrolyze the starch to dextrin. Dextrin solutions are less viscous hence, the starch gel is liquefied. Dextrin consists of short glucose chains, and small amounts of glucose and maltose (Kearsley and Dziedzic 1995). Dextrin can be further hydrolyzed to glucose by adding amyloglucosidase and this stage is called saccharification.

α-amylase reacts endogenously with α-1,4 glucosidic linkages of polysaccharides to produce oligosaccharides whereas amyloglucosidase hydrolyzes exogenously the non-reducing end α-1,4, α-1,6 and α-1,3 glucosidic linkages of oligosaccharides to produce glucose (Pazur and Ando 1960).

Lignocellulose based feedstocks include wood residues and agricultural residues such as sugarcane bagasse. Lignocellulose is composed of cellulose, hemicellulose and lignin (Sjöström 1993). Different wood and agricultural residues have different compositions of cellulose, hemicellulose and lignin. For example, sugarcane bagasse has a complex structure, composed of 40–50 % cellulose, 25 % hemicellulose and 25 % of lignin (Ferreira et al. 2010). The conversion of lignocellulose based materials into fermentable sugars is possible using thermal pre-treatment, followed by chemical or enzymatic hydrolysis (Hernández-Salas et al. 2009). Thermal pre-treatment aims at disorganizing and fractioning the lignocellulose complex into cellulose, hemicellulose and lignin. This pre-treatment can be integrated with an acid catalysed step in mild conditions for hydrolyzing the hemicellulose component, generating a pentose-rich liquid phase, which can be fermented by

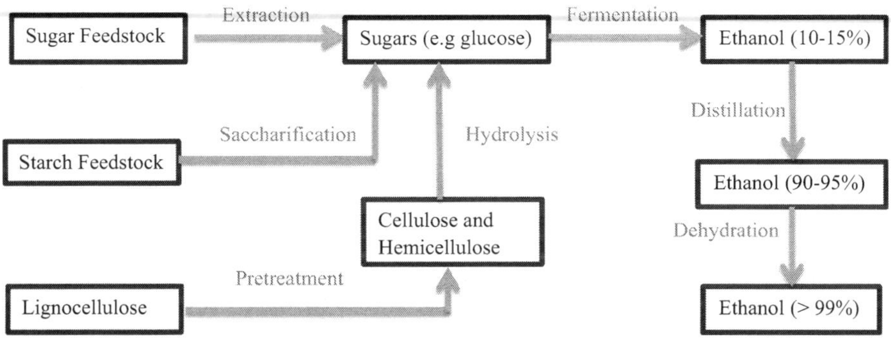

Fig. 20.1 Pathway to ethanol production using sugar, starch and lignocellulose feedstocks

appropriate microorganism (Ferreira et al. 2010). The cellulose hydrolysis stage can be carried out by chemical routes, such as acid hydrolysis, which is an eminent process, or by biological routes, using enzymes, which is a fairly new process and in development. The sugars derived from hemicellulose and cellulose are then fermented to ethanol using yeasts.

The solution produced from the process of fermentation is a dilute mixture of ethanol and water. Hence, the process of distillation is used to separate the ethanol component. During this process the separation of ethanol and water is made possible due to the fact that the boiling points of ethanol (78.5 °C) and water (100 °C) are different. As the ethanol obtained from this process still contains some water, the process of dehydration needs to be employed to obtain 100 % or absolute ethanol.

The pathway for ethanol production from different feedstock is shown in Fig. 20.1.

Overview of Resources

One of the potential sugar based resource to be considered for ethanol production in Fiji is sugarcane juice and molasses. The sugarcane industry had been Fiji's backbone in supporting the country's economy but in the recent years this industry has been in a state of decline.

The number of cane farmers/growers dropped from 22,179 in 2000 to 16,259 in 2011. The result has been a sharp drop in sugarcane production from 3,786,000 tonnes in 2000 to 2,096 tonnes in 2011 as shown in Fig. 20.2. The reason behind this decline since 2000 can be attributed to the expiry of cane-farming leases and other social factors. Secondly, the industry also suffered a second setback in 2009 when the Sugar Protocol was abolished (Singh 2012a). In the past 5 years, interest has arisen in the possibility of using sugarcane juice and molasses as feedstock for ethanol production.

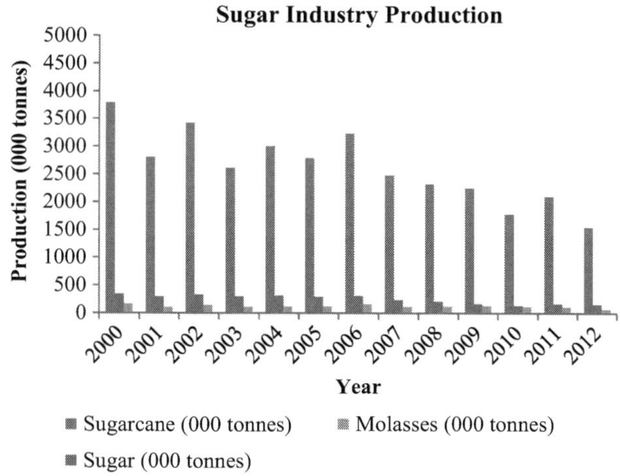

Fig. 20.2 Sugarcane, sugar and molasses production in Fiji, 2000–2012 (Bureau of Statistics 2013)

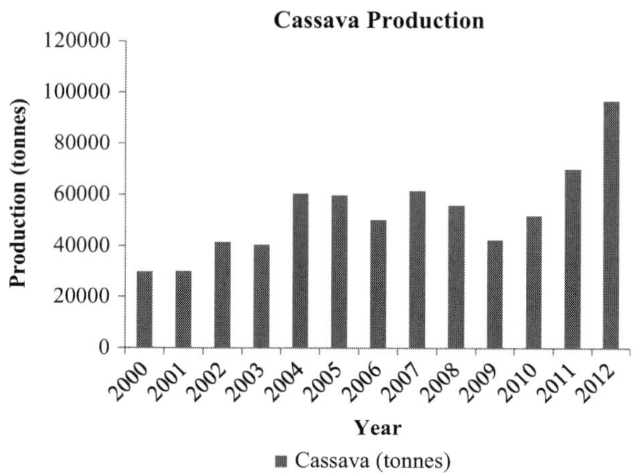

Fig. 20.3 Cassava production in Fiji, 2000–2012 (Bureau of Statistics 2013)

Another resource that has received considerable interest for ethanol production is cassava. In Fiji, cassava is predominantly grown for human consumption. There is almost no processing of cassava into dried form for human or animal use. Apart from local consumption as food, cassava is exported to Australia and New Zealand as frozen tubers. Cassava is grown in most parts of Fiji. As it is tolerant to a range of climatic conditions as well as growing in marginal land, limited effort is currently being placed in improving conditions for planting. There is minimum land preparation, weeding is hardly done and limited fertilizer applied. In 2007, the cassava yield was 13.80 t ha^{-1} (SOPAC 2009). This yield can be increased with sustainable cultivation practices and also by identifying high yielding varieties. Cassava research in Fiji is mostly done by the Koronivia Research Station and other stations of the Ministry of Agriculture, Fisheries and Forest. There has been a gradual increase in cassava production since year 2000 and this can be seen in Fig. 20.3.

Table 20.1 Comparison of ethanol yield made from various energy crops

Crop	Yield (t ha^{-1} $year^{-1}$)	Conversion rate to sugar or starch (%)	Conversion rate to ethanol (L t^{-1})	Ethanol yield (kg ha^{-1} $year^{-1}$)
Sugarcane	70	12.5	70	4,900
Cassava	40	25	150	6,000
Carrot	45	16	100	4,500
Sweet sorghum	35	14	80	2,800
Maize	5	69	410	2,050
Wheat	4	66	390	1,560
Rice	5	75	450	2,250

Source: (Rao 1997)

The Fiji Government has plans to produce bioethanol from agricultural sources available in Fiji, including sugarcane, molasses and cassava. It has been pointed out (Rao 1997) that cassava is one of the best crops to be used for bioethanol production. The ethanol yield of cassava per unit land area is higher than any other known energy crop as seen in Table 20.1. In addition, it is much cheaper to set up a cassava ethanol factory because of lower investment and the processing technology is much simpler due to the special characteristics of starch (Wang 2002). The net cost of cassava ethanol can be lowered through the production of useful by-products from different parts of cassava plant (Wang 2002).

However, since cassava is primarily produced in Fiji for food by the people, an approach needs to be found that would balance out the use of agricultural land for food and fuel. The use of food crops for fuel usually drives up the prices of these crops. For this reason governments in many countries are now ensuring that biofuels do not increase the price of staple foods. The Fiji Government has dismissed the threat to food security on the grounds that more than enough land is idle in Fiji according to Food and Agriculture Organization (FAO) 2006 figures (SOPAC 2009). However, no comprehensive study seems to have been carried out of the nature and amount of this additional resource. In particular, a detailed quantitative assessment is required of the various categories of marginal land, and their availability for use.

Table 20.2 considers the coconut, cassava and sugar cane production for 2012 for all PICs. It is seen that a vast difference exists amongst the countries in their feedstock production. Apart from Fiji and Papua New Guinea, the PICs either do not plant sugar cane or the amount is negligible. Similarly, only the Cook Islands, Federated States of Micronesia, Fiji and Tonga plant cassava in large quantities. All the PICs produce coconut in significant amounts.

Table 20.2 Coconut, cassava and sugar cane production in PICs for 2012 (FAOSTAT 2014)

Country	Coconut (tonnes)	Cassava (tonnes)	Sugar cane (tonnes)
American Samoa	7,000	–	–
Cook Islands	2,000	1,000	–
Federated States of Micronesia	55,000	9,000	–
Fiji	210,000	70,000	1,620,000
Guam	52,000	–	–
Kiribati	170,000	–	–
Nauru	2,600	–	–
Niue	3,300	–	–
Papua New Guinea	900,000	–	340,000
Republic of Marshall Islands	45,000	–	–
Samoa	180,000	–	–
Solomon Island	408,000	–	–
Tokelau	4,300	–	–
Tonga	60,500	7,500	–
Tuvalu	2,100	–	–
Vanuatu	400,000	–	–

Production Potential

Biofuels can play a fundamental role in the transportation sector of Fiji and the other PICs. The transportation needs of Fiji can be satisfied by using biodiesel and its blends with diesel for Compression Ignition (CI) engine vehicles or by using bioethanol and its blends with petrol for SI engine vehicles.

Fiji has the necessary resources for the production of bioethanol (Singh 2012a). The possible feedstock identified over the past few years has been sugarcane and cassava. When considering sugarcane, there are two options: either to use sugarcane juice, or molasses for ethanol production. In a presentation made by the Director of Fiji Department of Energy (FDOE) (Nakavulevu 2011) during a workshop organized by the International Renewable Energy Agency, it was indicated that Fiji would require 7,466,242 L of ethanol for blending with petrol for the production of E10. His presentation also indicated that the 100,000 tonnes of molasses that Fiji Sugar Corporation produces could yield 34 ML of ethanol annually, of which 8 ML would be needed for blending to produce the required volume of E10 for Fiji while the rest could be exported. For developing the biofuel or bioethanol industry in Fiji it is important to have the available resources. These include not only feedstock plantations but also available land area for the further development of the industry. Fiji has about 1.8 million ha of available land, of which less than 19 % is available for agriculture. Since the portion of the land that is arable is ostensibly quite small, one would need more assurance, to substantiate the FAO claim reported above that the extra land for biofuels development is indeed available.

Table 20.3 Total fuel imports for Fiji (2011)

Fuel type	Total imports (L)[a]
Motor spirit	49,404,283
Automotive diesel fuel	33,922,843
Industrial diesel fuel	235,182,797
Residual fuel oil	20,318,647
Total	338,828,570

[a]The total imports is the retained imports in Fiji (it is less the re-exported value). *Source*: Nakavulevu (2011)

Table 20.4 Bioethanol yield from different feedstocks

Crop	Global/ Estimates	Crop yield ($t\ ha^{-1}$)	Conversion efficiency ($L\ t^{-1}$)	Bioethanol yield ($L\ ha^{-1}$)
Sugar beet[a]	Global	46	110	5,060
Sugar cane[a]	Global	65	70	4,550
Cassava[a]	Global	12	180	2,070
Maize[a]	Global	4.9	400	1,960
Rice[a]	Global	4.2	430	1,806
Wheat[a]	Global	2.8	340	952
Sorghum[a]	Global	1.3	380	494
Molasses[b]	Australia		270	
Cassava[c]	Fiji		102–138	

Sources: [a]Rajagopal et al. (2007), [b] Deverall (2005), [c]Bijay (2012)

Bioethanol is a promising alternative fuel for Fiji, with the availability of more than one feedstock to consider, as well as the potential for improvement in the required human capacity for the industry and interest by government and private sector. With proper planning and development of the biofuel industry, Fiji could improve its overall energy situation significantly.

The actual data for current fuel usage in Fiji is not that readily available. However, the figures given in Table 20.3 which were used by Director of Fiji Department of Energy (Nakavulevu 2011) can be used to make estimates on the amount of E10 required to replace some of the petrol that is being imported in Fiji. Based on the raw materials that can be produced in Fiji, namely sugarcane, molasses and cassava, an estimate can be made of the total amount of ethanol that could be produced. The estimate requires the total tonnage of feedstock available, and the volume of fuel that can be produced per tonne of the feedstock (litres per tonne). The latter information is shown in Table 20.4.

Firstly, if all the sugarcane produced in Fiji in 2012 (1,546,000 tonnes) is used for ethanol production and using the conversion factor of 70 L of ethanol per tonne of sugarcane as shown in Table 20.4, it is expected that 108.22 million litres

(ML) of ethanol will be produced. This is equivalent to 219 % of the total fuel imports into the country.

On the other hand if all the molasses produced in 2012 (67,000 tonnes) were to be used for ethanol production then 18.09 ML of ethanol will be produced using the conversion factor as shown in Table 20.4. This would replace 37 % of fuel imports in the country.

If cassava is to be used as feedstock and using the conversion factor for Fiji of 102–138 L t^{-1}, then we expect to produce about 9.87–13.36 ML of ethanol if all the cassava produced in 2012 was used. This will result in reduced in fuel imports by 20–27 %.

Biodiesel

The use of vegetable oils to run a diesel engine is not a new concept. This has been tried ever since the diesel engine was first made. The greatest advantages of vegetable oils are that they are obtained from seeds of various plants. In view of this, researchers have started showing renewed interest towards vegetable oils because of their advantages as a potential alternate fuel. Vegetable oils are renewable and eco-friendly in nature and at the same time, they can be easily produced in rural areas (Martin and Prithviraj 2011).

However, due to inherent high viscosity, vegetable oils pose problems such as poor fuel flow and atomization and constrain their direct use in engine without any modifications. Various techniques have been identified and tested to reduce the vegetable oil viscosity to the extent that they can be used as substitute fuels for diesel engines.

Biodiesel is the second most common biofuel used around the world (Ren21 2012). There are four adopted methods to reduce the viscosity of vegetable oils in order for it to be used in a compression ignition engine without any major modification to the engine. These include pyrolysis, blending with petroleum diesel, microemulsification and transesterification (Knothe et al. 2005; Ramachandran et al. 2013).

Production Science and Technology

In the transesterification process, triglyceride (vegetable oils) reacts with an alcohol (preferably methanol or ethanol) forming a methyl or ethyl-ester (biodiesel) as the core product and glycerol as a by-product (Bergmann et al. 2013; Nasir et al. 2013; Subramaniam et al. 2013; Knothe et al. 2005). Even though ethanol is the preferred alcohol in the transesterification reaction as ethanol can be produced renewably and has low toxic level, methanol is mostly utilized as it is cheaper than most other alcohols (Bezergianni and Dimitriadis 2013). This leads to a biodiesel which is predominantly a fatty acid methyl ester (FAME) (Bezergianni and Dimitriadis 2013; Bergmann et al. 2013; Nasir et al. 2013; Arbab et al. 2013).

A catalyst is used to enhance the reaction rate and increase the reaction yield (Silitonga et al. 2013). The catalyst can be either alkaline or acidic. For experimental purposes, commonly used alkaline catalyst are sodium hydroxide (NaOH), potassium hydroxide (KOH), carbonates and sodium or potassium methoxide whereas sulphuric acid (H_2SO_4), hydrochloric acid (HCl), phosphoric acid (H_3PO_4), and sulfonic acids are typical acidic catalyst (Santori et al. 2012). However many of these catalyst are not suitable for industrial purposes. Mostly KOH is used in the production of the biodiesel in large amounts. The concentration of the catalyst essential for the reaction ranges from 0.1 to 1.5 % of the vegetable oil by weight (Santori et al. 2012). Kumar et al. (2010) prepared biodiesel with 0.75 % concentration of KOH and yielded a maximum ester formation of 97.8 %.

Excess alcohol is used to shift the equilibrium to the product side as the reaction is reversible (Singh 2009). Santori et al. (2012) mentioned that the molar ratio of alcohol: vegetable oil as 6:1. There are over 4,000 plant species from which vegetable oil can be derived (Santori et al. 2012) but their yield per hectare varies drastically. To avoid the food versus energy conflict non-edible oil sources are preferred. The characteristics which can classify a particular vegetable oil as good source for biodiesel production are large oil content, good adaptation to growing conditions, low maintenance costs, regular and frequent maturations and environmentally friendly waste products (Mohammad et al. 2013). Vegetable oil to biodiesel conversion requires various stages and it can be summarized as shown in Fig. 20.4.

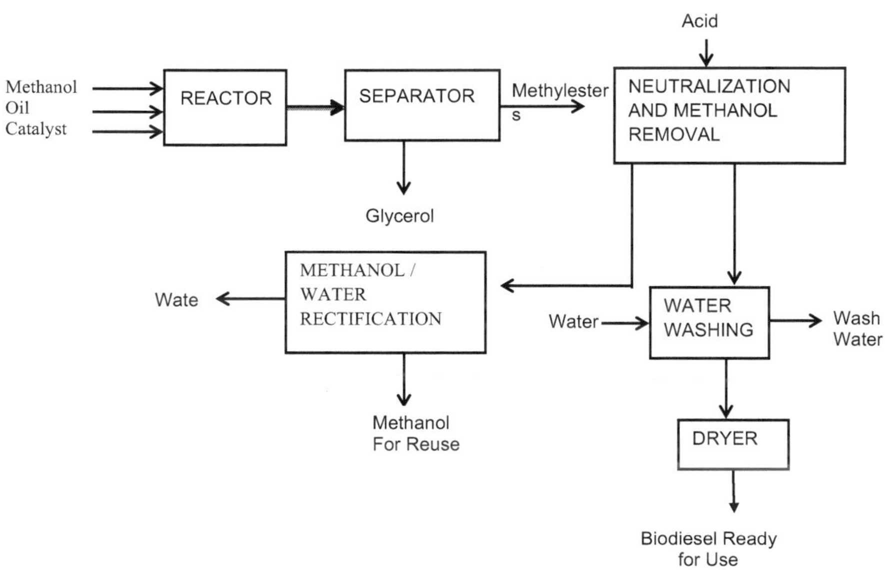

Fig. 20.4 Vegetable oil to biodiesel conversion (Knothe et al. 2005)

Overview of Resources

The possible indigenous feedstock in Fiji and the PICs that can be considered for biodiesel production include coconut oil which is quite readily available. Other potential vegetable oils that can be considered for biodiesel production include pongamia, castor and jatropha.

Coconut palms are grown throughout the tropics and are widely called the "tree of life" for their important role in smallholders' livelihoods as a direct source of cash income, nutrition and materials (Warner et al. 2007). However, the paramount economic benefit to coconut producers has come from drying the coconuts into copra for further processing into copra oil (PARDI 2011).

In Fiji, over the years there has been a decline in the production of copra which has eventually led to a reduction in the coconut oil production as seen in Fig. 20.5. The decline can mostly be attributed to the decreasing world prices and low return on labour leading to farmers ceasing copra and coconut oil production. Other reasons include: a large representation of senile trees with low productivity; poor delivery of services from current copra buyers; product range restricted to copra production only; slow rate of replanting; high labour and freight costs; and irregular shipping services for the outer islands.

Fiji possesses the required potential for replacing expensive imported fuels with locally produced, environmentally friendly fuel from coconut oil. There is also a shift to transform the industry from the traditional copra production to one where coconut farmers are involved in value added products as well, such as health food and cosmetics. Currently, companies involved in processing coconut include; Punjas (coconut oil), Pure Fiji (oil, soap, lotion-cosmetics), Makosoi (cosmetics), Organic (cosmetics), Fiji Department of Energy (Biofuel), Niu Industry (Biofuel), Pacific Green (Timber, Furniture), and Food Processors (coconut cream).

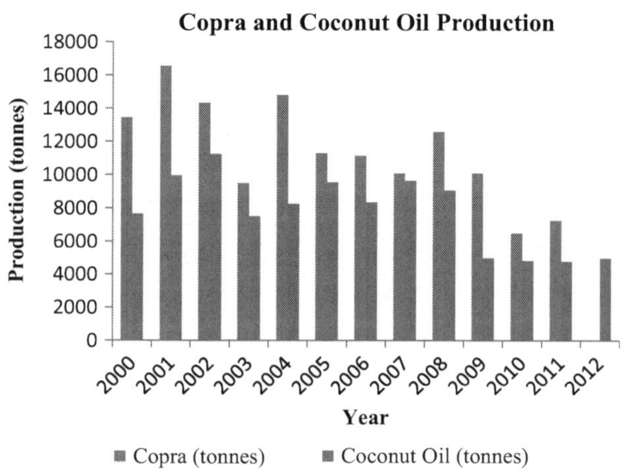

Fig. 20.5 Copra and coconut oil production from 2000 to 2012 in Fiji (Bureau of Statistics 2013)

Copra production in 1977 was as the highest at 30,644 tonnes, producing a total of 18,502 tonnes of coconut oil (Bureau of Statistics 2013). An effective way of returning the industry to these previous levels would be to replace senile trees with young higher yielding trees; increase coconut production through proper farm maintenance, efficient production of coconut oil through decentralized small scale milling units; establishment of biodiesel plant in Fiji producing fuel that will add value to coconut and save foreign exchange spent on import of light diesel.

Non-edible oils that are being explored for biodiesel production in Fiji include jatropha, pongamia and castor oil. Biofuels International (n.d.), a privately funded initiative, has been focused on the development of pongamia and jatropha as the world's new energy source. The project's first phase is currently underway in which FJ $174 million has been invested in the planting of *Pongamia pinnata*. Currently, Biofuel International has a quarter million pongamia trees growing in the province of Bua on the island of Vanua Levu in Fiji and according to the Chief Executive Officer, production of fuel will start next year (FBC 2014). Apart from pongamia and Jatropha, Biofuels International and Bio-Agro-tech plan to plant special patented castor seeds throughout the planned plantations in Fiji and Tonga.

Production Potential

Biodiesel is considered a good substitute for petroleum diesel either in pure form or blended with petroleum diesel. Currently, the Fiji government has approved the use of 5 % biodiesel blended with petroleum diesel in automobiles. Furthermore, the Fijian Government's biofuel outer island programme include use of blends consisting of 20 % coconut oil and 80 % petroleum diesel in diesel generators with no modification and use of 100 % coconut oil with modified diesel generators.

If automobiles were to switch to B5 (5 % biodiesel with 95 % petroleum diesel) then the total automotive diesel fuel import would be reduced to 33,922,843 L (see Table 20.3). The 4,977 tonnes (approximately equal to 5.4 ML assuming the density of coconut oil is 0.925 g mL^{-1}) coconut oil produced in 2012 will yield 4.32 ML of biodiesel. This is assuming that 1 L of coconut oil will produce 0.8 L of biodiesel (Beer et al. 2007). This will amount to 13 % of the total automotive diesel imports per year in Fiji.

If all the 4.32 ML of biodiesel is used to make B5 then a total volume of 86.4 ML will be produced. The current automobile petroleum diesel required for Fiji is approximately 34 ML from Table 20.3. Thus, this is 254 % of the total needs for Fiji. The additional B5 can be used for industrial purposes for running Compression Ignition engines, or exported.

The current major fiscal incentives available for renewables include the following (Nakavulevu 2011):

- The diesel which is used for blending with biodiesel has a reduced duty of only \$0.05 L^{-1} compared to the normal duty of \$0.18 L^{-1}; and
- A 10 year tax holiday is available to a taxpayer undertaking a new activity in processing agricultural commodities into biofuels. This has been approved for the period extending from 1 January 2009 to 31 December 2014. In order to qualify, the taxpayer must have; (1) a minimum level of investment of FJ \$1,000,000; and (2) employ at least twenty local employees or more for every income year.

Technology, Human Capacity and Other Pre-requisites for a Biofuels Industry

Fiji has recently made significant strides in developing its human capacity for the development of renewable energy resources, especially in the biofuels area. As discussed by Singh (2012a), the Fiji Department of Energy (FDOE) has established biofuels units with staff experienced in biofuels research. The local universities have started producing graduates with scientific training in biofuels, and the FDOE's biofuel unit has developed a biofuels standard for Fiji. In addition, legislation is now in place to enforce these standards. This includes the use of B5 in compression ignition (CI) or diesel engine vehicles and 10 % ethanol with 90 % petrol (E10) in Spark Ignition (SI) engine, or petrol engine vehicles. Currently, the use of B5 and E10 are voluntary as the lack of production capacity is preventing the mandatory status.

Recently, there has also been significant interest by the private sector in the biofuels industry. This has mainly resulted from the current incentives offered by the Government of Fiji. The plant, machinery and equipment for the initial establishment of the factory and the chemicals required for biofuel production do not incur any fiscal duty and only require a 15 % Value Added Tax (Nakavulevu 2011).

As indicated earlier in this chapter, there are other factors, beyond the elementary considerations above, that play a deciding role in the ultimate success of any new venture in developing nations such as the PICs. The situation is best seen via a comparative analysis. The issue of setting up any new industry in a developed nation is a relatively simple one as compared to the case of developing nations. Developed nations have both the technological infrastructure as well as the science base to support the new technology (Singh 2012b). Developing nations on the other hand are usually deficient in these aspects, the problem being most acute in the case of the PICs. Thus to set up a new industry, developing countries (particularly from the Pacific) must do a lot more. They are usually forced to take a piece-meal approach to the development, identifying each deficiency and resolving it on an individual basis. This is due to the scarcity of available human and technical capacity in these countries. The development process can be painstaking and morally debilitating.

To consider a relevant example, government departments rely on scientific advice for the development of the appropriate fuel standards for new fuels. In developed nations, such expertise is easily available either in-house or in the numerous universities and technical institutions available in the country. In the case of the PICs, such advice is severely limited by the scarcity of the appropriate advisors.

To put the argument more simply, the development of any new industry must necessarily involve an institutional learning curve where all stakeholders, including the business enterprise, the government as well as the users learn to adapt and adopt the new technology. For the PICs, this learning curve is a lot steeper (often prohibitively so) than those for the developed nations. The PICs must learn to live with this predicament.

Future Prospects for a Biofuels Industry for Fiji and the PICs

Worldwide liquid biofuels make a small but increasing contribution to fuel usage (REN21 2012). In 2011 biofuels provided about 3 % of global road transport fuels and is expected to rise to 27 % by 2050 (IEA 2011). Therefore, it is timely for developing countries like Fiji to consider developing their biofuel industry in earnest.

The prospects and challenges for this endeavour have already been highlighted. Nevertheless, in the case of Fiji, with the relevant biofuel policies already put in place by FDOE, the country should start looking forward to the sustainable production and use of biofuels for transportation. According to the current biofuel regulations, SI engine cars later than the 1990 model in Fiji can use a blend of up to 10 % ethanol in petrol without engine modifications. For models older than 1990, modification to the carburetor will be required before ethanol-petrol blends with 5–10 % ethanol can be used (ANFAVEA 2005). This mainly includes the material of the body and cover of the carburettor which cannot be aluminium or Zamak and also any component of Nylon, which will have to be replaced if unprotected. As most fleets in Fiji are currently 1990 models and later, the compatibility issue should not arise for blends containing a maximum of 10 % ethanol (E10).

In order to use higher ethanol-petrol blends in Fiji, certain modifications to engines will have to be made so that the benefits associated with the higher oxygen content could be utilized, which include improved fuel efficiency and reduced emissions. Flex-fuel, or flexible fuel vehicles (FFVs), are a solution to using higher blends of ethanol-petrol. These have engines that can alternate between two sources of fuel, including petrol and bioethanol or petrol and natural gas (The Royal Society

2008). FFVs for petrol and bioethanol have been used extensively in Brazil and to some extent in USA and Sweden and offer many advantages (Joseph 2007).

In Fiji, there are currently no moves to introduce FFVs, with the focus being mostly on introducing blends up to E10 in the existing vehicle fleets. However, this might change should the bioethanol industry develop and the production of bioethanol increases. In order to ensure sustainability, with increasing bioethanol use there needs to be parallel and compatible development of engine technologies, bioethanol feedstock and feedstock conversion technology.

As some of the crops being considered for biofuel are primarily produced for food in Fiji, a suitable compromise needs to be established that would balance out the use of agricultural land for food and fuel. The use of food crops for fuel usually drives up the prices of these crops. Although the Fiji Government has dismissed this threat, indicating that there is sufficient unused land, there is still a need to take necessary precautions to avoid a food versus fuel crisis in Fiji. Setting aside land for food production is one strategy. However, governments need to make a national-level decision as to what extent staple crops should be used for biofuel production. Clearly, there are several measures that need to be taken in order to develop bioethanol production on a sustainable basis. These should include the proper coordination and integration of national policies and the conduct of feasibility studies which evaluate the opportunities for biofuel production from different feedstock.

On the regional scale, judging from the availability of feedstock alone, Papua new Guinea (PNG), Solomon Islands, Samoa, Vanuatu and Kiribati produce enough coconuts to meet the requirements of large-scale biodiesel production. However, none of these countries are currently producing the biofuel. In the case of PNG, there is a preference to use blends of straight vegetable oil (SVO) with petroleum diesel for both land and sea transportation. An ethanol cassava project that was to ensue from a public-private partnership between the PNG government and the Changhae Tapioka company of South Korea, under which 13,700 ha of land was to be planted with cassava to provide feedstock for an ethanol plant, has evidently failed (SABL 2014).

It will certainly not be feasible to replace all fossil fuel imports in Fiji and the PICs with biofuels for transportation for some time. However, an overall reduction in fossil fuel imports and the concomitant results on the country's GHG emissions would still be possible through the use of biofuel blends in the vehicle energy mix of Fiji (Singh 2012a). In order for Fiji to reduce its petroleum import, drastic measures may be required. Having a large agro-based economy, Fiji has a biofuel advantage due to its rich endowment of natural resources.

The simplest and fastest way of expanding the use of biofuels in vehicles is to use blends on vehicles already on the road, without the need for modifying engines. Developing countries that currently have a limited capacity to produce cost-efficient bioethanol and biodiesel with good energy and environmental balances can diversify their liquid fuels options by importing biofuels from regions with favorable conditions for biofuel production. Alternatively, they can look towards the adoption of new technologies, and their adaptation for the local requirements.

New Technologies

One major disadvantage of biodiesel is its cost of production, which prevents its small-scale production. In Fiji, for instance, it has not been viable to produce biodiesel in volumes suitable for local consumption even with the government incentives mentioned above in place.

A simple solution that is being tried in Fiji is to use CNO-petroleum diesel blends in an effort to reduce the fossil fuel import burden of the country. However, such fuels are based on compounds containing oxygen, which give rise to issues relating to fuel standards.

Petroleum diesel, such as the number two diesel used by automobiles, is made up of pure hydrocarbons (i.e., molecules that contain only hydrogen and carbon in them). Thus any new fuel for diesel engines containing oxygen, such as the triglyceride-based blends containing CNO, are automatically disqualified from satisfying the internationally-accepted diesel fuel standards. Other issues relate to the average size of the molecules making up the fuel.

A solution has been found in the form of hydro-deoxygenated fuel (HDO) (Singh 2013) or hydrotreated vegetable oil (HVO) (Mikkonen et al. 2012). These new diesel engine fuels are produced in a process which removes the oxygen content of the oil used as feedstock, and are thus are better suited to meeting international diesel fuel standards.

Currently, a new production technique is being developed for the cheaper production of HVO from vegetable oils such as CNO which is readily available in the PICs. The technique is based on a micro-channel reactor for the production of the fuel, and possibilities that such a development may reach Fiji's shores in the not-too-distant future are not far-fetched.

Conclusions

- Fiji and most of the other PICs are committed to their UNFCCC obligations, while at the same time face the burden of heavy fossil fuel import bills.
- Fiji can produce enough feedstock for the production of bioethanol and biodiesel on a significant scale.
- Most PICs produce coconuts in sufficient quantities for the production of biodiesel.
- In the PICs the development of viable biofuel industries is hampered by barriers that extend beyond the usual requirements of the availability of resources, infrastructure, human capacity and legislative framework.
- Fiji is in a position to produce E10 and B5 in excess of its domestic requirements.
- New feedstock and technologies are now offering possibilities for the cheaper production of biofuels in the Pacific region.

References

ANFAVEA (2005) Ethanol fuel vehicular application technology. Presentation of Henry Joseph Jr. (Brazilian Automotive Industry Associations Energy and Environment Commission anfavea@anfavea.com.br) at CEPAL/CENBIO/USP Seminar São Paulo, 17 Aug 2005

Arbab MI, Masjuki HH, Varman M, Kalam MA, Imtenan S, Sajjad H (2013) Fuel properties, engine performance and emission characteristic of common biodiesels as a renewable and sustainable source of fuel. Renew Sust Energ Rev 22:133–147

Badger PC (2002) Ethanol from cellulose: a general review. In: Whippey A, Janick J (eds) Trends in new crops and new uses. ASHS, Alexandria, VA

Beer T, Grant T, Campbell PK (2007) The greenhouse and quality emissions of biodiesel blends in Australia, Report KS54C/1/F2.27. CSIRO, Aspendale, VIC

Bergmann JC, Tupinamba DD, Costa OYA, Almeida JRM, Barreto CC, Quirino BF (2013) Biodiesel production in Brazil and alternative biomass feedstocks. Renew Sust Energ Rev 21:411–420

Bezergianni S, Dimitriadis A (2013) Comparison between different types of renewable diesel. Renew Sust Energ Rev 21:110–116

Bijay P (2012) Viability of using cassava as feedstock for bio-ethanol production in Fiji. Presentation at the international conference on technology transfer and renewable energy, Mauritius, 21–22 June 2012

Biofuels Digest (2012) Brazil sets up $38 billion ethanol subsidy program to stimulate expansion. Biofuels Digest, 27 Feb 2012. Available http://www.biofuelsdigest.com/bdigest/2012/02/27/brazil-sets-up-38-billion-ethanol-subsidy-program-to-stimulate-expansion/ (date accessed 10 March 2014)

Biofuels International (n.d.) Available online http://www.biofuelsonline.net/ (date accessed 12 March 2014)

Bureau of Statistics (2013) Key statistics 2013. Bureau of Statistics, Suva

Colitt R, Nielsen S (2012) Brazil ethanol drive falters on domestic supply shortage. Bloomberg Businessweek, 13 March 2012. Available http://www.businessweek.com/news/2012-03-13/brazil-ethanol-slows (date accessed 16 April 2014)

Crutzen PJ, Mosier AR, Smith KA, Winiwarter W (2007) N_2O release from agro-biofuel production negates global warming reduction by replacing fossil fuels. Atmos Chem Phys Discuss 7:11191–11205

Deverall E (2005) The Fijian sugarcane industry – investing in sustainable technology. Oxfam briefing paper 77, Oxfam International, Oxfam

E4tech (2008) Biofuels review: advanced technologies overview for the renewable fuels agency. Available http://www.globalbioenergy.org/uploads/media/0805_E4tech_-_Advanced_technologies.pdf (date accessed 10 March 2014)

FAOSTAT (2014) FAO of the UN. Available http://faostat.fao.org/site/567/default.aspx#ancor (date accessed 7 Feb 2014)

FBC (2014) Biofuel project in the North. Fiji News, Fiji Broadcasting Commission, 26 Feb 2014

Ferreira V, Faber MO, Mesquita SS, Pereira N Jr (2010) Simultaneous saccharification and fermentation process of different cellulosic substrates using a recombinant Saccharomyces cerevisiae harbouring the β-glucosidase gene. Electron J Biotechnol 13(2):1–7. Available http://www.ejbiotechnology.info/content/vol13/issue2/full/1/ (date accessed 10 March 2014)

Hernández-Salas JM, Villa-Ramirez MS, Veloz-Rendón JS, Rivera-Hernández KN, González-César RA, Plascencia-Espinosa MA, Trejo-Estrada SR (2009) Comparative hydrolysis and fermentation of sugarcane and agave bagasse. Bioresour Technol 100(3):238–1245

Hofstrand D (2009) Brazil's ethanol industry. AgDM Newsletter, January 2009. http://www.biofuelsonline.net/ (date accessed 10 March 2014)

IEA (2004) Biofuels for transport: an international perspective. International Energy Agency, Paris, France

IEA (2011) Technology roadmap: biofuels for transport. International Energy Agency, Paris

IPCC (2014) IPCC Working Group II contribution to the fifth assessment report. In: Climate change 2014: impacts, adaptation and vulnerability, March 2014. Available online http://www.ipcc.ch/report/ar5/wg2 (date accessed 24 May 2014)

Johnston P (2012) Promoting regional cooperation for enhanced energy security and the sustainable use of energy in Asia and the Pacific: Pacific perspective on the challenges to energy security and the sustainable use of energy. ESCAP high level policy dialogue, Tanoa Hotel, Nadi, 8–9 Oct 2013

Joseph H (2007) The vehicle adaptation to ethanol fuel. Presentation to the Royal Society International biofuels meeting held in London, April 2007. Available online www.royalsoc.ac.uk/downloaddoc.asp?id=4248 (date accessed 13 April 2014)

Kearsley MW, Dziedzic SZ (eds) (1995) Handbook of starch hydrolysis products and their derivatives. Blackie Academic and Professional, New York

Knothe G, Gerpen JV, Krahl J (eds) (2005) The biodiesel handbook. AOCS Press, Champaign, IL

Krishna SH, Reddy TJ, Chowdary GV (2001) Simultaneous saccharification and fermentation of lignocellulosic wastes to ethanol using a thermotolerant yeast. Bioresour Technol 77 (2):193–196

Kumar AKN, Hasan MM, Chaudhary R, Siddiquee AN, Khan ZA (2010) Production of biodiesel from thumba oil: optimization of process parameters. Iran J Energ Environ 4:352–358

Lee CG, Kim CH, Rhee SK (1992) A kinetic model and simulation of starch saccharification and simultaneous ethanol fermentation by amyloglucosidase and Zymomonas mobilis. Bioprocess Eng 7:335–341

Martin M, Prithviraj D (2011) Performance of pre-heated cottonseed oil and diesel fuel blends in a compression ignition engine. Jordan J Mech Ind Eng 5(3):235–240

Mikkonen S, Hartikka H, Kuronnen M, Saikkonen P (2012) HVO, hydrogenated vegetable oil—a premium renewable biofuel for diesel engines. Nest Oil Proprietary Publication, Nest Oil, Keilaranta

Mohammad M, Hari TK, Yaakob Z, Sharma YC, Sopian K (2013) Overview on the production of paraffin based-biofuels via catalytic hydrodeoxygenation. Renew Sust Energ Rev 22:121–132

Nakavulevu P (2011) Promoting renewable energy policies in Fiji. Presentation at workshop organized by IRENA, accelerated renewable energy deployment on islands with emphasis on Pacific Islands, Sydney, 26–28 Oct 2011

Nasir NF, Daud WRW, Kamarudin SK, Yaakob Z (2013) Process system engineering in biodiesel production: a review. Renew Sust Energ Rev 22:631–639

OECD (2011) Brazil's biofuel sector: what future? OECD Observer, No. 287. Available http://www.oecdobserver.org/m/fullstory.php/aid/3748/Brazil_92s_biofuel_sector:_What_future_.html (date accessed 13 March 2014)

Pacific Agribusiness Research and Development Initiative (PARDI) (2011) Coconut chain review. Available online www.adelaide.edu.au/global-food/documents/pardi-cocoa-chain-review-oct-2012.pdf (date accessed 10 April 2014)

Pazur JH, Ando T (1960) The hydrolysis of glucosyl oligosaccharides with α-D-(1→4) and α-D-(1→6) bonds by fungal amyloglucosidase. J Biol Chem 235:297–302

Rajagopal D, Sexton SE, Roland-Holst D, Zilberman D (2007) Challenge of biofuel: filling the tank without emptying the stomach? Environ Res Lett 2:044004. doi:10.1088/1748-9326/2/4/044004

Ramachandran K, Suganya T, Gandhi NG, Renganathan S (2013) Recent developments for biodiesel production by ultrasonic assist transesterification using different heterogeneous catalyst: a review. Renew Sust Energ Rev 22:410–418

Rao PJM (1997) Industrial utilization of sugarcane and its co-product, Indian commission of sugar industry development. ISPCK, New Delhi

REN21 (2012) Renewables 2012 global status report, REN21 Secretariat, Paris

SABL (2014) SABL Commission of inquiry report 1. Available online http://pngexposed.wordpress.com/2013/11/25/sabl-case-study-no-1-puka-temu-and-the-changhae-tapioka-cassava-project/ (date accessed 20 May 2014)

Santori G, Nicola GD, Moglie M, Polonara F (2012) A review analyzing the industrial biodiesel production practice starting from vegetable oil refining. Appl Energy 92:109–132

Shen Y, Zhang Y, Ma T, Bao X, Du F, Zhuang G, Qu Y (2008) Simultaneous saccharification and fermentation of acid pretreated corncobs with a recombinant *Saccharomyces cerevisiae* expressing β-glucosidase. Bioresour Technol 99(11):5099–5103

Silitonga AS, Masjuki HH, Mahlia THI, Ong HC, Chong WT, Boosroh MH (2013) Overview properties of biodiesel diesel blends from edible and non-edible feedstock. Renew Sust Energ Rev 22:346–360

Singh PJ (2009) Preparation, characterisation and engine performance of coconut oil based hybrid fuels. MSc thesis, School of Engineering and Physics, The University of the South Pacific, Suva

Singh A (2012a) Biofuels and Fiji's roadmap to energy self-sufficiency. Biofuels 3(3):269–284

Singh A (2012b) Renewable energy in the Pacific island countries: resources, policies and issues. Manag Environ Qual 23(3):254–263

Singh A (2013) Fiji's biofuel industry: perched for a soaring takeoff? Biofuels 4(6):587–589

Sjöström E (1993) Wood chemistry: fundamentals and applications. Academic, New York

SOPAC (2009) Potential for liquid biofuels in Fiji. SOPAC miscellaneous report 677

Subramaniam D, Murugesan A, Avinash A, Kumaravel A (2013) Bio-diesel production and its engine characteristics—an expatiate view. Renew Sust Energ Rev 22:361–370

Sukumaran RK, Singhania RR, Mathew GM, Pandey A (2009) Cellulase production using biomass feed stock and its application in lignocellulose saccharification for bio-ethanol production. Renew Energ 34(2):421–424

The Royal Society (2008) Sustainable biofuels: prospects and challenges. Policy document 01/08. Available http://royalsociety.org (date accessed 9 March 2014)

UNFCCC (2014) UNFCCC Non-Annex 1 national communications. Available online http://unfccc.int/national_reports/non-annex_i_natcom/items/2979.php (date accessed 25 May 2014)

Wang W (2002) Cassava production for industrial utilization in China-present and future perspectives. In: Howeler RH (ed) Cassava research and development in Asia: exploring new opportunities for an ancient crop. Proceedings of seventh regional workshop held in Bangkok, 28 Oct–1 Nov 2002

Warner B, Quirke D, Longmore C (2007) A review of the future prospects for the world coconut industry and past research in coconut production and product. The final report for PLIA/2007/019, Australian Centre for International Agricultural Research (ACIAR)

Weir T, Virani Z (2010) Three linked risks for development in the Pacific islands: climate change, natural disasters and conflict. PACE-SD occasional paper no. 2010/3. Available online www.usp.ac.fj/pace (date accessed 2 June 2014)

Wilson C (2014) Fiji leads Pacific region on climate adaptation efforts. IPS-Inter Press Services. Available online http://www.ipsnews.net/2014/05/fiji-leads-pacific-region-climate-adaptation-efforts/ (date accessed 2 June 2014)

World Watch Institute (WWI) (2007) Biofuels for transportation: global potential and implications for sustainable agriculture and energy in the 21st century. Available online www.worldwatch.org/node/4078 (date accessed 26 March 2014)

Chapter 21
Towards Low Carbon Hotels in the Pacific Region: A Study of Energy Consumption and Efficiency in Hotels Using Models Based on Energy Performance Indicators

Krishneel Prasad and Anirudh Singh

Abstract Low carbon tourism is becoming an increasingly important feature in the sustainable development of the tourism industry, within the Pacific Island Countries (PICs). This growing interest is due at least in part to the desire for these nations to contribute meaningfully to the fight against climate change. One way of reducing carbon emissions is through energy efficiency in hotels. This requires the study of the energy consumption of hotels via modeling techniques involving energy audits. While such studies have been carried out extensively for European and other hotels in the Northern Hemisphere, the PICs have not received similar attention. Because of the differences in the heating/cooling requirements, hotels in temperate regions will have energy consumption characteristics that may be quite different from those of the tropics. This chapter reports an analysis of the energy consumption characteristics of Pacific Island hotels using Fijian hotels as a case study. The study employs the results of detailed energy audits and modeling based on Energy Performance Indicators (EPIs) that permit the prediction of the energy consumption of hotels within specific architectural groupings and star ratings. The models allow the investigation of energy savings possible through retrofitting and other energy-saving techniques.

Introduction

The demand for energy has been increasing over the years. The current global situation is such that there is a scarcity of energy resources with an escalation of fuel prices and increase in climate change. Hence it has been recognized that energy efficient technologies and renewable energy paves way for a sustainable environmental and socio-economic development. Figure 21.1 shows the global energy

K. Prasad (✉) • A. Singh
School of Engineering and Physics, University of the South Pacific, Laucala Campus, Suva, Fiji
e-mail: prasad_ki@usp.ac.fj; singh_ag@usp.ac.fj

W. Leal Filho (ed.), *Climate Change in the Asia-Pacific Region*, Climate Change Management, DOI 10.1007/978-3-319-14938-7_21

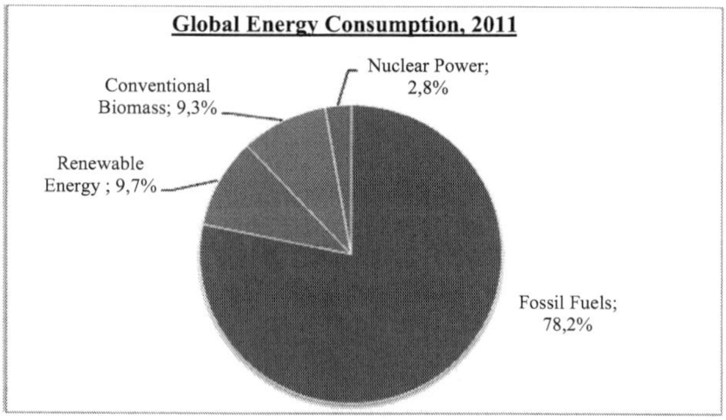

Fig. 21.1 Global energy demand and consumption (REN21 Global Status Report 2013)

consumption from the different energy sources in as reported in 2011 by the global status report from REN21. Fossil fuels have continued to be the major source of energy generation in the world today. The consumption of huge volumes of fossil fuels emits a greater degree of greenhouse gases (GHGs) directly into the earth's atmosphere.

The tourism sector is a thriving industry in many Pacific island countries. It benefits these small developing countries contributing to their economic growth, having improved infrastructure, bringing in foreign currency and creating employment for many people.

On the other hand, there is an increase in energy demand to provide services to the tourists. Transportation is a sub sector of tourism that causes around 75 % of the CO_2 emissions including aviation as a major contributor of 40 % while the accommodation sector represents about 20 % of emissions from tourism activities (Scott et al. 2008).

Tourism is one of the largest economic sectors in the Pacific Island Countries (PICs), contributing significantly to all these island states. The number of tourist arrivals in Fiji increased by 17 % from 2007 to 2010 (Statistics 2012). These PICs are almost exclusively dependent on imported fossil fuels for their energy needs for power generation and transportation. The complex fuel logistics and distribution to these island states add to the high risk of volatile price of fuel supply and energy costs. The dominance of high energy costs adds particular concern to the imbalance of the small regional economies in terms of limited investments in infrastructure, key service providers and other essential items (Linus Mofor 2013).

The tourism industry in small island states of the Pacific is dependent on the environmental resources offered by the surrounding ecosystems (IPCC AR5 WG II 2014a). Hotels and resorts near the coastal shoreline is a very common sight in PICs. These tourist destinations are very prone to the direct and indirect impacts of climate change, including coastal erosion, siltation, and sea level rise, disturbance to wild life habitats, coral degradation, and extreme weather events such as flooding. These issues further impact the shoreline and coastal environment as it reduces the recreational opportunities for tourists. It inhibits the activities such as deep sea diving, snorkeling fishing, boating which are some common leisure interests looked forward to by the tourists during their stay at the resorts (IPCC AR5 WG II 2014b).

This paper aims to investigate the energy consumption in hotel buildings in the Pacific, taking two hotels from Fiji as case studies. It compares the energy use between similar areas of the hotels studied. It also evaluates the need for energy efficiency by using energy performance indicators to benchmark the level of performance in terms of energy consumption. Recommendations to further enhance energy efficiency are also discussed.

The following sections begin with a literature review giving some background knowledge of the energy consumption in hotel buildings on the global scale and key information from similar studies by other authors. This is followed by a description of the methodology, describing the analytical processes followed to carry out this study. The outcome and results of the study are then presented and discussed. We conclude this chapter by summarizing the outcomes of the study.

Literature Review

Energy Sources in a Hotel Building

The use of conventional energy sources such as oil and fuels also adds to the aggravating problem of environmental pollution emitting toxic gases directly into the atmosphere (Santamouris et al. 1996). In Fiji, The Fiji Electricity Authority reports 58,996 metric tons of fuel consumption for the year 2012, costing the company FJ$105,136,000 (FEA 2012). For instance (Jaber and Probert 2002) state that the an equivalent of 80 % of total conventional fuel had been consumed for electricity generation in Jordan, in the Middle East.

Using fossil fuels result in large amounts of greenhouse gases emitted into the atmosphere. The building sector in the European Union contributes to an alarming 36 % of CO_2 emissions from 40 % of energy consumption as reported by (Goncalves et al. 2012). A clear example has been identified by (Wang 2012) which reports 21 % of carbon dioxide emitted by the tourism sector in Taiwan.

A major sub-sector of the tourism industry is the hospitality sector providing services such as accommodation and leisure activities to tourists. Generally hotels use different types of energy to provide its services including electricity, (LPG) liquefied petroleum gas, diesel fuel and natural gas, etc. In its primary form, electricity is used for a variety of activities and facilities, such as, lighting, kitchen appliances, lifts/escalators, heating ventilation and air conditioning (HVAC). The hotels studied in this paper consumed energy in the form of electricity and liquefied petroleum gas.

Energy consumption is also dependent on the type of building. The building's features such as the building envelope and façade as well as the operational characteristics contributes variably to the total energy consumption (Jaber and

Probert 2002; Deng and Burnett 2000). For instance, the different materials used in construction, maintenance, lighting, and the heating and cooling systems. The design of a building can also include its orientation, insulation and the use of natural ventilation and daylight wherever possible as a means of energy reduction. Hotels have higher energy consumption than other residential and commercial buildings to provide services to their occupants or guests, for instance, bars, restaurants, spacious lounges conference rooms, gym and swimming pools are a few facilities to name. In fact, hotels have to be operated on a 24 h basis, meaning that energy is being consumed continuously, as compared to other commercial and office buildings (Xin et al. 2012; Deng and Burnett 2000).

Analysis of Energy Consumption in Hotels

A number of studies to evaluate the annual energy use per unit area of the hotel buildings have been noted. The energy performance of hotel buildings can be analyzed by the use of some simple indicators (Goncalves et al. 2012). Some of the indicators used as part of their study include:

- Annual primary energy demand per square meter,
- Primary energy ratio (PER) and,
- Energy efficiency

Electricity has been identified to dominate the energy consumption in the case of hotel buildings in Taiwan noting an average of 84 % of the total energy (Wang 2012). Wang showed significance of the Pearson correlation coefficients of the annual energy consumption with the Energy use index (EUI). Together with that, using regression analysis of multiple variables, it was possible in predicting annual energy consumption.

Measures to Reduce Energy Consumption

A research in China focusing on the establishment of energy consumption quota gave positive results to promote energy efficiency (Xin et al. 2012). The implementation of an Energy Conservation Building Code (ECBC) on six different types of buildings in the city of Jaipur, India estimated to bring about 17–42 % of energy savings (Tulsyan et al. 2013). The authors report an energy reduction of 100.94 kWh/m^2 per year for a five star luxury hotel in the city area.

The directive 2002/91/EC in the European Union has been instrumental to improve the buildings energy performance while the directive 2010/31/EU with its streamlined provisions and adoption of new methodologies is aimed to strengthen the performance of energy in European buildings (Goncalves et al. 2012). A 20 % energy conservation reported by (Santamouris et al. 1996) is possible via retrofitting techniques and employing new energy efficient materials and systems in Hellenic hotels. The energy consumption data was collected from 158 Hellenic hotels during an energy audit of buildings of the National Energy Program.

Proper and timely maintenance of equipment and technology are crucial in optimizing their energy efficiency. For instance, an energy saving of 37 % was noted at the London Marriott County Hall in the United Kingdom (World Tourism Organization 2008). This success of energy conservation was achieved by cleaning the filters and coils of the air-conditioning units. The use of new technology by (Yu and Chan 2010) report a reduced annual electricity consumption of 8.6 % and reduced operating cost of 9.9 % with the use of water-cooled chiller system against their current and conventional counterpart.

Renewable Energy for Energy Efficiency

Renewable energy sources including solar energy, wind energy, hydropower, geothermal power and bio-fuels are regarded as clean sources (REN21 2012). The global investment in renewable energy technologies has been increasing over the years (see Fig. 21.2).

The use of renewable energies can be beneficial towards improving the energy efficiencies by reducing the electricity requirements and hence lowering the costs of energy consumption. Khemiri estimated an optimal potential of energy savings of 50 % in a Mediterranean hotel via the installation of renewable sources of energy production (Khemiri and Hassairi 2005).

To reduce the energy consumption, both energy efficiency and renewable energy can be incorporated. This is evidently clear in the following areas of energy consumption:

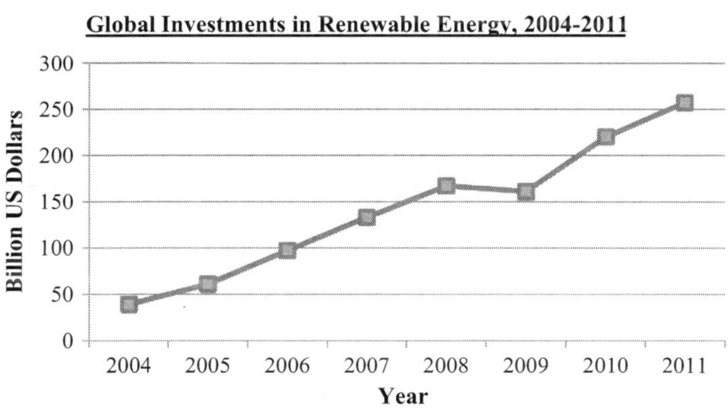

Fig. 21.2 New investments in renewable energy from 2004 to 2011 globally (REN21 2012)

- **Electrical Services**
  This can be achieved by the use of highly efficient refrigerators, pumps and lighting (such as compact fluorescent lamps and LED lights) and other appliances. With appropriate design for building envelope, daylight can be used to provide the illumination inside building.
- **Heating and cooling**
  The heating and cooling requirements are very common in all parts of the world. Heating constitute to space heating in colder climates and can be met by passive solar designs. Solar collectors are also used in domestic hot water systems. Cooling is mainly required in tropical and hot zones. Building designs with appropriate shading and using ventilation techniques, the cooling load can reduced from the use of air conditioning systems.
- **Industrial Processes**
  The industrial processes require energy in the form of heat, electricity and mechanical operations. These energy requirements can be met by using rooftop solar photovoltaic systems, wind power and biomass technology.
- **Transportation**
  The use of renewable energy has been implemented in the transportation sector. This is in the form of bio-fuels and electric vehicles (battery operated or hybrid) and the renewably generated hydrogen for fuel cells. The electric vehicles and those operating on hydrogen fuel cells boast a higher efficiency in terms of energy requirement per driven distance.
  The use of renewable energy reduces the emission of carbon dioxide and other greenhouse gases into the earth's atmosphere. A shift towards renewable energy confirms a 45 % CO_2 emission reduction in Balearic Islands hotels from the increase of 76.6 % between 1990 and 2010 (Rosselló-Batle et al. 2010).

Methodology

An analysis of the energy consumption of the case studies was done. To begin with, an energy audit of the hotels was carried out with estimates of energy consumption and used normalized ratios to determine the energy savings via energy efficiency. This section outlines the model used to determine the possible energy savings via retrofitting.

Energy Consumption

To assess the electrical energy used by the hotel, the utility data was obtained from the hotel administration. The electricity bill data given in dollars was converted to energy use in kilowatt hours (kWh) using the tariff rates of the electricity utility company.

$$Energy\ Usage, (kWh)W_O = \frac{Energy\ bill(\$)}{tariff\ rate\left(^{\$}/_{kWh}\right)} \tag{21.1}$$

Energy Audit

The energy audit of the hotels was achieved by noting all the electrical devices, their power ratings and the estimated duty cycle of the operation of the equipment for a typical day. The energy consumed was calculated using the formula:

$$E_\sigma = \sum_1^n V \times I \times t \tag{21.2}$$

E_σ is the sum of energy consumed in an area of the hotel by the number of electrical devices operated for the period of time, t (in seconds) multiplied by its power[1] given by the product of voltage (V) and the current, I (A) drawn by the equipment.

Normalized Energy Consumption from the Estimated Energy Audit

The estimated energy values of different areas such as guest rooms, conference rooms, kitchen, restaurant & bar, reception lounge, administration etc. were normalized using the actual consumption as per the utility bill. To yield:

$$E_{\sigma'} = \frac{E_\sigma}{\sum E_\sigma} \times W_O \tag{21.3}$$

$E_{\sigma'}$ is the normalized energy use in an area and W_O is the actual energy from bill. Therefore, the sum of all normalized values of the estimated energy consumption from the estimated energy audit should be equal to actual energy being consumed by the hotel.

[1] The rated power of the device(s) is taken if indicated. Power is given in Watts (W). Power = Voltage × Current.

$$\sum_{i}^{n} E_{\sigma'} = W_O \tag{21.4}$$

The energy consumption by the hotel was analyzed into the three arbitrarily chosen sectors of Production, Service and Management as follows:

- Production (*Guest rooms, conference rooms, restaurant and bar, business center and internet services*)
- Service (*Kitchen, hot water systems, laundry and swimming pool*)
- Management (*Administration & reservations, reception area, maintenance workshops*)

Retrofit and Recalculate

One way of improving energy efficiency (EE) is by retrofitting, that is, by replacing the current equipment and appliance with those of a higher efficiency rating and a low energy consumption.

The energy consumption was recalculated after retrofitting the current devices and using other efficiency measures to yield a more energy efficient consumption.

$$E_{eff} = \frac{E_{eff.\sigma}}{\sum E_{eff.\,\sigma}} \times W_O \tag{21.5}$$

The sum of the energy efficient E_{eff} values is equal to the new improved energy bill, W_E.

$$\sum_{i}^{n} E_{eff} = W_E \tag{21.6}$$

W_E, the new efficient total energy consumption is less than W_O, the current total energy consumption by the hotel. The difference between W_O and W_E gives the possible energy savings.

$$E_{saved} = W_O - W_E \tag{21.7}$$

Estimated Fuel (Diesel) Savings

1 kWh = 3.6 MJ, therefore, using this factor, the energy saved in kWh can be converted to Mega Joules.

$$E_{saved}(MJ) = 3.6 \times E_{saved}(kWh) \tag{21.8}$$

The calorific value for diesel is accepted as 38 MJ/L. Using this figure, the ideal volume of diesel saved as a result of energy saved can be calculated using equation 21.9.

$$Volume_{Diesel\ saved} = \frac{E_{saved}(MJ)}{Calorific\ Value_{Diesel}\left({}^{MJ}/_{L}\right)} \tag{21.9}$$

The volume of diesel saved also depends on the operational efficiency of the diesel generators. If the efficiency of the generators is known, a more accurate value of the diesel savings can be calculated. After determining the volume of the diesel, the equivalent mass of diesel can be calculated using the density as 0.832 kg/L.

For instance;

In the case of hotel 1, from Table 21.3, the energy saved per annum after retrofitting is 52,019.47 kWh which equates to 187,270.092 MJ. Therefore, using the calorific value of diesel as 38 MJ/l, the volume of diesel in liters consumed can be calculated. Hence, 4928.16 L of diesel can be ideally saved. Additionally, there are always some losses in the diesel generator system since the generators have an efficiency rating. Thus, the power input to a system is always greater than the useful power output. Assuming, the generator is operating at an efficiency of 55–75 %, thus the total volume of diesel saved is approximately between 7,638.65 and 8,624.28 L. Similarly, taking the same efficiency level for hotel 2, the volume of diesel saved is between 35,270.85 and 39,821.93 L.

Using the density of diesel as 0.832 kg/L, the mass of diesel in tons is recorded in Table 21.2 as amount of diesel consumed equivalent to the average energy consumption. The energy savings via retrofitting and its diesel equivalent has been shown in Table 21.3.

Calculating the Carbon Emissions

Common fuels used for energy production are presented in Table 21.1. It also shows the specific carbon dioxide emission per unit of energy from these fuels.

To calculate the carbon dioxide emission, the following formula is used.

$$Q_{CO_2 = \frac{C_f}{h_f} \times \frac{C_{CO_2}}{C_m}} \tag{21.10}$$

Where;

Q_{CO_2} = the specific CO_2 emission (kg_{CO2}/kWh)

C_f = specific carbon content in the fuel (kg_C/kg_{fuel})

hf = specific energy content (kWh/kg_{fuel})

Cm = specific mass carbon (kg/mol Carbon)

C_{CO_2} = specific mass carbon dioxide (kg/mol CO_2)

Table 21.1 Some common fuels and their specific energy content with CO_2 emission levels

Fuel	Specific carbon content C_f (kg_C/kg_{fuel})	Specific energy content (kWh/kg_{fuel})	Specific CO_2 emission (kg_{CO2}/kg_{fuel})	Specific CO_2 emission Q_{CO2} (kg_{CO2}/kWh)
Coal (bituminous/anthracite)	0.75	7.5	2.3	0.37
Gasoline	0.9	12.5	3.3	0.27
Light oil	0.7	11.7	2.6	0.26
Diesel	0.86	11.8	3.2	0.24
LPG—liquid petroleum gas	0.82	12.3	3	0.24
Natural gas, methane	0.75	12	2.8	0.23

Source: http://www.engineeringtoolbox.com/co2-emission-fuels-d_1085.html (accessed: 14/4/2014)

Outcomes of the Study

Energy Consumption in the Case Studies

Table 21.2 shows the electrical energy consumption, electricity cost, the approximate equivalent of diesel for power generation and the associated carbon emission for the two hotels studied. These approximations are true if all the power generation was from diesel alone. As the current diesel/Renewable Energy ratio in the FEA generation mix is approximately 50 %, one would expect that the carbon emissions are reduced by the same factor. Hence, based on the annual energy consumption and assuming 100 % diesel generation, the carbon footprint of the two hotels were calculated as 77 tons CO_2 per annum and 294 for hotel 1 and 2 respectively.

The annual energy consumption and the competition in the demand for services in the hotels have been noted to be on an increase. Figure 21.3 clearly shows the trend of the increasing monthly energy use over the course of 2009–2011.

The annual energy consumption was also analyzed in terms of the three sectors of Production, Service and Management referred to in Sect. 3.3 above.

Categorically the three sectors were defined as follows:

- **Production** is where the economic production can occur or money can be made;
- **Service** sector consists of amenities to provide services to the guests;
- While the **management** consists of the energy use in staff areas such as administration and reception that are used mostly by staff in assisting their hotel guests.

Figure 21.4 shows that the highest amount of energy is used by the service sector at 51 % of the annual energy Consumption followed by Production at 46 and 3 % is used by the Management and employees of the hotel.

Table 21.2 Electrical energy consumption by hotels 1 and 2

Case study	Hotel 1	Hotel 2
Average energy consumed (kWh) 2009–2011	322,705.85	1,226,850.18
Electricity cost FJ ($)	127,372.00	510,931.56
Equivalent liters of diesel consumed (maximum)	53,501.23	116,227.89
Tons of diesel equivalent	44.51	96,701.61
Equivalent carbon dioxide emissions (kg)	77,449.40	294,444.04
Carbon footprint (ton_{CO2})	77	294

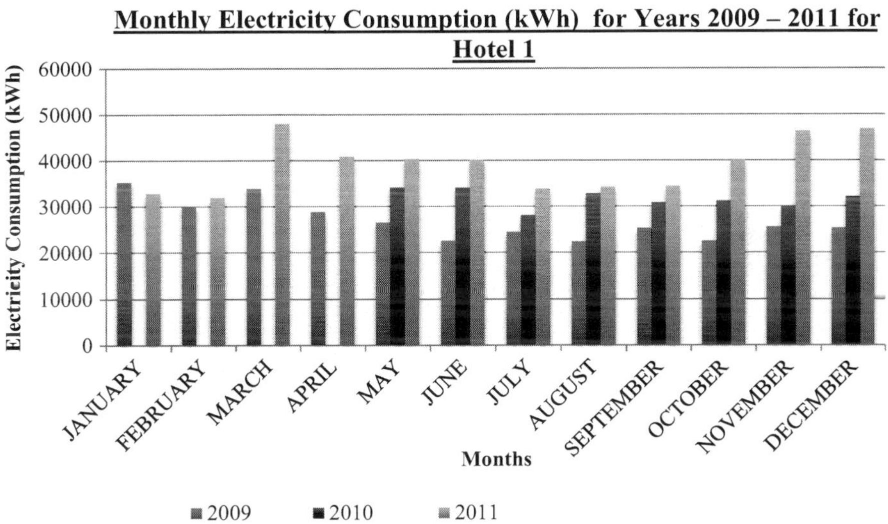

Fig. 21.3 The monthly electricity consumption of hotel 1 between 2009 and 2011

Energy Savings Due to Retrofits

Retrofitting involves replacing the current or conventional devices and appliances with those of a higher efficiency rating. Research and development continue to provide the current market with efficient systems that consume less energy and do the same or more work than the current and phased technologies (Han et al. 2010).

In this research, the retrofitting and the energy efficiency discipline were used to estimate the energy savings from the current energy consumption. For hotel 1, Fig. 21.5 clearly shows the consumption of energy in the various sections of the hotel with the decrease in energy consumption after applying the retrofitting techniques. Figure 21.6 shows the similar plot of energy consumption with the effect of retrofitting in hotel 2.

The estimated energy saved in the case studies of hotel 1 and hotel 2 is revealed in Table 21.3. According to the utility tariff rates of commercial and industrial category, the energy saved for hotel 1 is about 16 % while for hotel 2 is

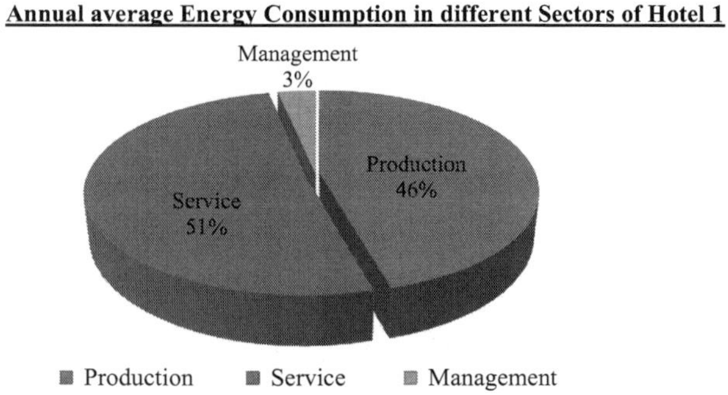

Fig. 21.4 Sector—wise analysis of the annual energy consumption in hotel 1 (see text for definition of the sectors)

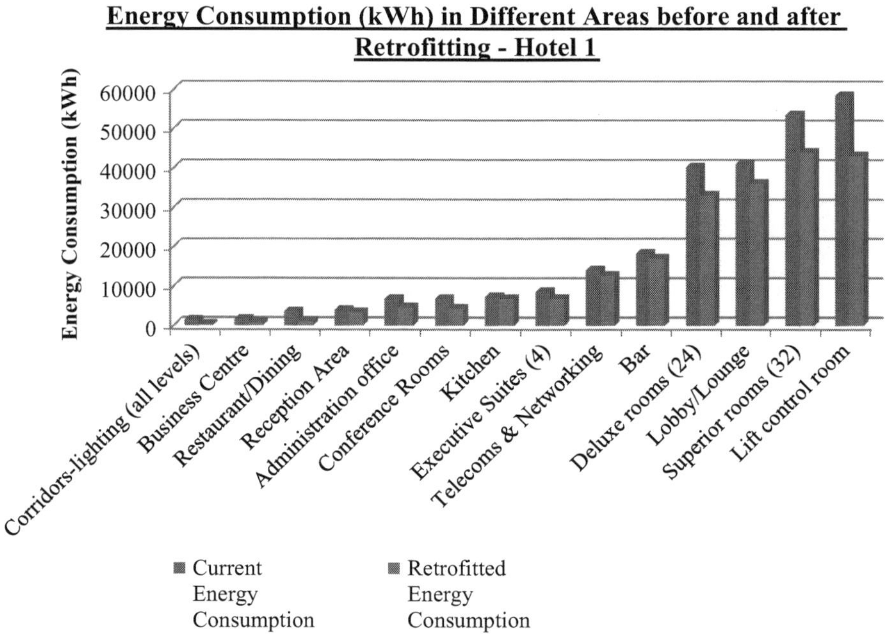

Fig. 21.5 The effects of retrofitting on energy consumption in different areas of hotel 1

approximately 19 % when compared to the annual energy consumption from 'business as usual'. Following the energy efficiency measures of retrofitting and the general energy discipline, the carbon emissions is estimated to drastically reduce by 12.48 tons in hotel 1 and 57.65 tons in hotel 2.

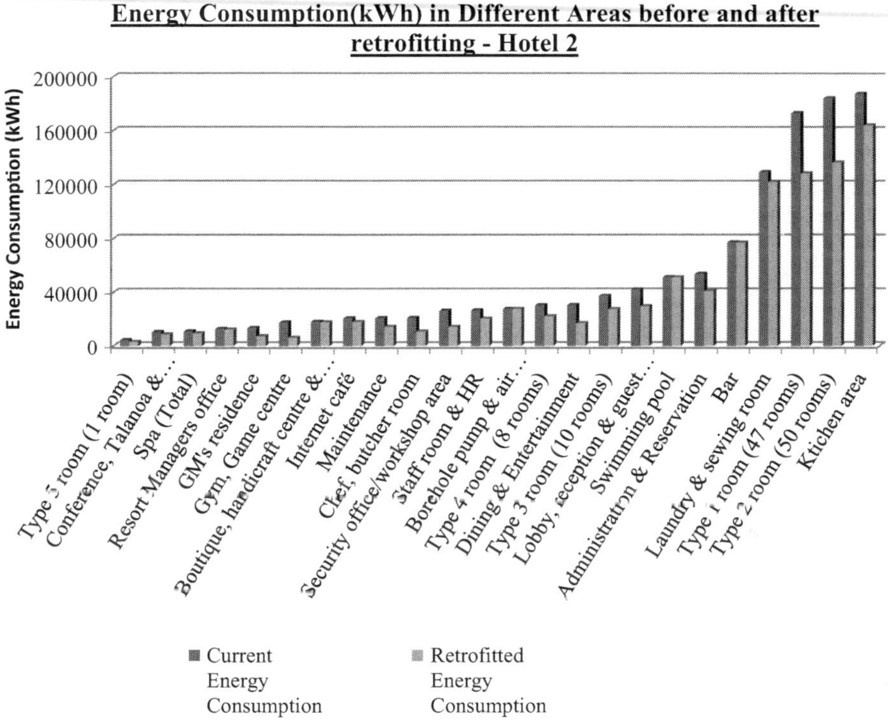

Fig. 21.6 Retrofitting techniques applied in hotel 2 showing the current energy consumption with the energy efficient retrofitted value

Table 21.3 Energy and cost savings per annum after retrofitting in the hotel case studies

Case study	Hotel 1	Hotel 2
Energy saved (kWh)	52,019.47	240,195.77
Electricity cost savings ($)	22,888.57	105,686.14
Liters of diesel saved (maximum)	8,624.28	22,755.39
Tons of diesel equivalent	7.17	18.93
Carbon dioxide emissions reduced (kg)	12,484.67	57,646.98

Development of EPIs

To evaluate the performance of energy consumption of the hotel buildings, simple energy performance indicators were developed. The EPIs act as baseline data which can be used to determine the performance of energy from one hotel to the other in terms of similar activities and areas. The first EPI was defined as energy consumption per unit area in kWh/m^2. EPI 1 was used to gauge consumption of energy by different hotels in similar areas like restaurant, kitchen, bar, administration and reception. Table 21.4 summarizes the results and shows the comparison of the energy consumption between the two hotels.

Table 21.4 Comparison of Energy consumption per unit area between similar areas of the two hotels

EPI 1 (kWh/m²/day)		
Area	Hotel 1	Hotel 2
Restaurant	0.1	0.6
Administration and reception	2.4	4.9
Kitchen	1.2	6.7
Bar	9.1	10.7

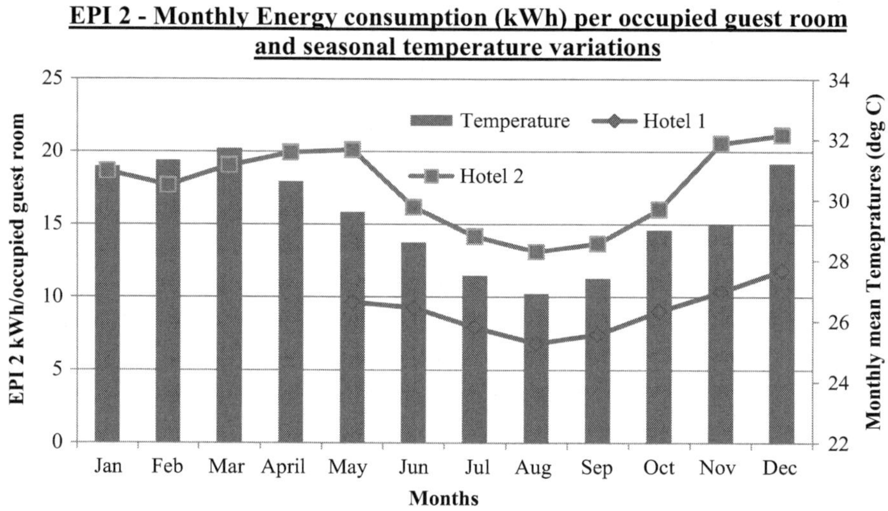

Fig. 21.7 shows the energy consumption in an occupied guest room in both the hotels with seasonal variation of temperature change

The second energy performance indicator (EPI) is energy consumption per occupied guest room. EPI 2 is generally used to compare energy consumption of guest rooms between different hotels (or different grades of rooms in the same hotel). In this instance the consumption is based on the guest's occupancy and behavior and is central to the core productivity. Figure 21.7 shows a graph of EPI 2 as energy consumption (kWh) on a particular day in each month. The average temperatures of each month were plotted and this showed very close correlations to the energy use.

Discussion

Fossil Fuels Used for Energy Generation in Fiji

In Fiji, the electricity utility company uses Industrial diesel oil and heavy fuel oil for their thermal power generation to meet the electricity demands. The Fiji Electricity Authority is also assisted by a few independent power producers (IPPs) for energy

Table 21.5 Average fuel prices of the fuels used for thermal power generation in Fiji

Average fuel price per year	Industrial diesel oil ($/t)	Heavy fuel oil ($/t)
2010	1,539	1,244
2011	2,079	1,542
2012	2,057	1,438

Source: FEA (2010, 2011,2012)

Table 21.6 Fuels consumed to meet the energy demands in Fiji

Fuel used for thermal power generation per year	Industrial diesel oil (IDO) (tons)	Heavy fuel oil (HFO) (tons)	Total fuel consumption (tons)	Total fuels cost ($' 000)	Electricity generated (GWh)
2010	60,113	27,483	87,596	126,756	383
2011	53,238	17,648	70,886	137,881	425
2012	30,694	28,302	58,996	105,136	467

production. The average energy generation mix in 2011 included 56 % of renewable sources from hydro (55 %) and wind (1 %), 40 % of diesel and heavy oil and 4 % from IPPs. The average fuel price per ton is indicated in Table 21.5.

The large quantities of fossil fuels used to cater to the demand of energy requirements in Fiji has been presented in Table 21.6 together with the total cost of the two fuels and the energy generated by the utility company (FEA 2010, 2011, 2012).

Comparison of EPI

A list of literature in Table 21.7 shows the energy performance indicators used in hotels from different climatic zones of the world. These values represent EPI 1, that is, the energy consumed per unit floor area (kWh/m^2) annually. Similarly, the energy performance of hotels 1 and 2 from the case studies had similar correlation to the hotels studied by other authors.

- 273 kWh/m^2 (M. Santamouris et al. 1996) in Hellenic Greece in 1996
- 564 kWh/m^2 (Deng and Burnett 2000) in Hong Kong in 1995,
- 446 kWh/m^2 (Pedro Goncalves et al. 2012) in Portugal in 2012,
- 427 kWh/m^2 (Priyadarsini et al. 2009) in Singapore in 2009,
- 389 kWh/m^2 (Onut and Soner 2006) in Turkey in 2005.

In this case study of hotels in Fiji,

- Hotel 1 was recorded with 426.64 kWh/m^2 and
- Hotel 2 recorded 537.83 kWh/m^2.

The hotels of this case study show that hotel 1 had an annual total energy consumption of 426.64 kWh/m^2 while hotel 2 consumed 537.83 kWh/m^2.

Table 21.7 Energy performance indicator (annual total energy consumption per unit area) from different parts of the world

Climatic zone (longitude)	Country of study	EPI—kWh/m^2/year	Journal author
45.25′N	Ottawa, Canada	612	Zmeureanu (1994)
22.3′S	Hong Kong	366	Lam and Chan (1994)
	Hong Kong	564	Deng and Burnett (2002)
1.3′N	Singapore	427	Priyadarsini et al. (2009)
22.9′N	Taiwan (Total of 4 individual hotel studied)	Hotel 1—280.1	Jen Chun Wang (2012)
		Hotel 2—237.7	
		Hotel 3—186.3	
		Hotel 4—143.6	
36.9′N	Antalya, Turkey	389	Onut and Soner (2005)
53.5′N	UK	368	Taylor et al. (2010)
40.2′N	Coimbra, Portugal	446	Goncalves et al. (2012)
	European Hilton hotels	364.3	Bohdanowicz and Martinac (2007)
	European Scandic hotels	285	
39′N	Hellenic, Greece	273	Santamouris et al. (1996)
18.1′S	Fiji		Case studies in Fiji (This paper)
	Hotel 1	426.64	
	Hotel 2	537.83	

Table 21.8 represents the simulation of energy analysis, the energy consumption per unit area using computer based software.

EPI 1 in Table 21.4 was used to represent the performance of energy consumption per unit area in similar areas for a day. Being a much larger hotel, hotel 2 had higher energy consumption than hotel 1. The kitchen area was noted to have a highest energy consumption ratio. This can be accounted by the fact that the hotel 2 with EPI 1 of 6.7 kWh/m^2 had large cooler rooms for storing foods for its guests; while hotel 1 had an EPI 1 of 1.2 kWh/m^2 being located in the urban center did not deem the requirement of such facility.

The EPI 2 plotted in Fig. 21.7 was clearly seen to be depicting the seasonal variation of the region being studied. Fiji experiences hot and humid temperatures normally between the months of November to May followed by cold and dry season from June to November. The outdoor air temperature variation in Fiji is almost as close to the energy consumption required in a guest room. This is because the highest energy consuming equipment in a guest room is the air conditioning unit. Similar studies exhibiting energy use in hotels with seasonal patterns was observed in Hong Kong by (Deng and Burnett 2000).

Results of retrofitting and following energy efficiency measures indicates the energy savings amount close to $22,888 for hotel 1 and approximately $105,686 for hotel 2 per annum. Moreover, the energy savings also reduces the emission of greenhouse gases into the atmosphere. As a result, the carbon footprint of the buildings is improved.

Table 21.8 Computer simulations give a set of result for EPI 1 in different countries

	Simulation by softwares		
Climatic zone (longitude)	Country of study	EPI—kWh/m^2/year	Journal author
	XENIOS project (simulation)		Dascalaki and Balaras (2004)
47′N	France	215	
39′N	Hellenic (Greece)	174	
43′N	Italy	280	
40.4′N	Spain	287	
~35′N	Mediterranean countries	174–287 (Simulated)	Elena and Balaras (2004)

Summary and Conclusion

Studies of energy consumption in hotels help to identify the areas where large energy consumption occurs. Applying energy efficiency measures then reduces the energy bill of the hotel, thereby reducing the need for expensive fuel imports and is also helping in climate change mitigation.

Through the energy audit study, the management is able to:

- Identify the level of energy consumption in all areas of the hotel,
- Identify any inefficient energy use and,
- Investigate the possibility of achieving energy savings by other ways of improving energy efficiency in the form of general energy discipline.

Other ways of improving energy efficiency can be improved by improving the building's insulation, increasing the albedo or the reflectance coefficient together with its orientation and thermal envelope.

Renewable energy sources such as photovoltaic and solar energy, wind energy and bio-fuel systems can be easily introduced as a cost effective measure towards energy efficiency (REN21 2012). These new technologies do not have any adverse environmental impact as compared to the levels of toxic greenhouse gases emitted by the expensive conventional fuel sources. Much recognition has been given to renewable energy sources as they are becoming much less expensive over the years and are beneficial for economic and environmental sustainability.

Thus, this study can be used by future stakeholders as a benchmark in developing efficient buildings in the PICs. Combined with other efficiency measures this will further promote energy conservation in the region. The move towards energy efficiency will promote environmental sustainability and improve the carbon footprint of the tourism sector.

References

Bohdanowicz P, Martinac I (2007) Determinants and benchmarking of resource consumption in hotels—case study of Hilton International and Scandic in Europe. Energy Buildings 39:82–95

Deng S-M, Burnett J (2000) A study of energy performance of hotel buildings in Hong Kong. Energy Buildings 31:7–12

FEA (2010) Annual report 2010. Fiji electricity authority. http://www.fea.com.fj/userfiles/files/ANNUAL%20REPORT%202010.pdf

FEA (2011) Annual report 2011. Fiji electricity authority. http://www.fea.com.fj/userfiles/files/ANNUAL%20REPORT%202011.pdf

FEA (2012) Annual report 2012. Fiji electricity authority. http://www.fea.com.fj/userfiles/files/ANNUAL%20REPORT%202010.pdf

Goncalves P, Gaspar AR, Silva MGD (2012) Energy and exergy based indicators for the energy performance assessment of a hotel building. Energy Buildings 52:181–188

Han HJ, Jeon YI, Lim SH, Kim WW, Chen K (2010) New developments in illumination, heating and cooling technologies for energy-efficient buildings. Energy 35:2647–2653

IPCC. Working Group II (2014a) IPCC assessment report 5 volume I: global and sectoral aspects. Chapter 5 coastal systems and low-lying areas. http://ipcc-wg2.gov/AR5/images/uploads/WGIIAR5-Chap5_FGDall.pdf

IPCC. Working Group II (2014b) IPCC assessment report 5 volume II: regional aspects. Chapter 29 small islands. http://ipcc-wg2.gov/AR5/images/uploads/WGIIAR5-Chap29_FGDall.pdf

Jaber JO, Probert SD (2002) Purchased-energy consumptions in Jordan's commercial and public-service sector. Appl Energy 71:31–43

Khemiri A, Hassairi M (2005) Development of energy efficiency improvement in the Tunisian hotel sector: a case study. Renew Energy 30:903–911

Linus Mofor MI, Wade H, Soakai A (2013) Pacific lighthouses renewable energy roadmapping for islands. IRENA

Onut S, Soner S (2006) Energy efficiency assessment for the Antalya Region hotels in Turkey. Energy Buildings 38:964–971

Priyadarsini R, Xuchao W, Eang LS (2009) A study on energy performance of hotel buildings in Singapore. Energy Buildings 41:1319–1324

REN21 (2012) REN21 renewables 2012 global status report. REN21 Secretariat, Paris

Rosselló-Batle B, Moia A, Cladera A, MartíNez VC (2010) Energy use, CO_2 emissions and waste throughout the life cycle of a sample of hotels in the Balearic Islands. Energy Buildings 42:547–558

Santamouris M, Balaras CA, Dascalaki E, Argiriou A, Gaglia A (1996) Energy conservation and retrofitting potential in Hellenic hotels. Energy Buildings 24:65–75

Scott D, Amelung B, Becken S, Ceron J-P, Dubois G, Gossling S, Peeters P, Simpson MC (2008) Climate change and tourism responding to global challenges. World Tourism Organization, Madrid

Statistics FBO (2012) Key statistics. Fiji Bureau of Statistics, Suva

Tulsyan A, Dhaka S, Mathur J, Yadav JV (2013) Potential of energy savings through implementation of energy conservation building code in Jaipur city, India. Energy Buildings 58:123–130

Wang JC (2012) A study on the energy performance of hotel buildings in Taiwan. Energy Buildings 49:268–275

World Tourism Organization (2008) Climate change and tourism—responding to global challenges. World tourism organization and the united nations environment programme

Xin Y, Shilei L, Zhu N, Wu W (2012) Energy consumption quota of four and five star luxury hotel buildings in Hainan province, China. Energy Buildings 42:250–256

Yu FW, Chan KT (2010) Economic benefits of optimal control for water-cooled chiller systems serving hotels in a subtropical climate. Energy Buildings 42:203–209

Chapter 22
Integrating Technologies, Measures and Policies for Climate Resilient Economy: A Case Analysis from Emerging Countries in the Asia-Pacific Region

Rajesh S. Kumar, Nity Nishant, Keshav Jha, N.K. Binu, and Balachandran Thampi

Abstract The Asia-Pacific region accounts for significant level of greenhouse gas emissions globally. The recent abrupt drastic changes in the general climate are highly relevant for the Asia-Pacific than other regions owing to its highly vulnerable societies. However, the region has significant potential to contribute to both mitigation and adaptation efforts in the context of climate change. The regional scenario which is already experiencing multiple socio-economic problems is further worsened due to climate change. Nevertheless, it is pertinent to note that several countries in the region have initiated and implemented large-scale mitigation and adaptation programs. However, there is an urgent need to augment regional capacity, strengthen institutional coordination and knowledge sharing on mitigation and adaptation policies, strategies as well as programs. These interventions can potentially influence the extent to which countries integrate climate change policies within their broader development frameworks. In the current review we describe

R.S. Kumar (✉)
Indian Forest Service (IFS), 103 MS Flats, Chanakyapuri, New Delhi 110021, India
e-mail: rskumarifs@gmail.com

N. Nishant
University of Delhi, New Delhi 110021, India
e-mail: ms.nity@gmail.com

K. Jha
ICLEI–Local Governments for Sustainability, South Asia, NSIC – STP Complex, Okhla Industrial Estate, New Delhi 110020, India
e-mail: keshavjha@live.in

N.K. Binu
College of Forestry, Kerala Agricultural University, Thrissur, Kerala, India
e-mail: nk_binu1@yahoo.com

B. Thampi
National Rainfed Authority, Planning Commission of India, Government of India, NASC Complex, Dev Prakash Shastri Marg, PUSA, New Delhi 110012, India
e-mail: kbthampi@gmail.com

W. Leal Filho (ed.), *Climate Change in the Asia-Pacific Region*, Climate Change Management, DOI 10.1007/978-3-319-14938-7_22

the climate change related threats and associated developmental risks in the Asia-Pacific region as well as analyze the sensitivity, potential resilience strategies, various adaptation and mitigation initiatives in the region in the broad context of the role to be played by the region to deal with climate change. In specifics, we attempt to develop a general typology for adaptation strategies framed upon emissions reduction technologies, practices, measures, national policies adopted in the countries in the region. The key economic sectors such as energy, buildings, and water supply systems are covered as part of this assessment through a comprehensive methodology to position the climate change threats in the backdrop of resilient economy. This framework makes use of nested flowcharts to demonstrate how, social and environmental forces interact to create situations vulnerable to sudden changes. Our work is expected to contribute the initiatives in developing climate change adaptation toolkit for Asia-Pacific countries which will be crucial in undertaking vulnerability assessments as well as in identification and prioritization of appropriate actions required to build resilience in the context of changing climate in the region.

Introduction

> The latest government-agreed science shows that the world can still combat climate change but only if nations raise their collective ambition to achieve a carbon neutral world in the second half of the century. We cannot play a waiting game where we bet on future technological miracles to emerge and save the day. We already have the finance and proven technologies needed to do what is required now. The challenge and the opportunity is to increase the speed and scale of action and to make full and comprehensive use of the tools and the levers of international cooperation. Everyone who can make a difference needs to be engaged: business and investors, cities and citizens as well as governments (Christiana Figures, Executive Secretary UNFCCC, 2014).

According to reports, climate change has impacts on changing weather patterns in the Asia-Pacific region, with consequences on life security and livelihoods of the inhabiting population. The time series based climatic data analysis suggests increase in the intensity and frequency of several extreme climatic events and hazards such as heat waves, tropical cyclones, prolonged dry spells, intense rainfall, tornadoes, thunderstorms, severe dust storms etc., in the region. The regional disaster risk scenario is also reported to get compounded due to its geographic and geological vulnerability for various natural hazards such as earthquakes, tsunamis and landslides etc., (UNESCO 2014). As per reports, countries in Asia, particularly Bangladesh, China, India, Indonesia, Japan, the Philippines and Vietnam along with island nations in the Asia-Pacific region are most vulnerable to rising sea levels due to the impacts of climate change. The impacts of the disasters caused by climate change even in just 2 years viz., 2010 and 2011; resulted in the displacement of about 42 million people in the Asia-Pacific region. As the region represents a complex regional scape of countries with varied levels of economic development fraught with several key developmental issues (such as high population growth, population density, wide prevalence of poverty, heavy dependency on natural resources for subsistence, low per capita income, purchasing power etc.); profounded concerns have been raised at various national, regional and international platforms about its potentials for development and implementation of climate change mitigation and adaptation measures. It has also been reported that the key economic sectors of the developing countries in the region are already experiencing the negative effects of climate variability and change as well (USAID 2013).

Regional Venerability Scenario

The forecasts based on climate modeling experiments indicate rising rainfall concentration over the Asia-Pacific region leading to greater rainfall events during summer monsoon period and reduction in winter rainfall leading to development of arid situations. The vulnerability scenario is further expected to worsen due to the projected global sea level rise of 3–16 cm by 2030 and 7–50 cm by 2070 with consequential effects on the regional sea level. The scenario raises further alarm as climate models forecast increases in temperature par order of 0.5–2 °C by 2030 and 1–7 °C by 2070 in the region with immediate consequences over the arid areas of northern Pakistan, India and Western China. Similar studies also forecast high probability for more intense tropical cyclones as well as changes in important modes and causes for climate variability such as El Niño-Southern Oscillation, in the region. In this context, it needs no emphasis that such forecasted impacts present huge amount of additional risks for the highly vulnerable communities in the region; which are already struggling to live with poverty. The reports also adduce that the Asia-Pacific region has already accounted for 91 % of the world's total human causality and 49 % of the world's total damages on account of natural disasters in the last century. In the backdrop of such a precarious scenario, the advent of climate change challenges potentially poses serious additional threats to communities who live in remote, marginal areas such as mountains, dry lands and deserts; areas with limited natural resources, communication, transportation networks, poor financial markets, high market imperfections and weak governance institutions (ADB). The extent and nature of various major risk factors operating in the region in the context of climate change vulnerability, adaptive capacity and associated economic impacts have been organized into a risk matrix and is given in Table 22.1, which puts the context much obvious.

Table 22.1 Model matrix depicting the operation of risk factors in the context of climate change vulnerability in the Asia-Pacific region

Sensitivity	**Exposure**
Agriculture based production systems—agrarian based economies Issues of access, availability of safe drinking water and water for sanitation Low water storage, diversion and delivery capabilities to deal with rainfall extreme events Susceptibility of settlement and infrastructure Vulnerability of low lying coastal areas and island for sea level rise and storm surge	Monsoon pattern and its influences El Nino- southern oscillation—cyclonic drought, extreme sea levels Cyclones-high winds, storm surge, extreme rainfall
Economic impacts	**Adaptive capacity**
Compounds the pressure on natural resources and the environment associated with rapid urbanization, industrialization and Loss of agricultural revenue Additional costs for managing water resources, coastlines, and disease and other health risk will be a drag on economic activity Insurance industry and the loss due to climate change Degraded, threatening to undermine food security	Varying levels of economical, managerial, technological and infrastructural enabling condition Limited abilities on account of technological and financial resources Continued rapid exploration and degradation of natural resources Investing limited financial resources to face short term development needs than long term sustainable development

Source: Partially adapted from IFAD (2014), http://www.ifad.org/

GHG Emission Scenario in the Asia-Pacific Region

According to reports, under a business as usual scenario, the Asia-Pacific often described as the global engine of economic growth; is projected to contribute approximately 45 % and 60 % of global energy-related carbon dioxide (CO_2) emissions by 2030 and 2100 respectively. A time series analysis of CO_2 emissions by some of the emerging economies in the region also prompts to actively pursue the available options for climate change mitigation and adaptation to build the required resilience. The pattern of per capita CO_2 emission over the period 2006–2010 shows the emission behavior of some of the emerging economies in the Asia-Pacific (Fig. 22.1). As it could be observed, the per capita emission is also strongly linked to the population size, nature and scale of economic activity, level of economic development etc. Nevertheless, the encouraging signs in regional mitigation scenario include voluntary pledges by ten countries to reduce their emissions by 2020 (UNEP 2012). Besides, several prudential steps have been taken by the countries in the region to initiate national strategies to meet mitigation requirements as well as economic growth targets.

In this context, it is worthwhile to mention that despite recent economic downturn, low carbon industry has experienced significant growth in the recent years and at times outperformed traditional sectors. Though Asia has demonstrated strong economic growth in recent years, it has been reported that such increased growth

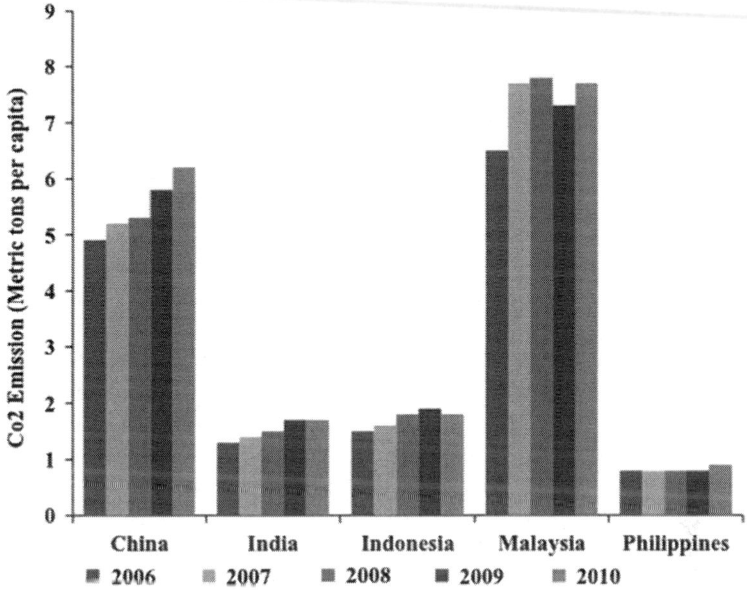

Fig. 22.1 Per capita CO_2 emission in some of the emerging economies in the Asia-Pacific. *Source*: World Bank (2014)

and trade created multiplied demands on resources as well has led to severe environmental damages in the region. For instance, concerns advanced as consequences of such growth pattern have been linked to high level of water shortages, emergence of new diseases, food insecurity issues as well as shifts in rain fall pattern. In this background, it is highly relevant to appreciate the vast commonalities existing in the Asia-Pacific region and to emphasize on areas of convergence for mitigation. Therefore, the climate change management strategies in the Asia-Pacific has to address the twin challenges of controlling high emission levels as well as high vulnerability to the impacts of climate change (EC 2012).

Adaptation: As a Way Out

It is a clear imperative that adaptation to the adverse effects of climate change is vital to respond to the current as well as future impacts of climate change and is among the key building blocks identified by the Bali Action Plan (Adaptation Forum 2013). However, successful positioning of adaptation strategies not only depends on governmental interventions, but also on efficient engagement of actors, stakeholders (including national, regional, multilateral, international organizations; public and private sectors, civil society etc.), besides effective management of adaptation knowledge. According to reports, though some countries in the Asia-

Pacific region have made significant strides in assessing climate vulnerabilities and developing adaptation strategies; the overall progress over the region has been limited. The limitations have been felt more particularly on implementation of adaptation options and strategies, which are ascribed to weaknesses in overall governance, processes for addressing adaptation within countries as well as among partners providing assistance (USAID 2013).

The emerging scenario in the domain emphasizes the need to track climate change adaptation progress especially with the institutionalization of adaptation fund. But capabilities to carry out effective adaptation tracking are potentially constrained by the complex nature of adaptation processes/strategies as well as due to the absence of indicators to measure such interventions. However, monitoring and evaluation of adaptation interventions could pave way for development of baseline scenarios and time series assessments of adaptation strategies and interventions. As per the domain literature, very little scholarship has been given to understand as to how to track adaptation despite the importance of such activities. The prevalence of such a scenario leads to limited and fragmented understanding of the state of adaptation. The current discourses indicate that approaches based on typological analysis are efficient in tracking the adaption globally at national level. However, the challenge is much complex as adaptation measures present a host of issues related to measurability, response period, potential for synergy beyond the political boundaries etc., compared to the mitigation efforts (Ford et al. 2013).

As reflected in the discussion, the availability of efficient adaptation reporting tools is critical for future adaptation tracking initiatives as well as to develop protocols to monitor progress which are requested at national and international levels. The overarching requirement of appropriate data sources has been much emphasized in the current literature as a sine qua non to facilitate the development and tracking of adaptation indicators. It is also reported that such data sources need to: *provide information on adaptation for a large number of countries to facilitate national-level comparative analysis, be systematically collected and follow standardized guidelines so comparison reflects real trends in adaptation as opposed to nature of data collection, provide sufficient detail for adaptations to be characterized, and be consistent in reporting over time and be collected at regular intervals. Currently, few existing data sources meet these requirements, and there has been little work examining how current data sources could be used to systematically track adaptation* (Ford et al. 2013).

In the above context, the current review was undertaken to develop an insight into the challenges for climate change mitigation as well as to develop a typology for adaptation measures undertaken in the emerging economies in the Asia-Pacific region. The typology is expected to find application in the, (1) development of base line scenarios (2) formulation of interventions for collective action across the different stake holders (3) exploration for synergies at trans-boundary level. The output of the current typological analysis could potentially contribute to address the current gaps identified in the domain apart from contributing to the development of an adaptation measurement toolkit for the use in the Asia-Pacific region, as there region shares many commonalities in terms of potentials as well as challenges.

Besides the study outcomes are expected to supplement the much needed critical scholarship to be developed for the region for collective policy development and action to deal with climate change.

The Methodology

In the current paper, an attempt has been made primarily to present a typological analysis of the various adaptation measures, polices and strategies pursued by the countries in the Asia-Pacific region by following a case study approach. Such a methodology was adopted to develop a generic typology to contribute the initiatives towards setting of baseline scenarios for adaptation measures and development of adaptation monitoring tools kits at regional, sub regional and national levels in areas collective action. The sampling frame for selection of the study countries included the emerging economies in the region. Accordingly six economies in the Asia-Pacific region viz., China, India, Malaysia, Philippines, Indonesia and Vietnam were considered for the analysis. The guiding considerations to fix the sampling frame included the current stage of economic development, geographic specifics, distribution of the various natural resources, contours of vulnerability to the impacts of climate change etc. In this context, we would also like to mention that our approach should not be construed as an intention to rank the countries or to identify a group of countries or a set of best practices/policies/programs. But, the guiding philosophy has been to develop a typology which can be used by the different stake holders to develop various adaptation measures and adaptation monitoring in the region.

The parameters considered for the adaptation typology included key sectors such as water resources, agriculture and animal husbandry, forestry, coastal and marine areas, health, energy and transportation etc. The required data were collected from the National Communications of the study countries to the United Nations Framework Convention on Climate Change (UNFCCC 2014). In order to develop an insight into the mitigation strategies, we have tried to comprehend the prevailing pattern by analyzing the country level cumulative mitigation potentials registered in the compliance market mechanism under the UNFCCC. Accordingly, mitigation potentials registered by the case study countries were abstracted from recognized Clean Development Mechanism registries and were considered for empirical analysis. As we already highlighted, the pattern analysis helps in direct understanding of the prevailing context of mitigation initiatives in the region as well as development and implementation of strategies for adaptation monitoring while fostering collective learning and action to deal with climate change in the region.

The typology and the tool kits so developed can also be used by different actors such as inter-governmental institutions, civil society; to apply analytical measures to compare and prioritize the various intervention requirements. It is reported in the current literature that potentials of the approaches to track adaptation interventions across the regions and sectors based on standardized baselines are highly valued

(Ford et al. 2013). The analysis presented in this work focuses deliberately on developing adaptation strategies—i.e., those "top-down" attempts geared towards framing "bottom-up" measures while allows for a higher level of comparability.

Results and Discussion

Climate Resilient Development

"Climate Change Adaptation involves adjustments in natural or human systems, in response to actual or expected climatic stimuli or their effects, moderating harm or exploiting beneficial opportunities" (Huttche 2013). It is widely recognized that there are several linkages between the varied requirements for climate change adaptation and the requirements of development in the countries in the Asia-Pacific region. In this context, it needs no emphasis that the regional developmental objectives need to be realized over a wide range of climatic conditions. Therefore adaption measures in the key domains of development such as poverty reduction, poverty elimination, water supply, sanitation, irrigation, agricultural and veterinary practices, flood control, transport, energy, health, education, integrated landscape management complement significantly to resilience enhancement in the region, in the wake of long term and uncertain impacts of climate change (ADB 2013). Although, the climate resilient development in the region is potentially challenged by various developmental, economic, social and political reasons; many countries have taken several measures towards climate change adaptation and resilience building. However, the case with least developed countries in the region present a different scenario altogether. Besides, it has also been reported that the progress of such initiatives has been limited in terms of implementation of various adaptation options and strategies (USAID). The reasons for such perceived limitations or shortcoming are reported to be linked with weaknesses in governance systems/ institutional arrangements, processes identified for the adaption, resource use and exploitation scenarios, level of investment etc.

The Vulnerability Conundrum

As we have already discussed, the countries in the region are fraught with complex networks of various, actors, policies, strategies, dynamic forces, operating in their landscapes. We have attempted to comprehend the climate change vulnerability conundrum through empirical analysis of the potentially contributing factors from the national communications of the study countries. The factors identified were arranged in a schematic matrix model to understand their interplay which leads to climate change vulnerability conundrum in the region. A schematic model depicting the interplay of these factors has been prepared and the same is given at

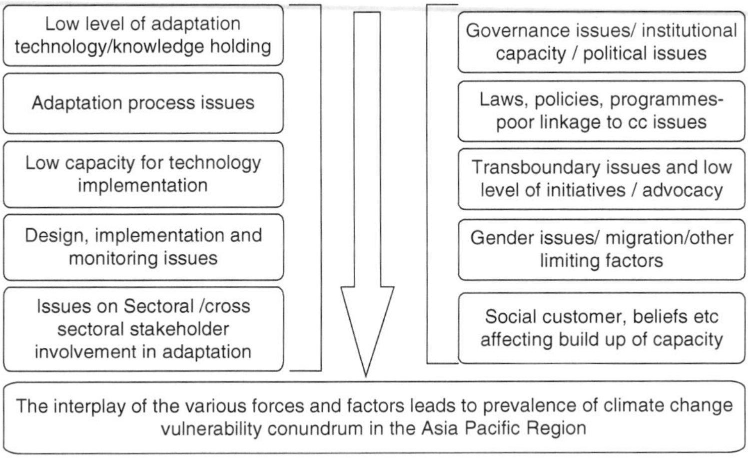

Fig. 22.2 A schematic model depicting the interplay of various forces and factors and the prevalence of climate change vulnerability conundrum in the Asia-Pacific Region. *Source*: Own elaboration with inputs from USAID

Fig. 22.2. It could be observed from the flow paths in the matrix that the complex scenario of climate change vulnerability in the region is contributed by the interplay of a spectrum of factors ranging from low level of adaptation technology, implementation capacity to stake holder participation, governance arrangements, institutional capacity, political issues, trans boundary issues, gender related aspects, social customs, etc.

The empirical analysis of the scenario supported by the schematic model reveals that the vulnerability conundrum in the region is contributed by various social, economic, political and resource management related forces/factors. It also prompted us to infer that these forces were to be balanced out to instill any efficiently operating adaptation strategy. Besides, we have found that the observations are further strengthening the arguments advanced in this direction by current discourses in the domain.

Mitigation Initiatives

The Cancun agreement has already prescribed for biennial update reports on mitigation measures undertaken by the developing countries. The international arrangements also stand for initiating efforts to develop and implement appropriate mitigation action. An analysis of the contours of development of the mitigation capacity in the region reveals that, the mitigation projects have much been advanced by the countries like China and India in the region, so also the other regional members. The details of the empirical analysis carried out for the study countries are presented in the Fig. 22.3. However, it is observed that the extent of development of mitigation initiatives/interventions are influenced by the economic

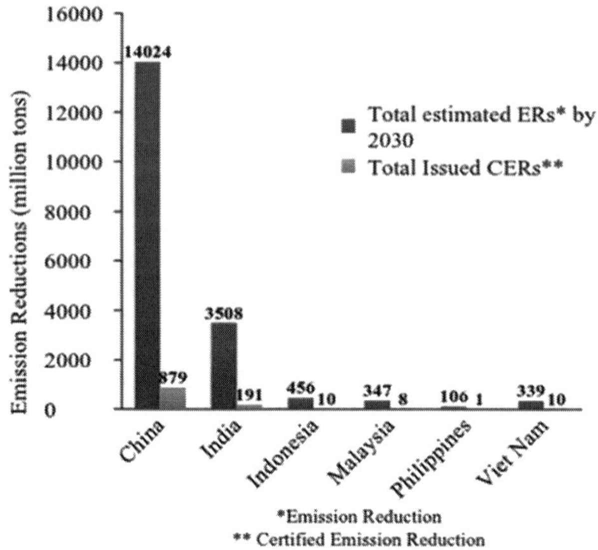

Fig. 22.3 Pattern of mitigation potentials registered by the study countries in the compliance markets domain. *Source*: IGES (2014)

development status, availability of markets, level of technical capacity available in the economies concerned along with the extent of self-emission regulations followed by the countries concerned.

Adaptation Typology

Adaptation refers to changes in processes, practices, and structures to moderate potential damages or to benefit from opportunities associated with climate change (UNFCCC 2014). In order to suggest a typology for the adaptation measures in the region, a comparative study of the various adaptation measures suggested/undertaken by the study countries was resorted to. The study profiled the adaptation interventions proposed across sectors such as water resources, agriculture and animal husbandry, forestry, coastal and marine areas, health, energy and transportation etc. The indicative typology can help the stakeholders to develop pragmatic adaptation plans/interventions/monitoring plans as well as complement in evolving trans boundary approaches to build adaption capabilities in the region. The indicative typology developed, based on the strategies adopted in the region is presented in the Table 22.2. The analysis also led us to infer that the adaptation strategies tend follow the contours of specifics of the countries while possessing several elements for replication and space for collective action in the region.

However, it may be observed that the implementation of various initiatives in building climate resilience is fraught with several issues. Though much capacity has been developed on greenhouse gas inventory on a regular basis, the overall scenario is challenged by constraints such as availability, reliability and variability of activity data and emission factors. There are reports of high variability because of

Table 22.2 An indicative typology for climate change adaptation measures in the Asia-Pacific

Sector	Adaptation measures
Water resources	Sustainable water resources development, integrated management of all river basins and regions based on social and economic development, ecosystems promoting water efficiency and conservation
	Construct new structures/reinforce/upgrade water resource infrastructure, existing system of river and sea dykes, flood diversion areas, flash flood relief channels, embankments for flood protection, saltwater intrusion barriers; artificial drainage structures for the low lying plain and coastal areas to improve operational efficiency and safety with design standards for flood mitigation
	Water-saving irrigation methods in agriculture, such as spray, and drip irrigation, modernize the observation and long range water resources forecasting network, seasonal and yearly forecasting for water resources and natural disasters (e.g., flooding, drought, salinity intrusion)
	Warning systems for flash floods and debris flows, with priority for mountainous areas and improved management of water resources with decision support systems
	Legal arrangements to promote sustainable water use and better water management which are crucial in adapting to climate change
	Improving storage efficiency by removing sediment from reservoirs and dams, eliminating losses from leakage and water theft
	Promote demand management practices to reduce per capita consumption of potable water by industrial, commercial and residential consumers, reducing wastage through behavioral changes and encouraging water harvesting for non-potable uses, water treatment and recycling for major water users, treated and un treated water supply/pricing
	Improve the efficiency of irrigation and other water uses that rely on non-potable sources such as rainfall and ground water as well as introduce low water use crops and farming practices
	Review design standards for flood risk management in all new infrastructure including water control structures, transportation structures and electrical, water and waste amenities to incorporate climate change factors
	Coordinate use of water basin/ground water source, strict implementation of existing forestry rules and regulations, water re allocation and compensation schemes to meet the demand of the sectors
	Water treatment and recycling, for major water users, treated and untreated water supply/pricing, urban flood prevention, rainwater collection and local processing systems
	Urban flood prevention, rainwater collection and local processing systems among others buildup of a flood/drought control mechanism, water-saving society, and unified management of water resources, etc.

(continued)

Table 22.2 (continued)

Sector	Adaptation measures
Agriculture and animal husbandry	Reforming regulations and agricultural institutions, development of information systems advocating and communication
	Development of agricultural insurance system related to drought, flood, pests and diseases and increasing subsides for water saving irrigation machines
	Improvement of water management, efficient irrigation, soil and fertilization management including organic fertilizer and development of carbon efficient farming systems
	Development of early, drought, salinity and inundation tolerant crop varieties, dynamic cropping calendar, flood and drought anticipation blue print, cropping pattern adjustment by preparation and dissemination of guidance and tools, sustainable farming practices, adjusting agricultural structure
	Prevent soil erosion, implement soil protection, preserve soil moisture and fertility levels, plant trees and form terraced fields on steep slopes to increase vegetation cover
	Expand fodder production and enhance storage, processing and utilization of animal feeds, build stables with adequate design, proper manure and water management systems
	Deployment of additional resources required to disperse technology and to train and build capacities of workers
	Implementing regional plans for competitive agricultural product layout high efficiency water saving irrigation and rain fed water efficiency farming systems
	Increasing subsides for improved crop varieties; actively developing intensive and standard livestock production and aquaculture development of an animal epidemic disease prevention system
	Recovery of grassland by reducing grazing intensity, managed grasslands; grassland ecological compensation mechanism, implementing protective measures, e.g., forage-livestock balance, grazing ban, rest, and rotational grazing, to restore natural grassland vegetation and to control grassland degradation
Forestry	Policies and regulations to reduce risk of land/forest fire, forest fire control and management program, community based fire management systems, early warning systems
	Strengthen sustainable forest management, prevent deforestation, development and increase forest coverage, afforestation and reforestation, enhance recovery and rehabilitation of degraded ecosystems to reduce climate change risks, integrated terrestrial ecosystem monitoring system at the national level
	Select and diversify plant species resistant to droughts, floods, pests and less prone to forest fires, genetic conservation plans and gene banks, seed center, botanical gardens, protected areas, captive breeding centers, rehabilitation centers for fauna, conservation corridors between forests
	Develop timber and non-timber product processing technologies and enhance timber use efficiency
	Coastal mangrove forest system restoration and development projects, plant protective dune forests (soft dyke's), forest development

(continued)

Table 22.2 (continued)

Sector	Adaptation measures
	components of sea dyke projects, reinforcing protection and management of wetland ecosystems to increase their resilience to climate change risks
	Support livelihood and improve living conditions for people living near forests, conversion of sloppy cropland into forestry
	Creating natural reserves in areas with high risks of climate change, reinforcing management and protection of terrestrial ecosystems, recovery and rehabilitation of degraded ecosystems to reduce climate change risks
Coastal and marine areas	Development of dikes equipped with polder system to protect hinterland, strengthening and elevating embankments, intensifying security services along coastlines, preventing saltwater intrusion, installing water pumps and
	Mangrove rehabilitation to increase soil surface and reduce wave energy destruction reduce erosion, sylvofishery, prohibition of mining of coral and beach sands
	Construction of adaptive infrastructure, focusing on developing new agricultural techniques and elevating houses above flood levels. Modification of drainage systems, specifications of minimum floor elevation and piling depth as well structural bracing for building code; allowing changes in land use such as conversion of agriculture land to aquaculture uses; prohibiting filling of wetlands, damming of rivers
	Natural resources, such as mangroves and coral reefs to be left to their natural processes to cope with sea level rise. Research is required on the adaptation and recovery mechanisms of coral beds to enable appropriate measures to be undertaken
	Withdrawal, which averts sea-level rise impacts by relocating inhabitants, enterprises and services towards safer areas
	Defensive engineering measures against inundation, tidal flooding, storm surges, and effects of waves on infrastructure, soil erosion and loss of natural resources such as mangroves. Consequently, hard measures such as sea-walls and groins and soft measures such as beach nourishment and wetlands/mangroves creation are possible adaptation measures
	Coastal reforestation to develop optimal planting methods and the creation of robust coastal forests that can strengthen the stability of coastlines and contribute to biodiversity enhancement
	Improving relevant laws, regulations and policies to constantly enhance marine ecosystem conservation and rehabilitation
Health	Improvement of the disease ecology surveillance system and development of early warning system for outbreaks. Empowering community health services systems for prevention and control of diseases, public health emergency response system
	Increasing political awareness of the effects of climate change on public health, enhancement of capacity building for the government, private sectors, and civil society institutions related to the prevention and mitigation of public health aspects affected by climate change
	Conducting research and developing methods for epidemiology and medicine to break the chains of disease transmissions, climate change

(continued)

Table 22.2 (continued)

Sector	Adaptation measures
	impacts on human health, and pursing scientific knowledge outreach and training
	Prevention and eradication of contagious and vector-borne diseases affected by climate change, disease control and prevention system and a health supervision and law enforcement system
	Strengthen residential planning, taking into account natural disasters impacts for vulnerable areas such as coastal zones, steep hills and mountains and fluvial low-lying areas, build capacity for rural health care institutions in disaster prone areas
	Mitigate the urban heat island effect by replacing concentrated metropolises with satellite cities, decrease outdoor activities, hard labour in particular, around the peak of the hot seasons during heat waves
Energy and transportation	Mainstream climate change issues into energy and transportation development strategies and plans. Guarantee energy security, transportation safety and sustainable development
	Improve energy efficiency in the exploitation, production, distribution and consumption of energy
	Elevate and renovate structures in the energy and transportation sectors in areas vulnerable to sea-level rise and flooding
	Reinforce transportation infrastructure, power transmission towers in flood prone areas, strengthen roads and rail networks in the mountains and areas with sloping terrain
	Construct bridges to replace submerged intersections and spillways to guarantee an uninterrupted transportation network

Source: MOEF (2012), MOE (2012), PRC (2012), IACCC (2000), MONRE (2011), and MNRE (2010)

continuous updates; default factors utilized in the inventory; which may put the context deviated from the ground reality. As regards the mitigation efforts, the scenario is vexed with issues related to technological affordability e.g., utilization of renewable in power production, determining full potential of the wastes sector; which represent considerable abatement opportunities. However strengthened arrangements are needed to implement the integrated abatement strategies to resolve the issues to great extent as well as to realize the expected outcomes.

Conclusions

It is much obvious that the impacts of climate change are already visible in different degrees in different parts of the Asia-Pacific region. The Cancun Adaptation Framework has clearly identified the need for enhanced action on adaptation, vulnerability reduction, integrating best science, multilevel institutional arrangements, engagement of all potential actors, financial and

(continued)

implementation arrangements to build resilience in the developing country members. There are several technologies, measures and policies already available for building climate resilient economies in the region. During the review, we could observe the presence of several such initiatives on the adaptation front, being practiced by the counties; indeed with their contours aligned to the country specifics. Nevertheless there exist several elements of generality and wide scope for regional, international co-ordination and collective action.

The typology of the adaption measures suggested in the current reviews primarily based on the strategies adopted by the case study countries and presents a non-exhaustive vertical spectrum of interventions. However it may be further calibrated to the requirements of the stakeholders for development and implementation of various adaptation measures. The outcome of the paper is expected to contribute to pilot/concept level initiatives to develop adaptation tool kit for Asia-Pacific. The calibration of such a tool kit is crucial to undertake vulnerability assessments, identification, prioritization of appropriate actions required for development of baseline data bases for adaptation monitoring. The outputs of the review can inter alia find useful in providing information to the policy makers and developers, in the regional context as well. We also acknowledge that the analysis of local sectorial adaptation activities is beyond the scope of this report; this will have led to some relevant trends and activities being overlooked. However, it is also suggested that more investigations and analysis are required to support the initiatives; as the study region presents a complex matrix in terms of economic, social, cultural, technological advancements.

References

Adaptation Forum (2013) Asia-Pacific climate change adaptation forum-2013. Archived on 27 March 2014. http://www.asiapacificadapt.net/adaptationforum2013/

ADB (Asian Development Bank) (2013) Promoting climate-resilient development. Archived on 20 March 2014. http://www.adb.org/themes/climate-change/climate-resilience

EC (European Commission) (2012) Engaging with Asia Pacific on climate change. Archived on 25 May 2014. http://ec.europa.eu/clima/events/articles/0052_en.htm

Figures C (2014) Science underlines urgency to act towards a carbon neutral world, United Nations Climate Change Secretariat. Archived on 27 March 2014. http://svmrevproxy01.unfccc.int/files/press/press_releases_advisories/application/pdf/pr20141304_ipccwg3.pdf

Ford JD, Berrang-Ford L, Lesnikowski A, Barrera M, Heymann SJ (2013) How to track adaptation to climate change: a typology of approaches for national-level application. Ecol Soc 18(3):40. Archived on 25 May 2014. doi:10.5751/ES-05732-180340

Huttche C (2013) Building climate change resilience for the Asia Pacific transport sector-climate proofing with adaptation assessment. Archived on 2 March 2014. http://climatechangechallengescities.files.wordpress.com/.../building-climate-ch

IACCC (Inter-Agency Committee on Climate Change, Philippines) (2000) The Philippines initial national communication on climate change. Archived on 27 March 2014. http://unfccc.int/resource/docs/natc/phinc.pdf

IFAD (International Fund for Agricultural Development) (2014) Climate change impacts in the Asia/Pacific region. Archived on 27 March 2014. http://www.ifad.org/events/apr09/impact/pacific.pdf

IGES (Institute for Global Environmental Strategies) (2014) GHG emissions/market database. Archived on 27 March 2014. http://www.iges.or.jp/en/climate-energy/mm/publication.html

MNRE (Ministry of Natural Resources and Environment, Socialist Republic of Vietnam) (2010) Vietnam's second national communication to the United Nations framework convention on climate change. Archived on 27 March 2014. http://unfccc.int/resource/docs/natc/vnmnc02.pdf

MOE (Ministry of Environment, Republic of Indonesia) (2012) Indonesia's second national communication under the United Nations framework convention on climate change. Archived on 27 March 2014. http://unfccc.int/resource/docs/natc/indonc2.pdf

MOEF (Ministry of Environment and Forests, Government of India) (2012) India second national communication to the United Nations framework convention on climate change. Archived on 27 March 2014. http://unfccc.int/resource/docs/natc/indnc2.pdf

MONRE (Ministry of Natural Resources and Environment, Government of Malaysia) (2011) Malaysia's second national communication (NC2) to the United Nations framework convention on climate change. Archived on 27 March 2014. http://www.nre.gov.my/Malay/Alam-Sekitar/Documents/Penerbitan/SECOND%20NATIONAL%20COMMUNICATION%20TO%20THE%20UNFCCC%20(NC2).pdf

PRC (People's Republic of China) (2012) Second national communication on climate change of the People's Republic of China. Archived on 27 March 2014. http://www.ccchina.gov.cn/archiver/ccchinaen/UpFile/Files/Default/20130218145208096785.pdf

UNEP (United Nations Environment Programme) (2012) Summary for Asia and the Pacific region on the eve of Rio +20. Archived on 27 March 2014. http://www.unep.org/geo/pdfs/geo5/RS_AsiaPacific_en.pdf

UNESCO (United Nations Educational Scientific and Cultural Organization) (2014) Addressing climate change education for sustainable development. In: Experts meeting on climate change education for sustainable development in Asia and Pacific, Manila. Archived on 27 March 2014. https://en.unesco.org/events/experts-meeting-climate-change-education-sustainable-development-asia-and-pacific, 2014)

UNFCCC (United Nations Framework Convention on Climate Change) (2014) Non-Annex I national communications. Acting towards a carbon neutral world, UNFCCC, Bonn. Archived on 24 May 2014. http://unfccc.int/national_reports/non-annex_i_natcom/items/2979.php

USAID (United States Agency for International Development) (2013) Proceedings of second annual meeting. In ADAPT Asia-Pacific, Nadi, 3–4 June 2013. Archived on 24 May 2014. http://adaptasiapacific.org/forum/2013-annual-meeting

World Bank (2014) World development indicators. Archived on 27 March 2014. http://databank.worldbank.org/data/views/reports/tableview.aspx?isshared=true&ispopular=series&pid=18